BYDALES SCHOOL

MODULAR SCIENCE
Chemistry

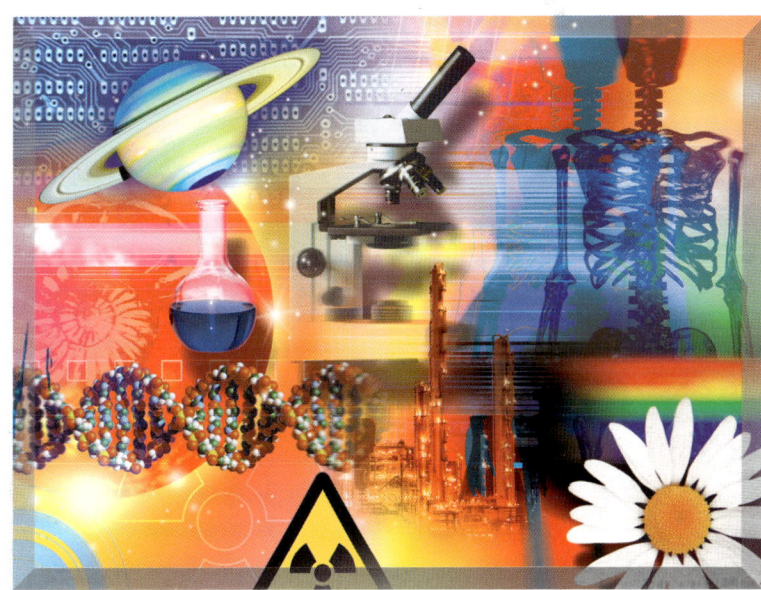

AQA Modular Science Higher Tier

Paul Ingram
RoseMarie Gallagher

OXFORD
UNIVERSITY PRESS

Great Clarendon Street, Oxford OX2 6DP

Oxford University Press is a department of the University of Oxford.
It furthers the University's objective of excellence in research, scholarship,
and education by publishing worldwide in

Oxford New York

Auckland Bangkok Buenos Aires Cape Town Chennai
Dar es Salaam Delhi Hong Kong Istanbul Karachi Kolkata
Kuala Lumpur Madrid Melbourne Mexico City Mumbai Nairobi
São Paulo Shanghai Taipei Tokyo Toronto

Oxford is a registered trade mark of Oxford University Press
in the UK and in certain other countries

© RoseMarie Gallagher and Paul Ingram 2001

The moral rights of the author have been asserted

Database right Oxford University Press (maker)

First published 2001

All rights reserved. No part of this publication may be reproduced,
stored in a retrieval system, or transmitted, in any form or by any means,
without the prior permission in writing of Oxford University Press,
or as expressly permitted by law, or under terms agreed with the appropriate
reprographics rights organization. Enquiries concerning reproduction
outside the scope of the above should be sent to the Rights Department,
Oxford University Press, at the address above

You must not circulate this book in any other binding or cover
and you must impose this same condition on any acquirer

British Library Cataloguing in Publication Data

Data available

ISBN 0 19 914816 3

10 9 8 7 6 5 4 3 2

Typeset in Palatino by AMR, Bramley

Printed in Spain by Edelvives

Acknowledgements

The publisher would like to thank the following for their permission to reproduce copyright material:

p8 Corbis/Wild Country; **p10** Powerstock/Zefa (top), OUP (bottom left and bottom middle), Bridgeman Art Library (bottom right); **p11** Science Photo Library/E. Willard Spurr/US Library of Congress; **p12** OUP (all); **p13** OUP; **p14** Andrew Lambert (top left, top middle and top right), Bridgeman Art Library (bottom **p15** James Davis Travel Photography (top), Bridgeman Art Library (bottom); **p18** Royal Mint (top), Thermit Welding UK (bottom); **p19** OUP; **p20** Barnaby's Picture Library/S.E. Barnes (top), Science Photo Library/NASA (bottom); **p21** GSF Picture Library (top left), Geological Museum (top middle and top right), Barnaby's Picture Library (bottom); **p22** SPL/D. Nunuk; **p24** Leslie Garland Picture Library (top left), OUP (top middle, top right, bottom left and bottom right), Bristock/IFA (bottom middle); **p26** Science Photo Library /Crown Copyright/Health & Safety Laboratory (top), Camera Press (bottom); **p30** Jonquiere Analytical Centre (top left), Alcan Aluminium (top middle, top right, bottom left and bottom middle), Science Photo Library/NASA (bottom right); **p31** Alcan Aluminium (top); **p32** Powerstock/Zefa/R. Maclennan (top), OUP (bottom); **p34** OUP (top left), Robert Harding/R. Estall (top middle), Bridgeman Art Library (top right), J Allan Cash (bottom); **p35** Heinz (top), Alcan Aluminium (bottom); **p37** Werner Forman Archive/British Museum London (top left), Werner Forman Archive/Royal Museum of Art & History Brussels (top right), Science Photo Library/S. Terry (bottom); **p38** OUP (all); **p39** OUP; **p41** OUP; **p44** OUP (top), NHPA/S. Dalton (middle), Hutchison/R. Aberman (bottom); **p45** James Davis Travel Photography (top left), OUP (top right), Friends of the Earth Trust (bottom); **p52** NASA; **p56** Ecoscene/Leeney (top left), Ecoscene/Alan Towse (top right); **p57** Eye Ubiquitous/Steve Brock; **p58** BP; **p59** Aviaton Photographs International, Andrew Lambert (middle), Colorsport (right); **p62** BP/John Rae; **p63** Exxon Mobil; **p64** OUP; **p65** Science Photo Library/P. Menzel; **p66** OUP; **p67** OUP; **p68** Eye Ubiquitous/P. Sehews (left), Symphony Environmental (right); **p69** Ecoscene/A. Gazidis (top left); **p71** Ecoscene/R. Nichol (top), Eye Ubiquitous (bottom); **p73** Eye Ubiquitous/B.Battersby (left), Environmental Images/S.Morgan (right;) **p74** Science Photo Library/NASA (left and right); **p77** Science Photo Library/Jean-Loup Charmet; **p78** Mary Evans Picture Library; **p79** Powerstock/Zefa (left), Ecoscene/Rod Gill (right); **p80** Powerstock/Zefa; **p81** Science Photo Library/Peter Ryan (top), Oxford Scientific Films/P. Parks (bottom); **p83** Tony Waltham Geophotos (all); **p85** Oxford Scientific Films/Paul Franklin (top), Oxford Scientific Films/Nigel Dennis (bottom); **p87** Tony Waltham Geophotos (both); **p88** Science Photo Library/Alfred Pasieka (top), Ecoscene/Sally Morgan (bottom left), Tony Waltham Geophotos (bottom right); **p89** James Davis Travel Photography (top), Geoscience Features (bottom); **p90** Tony Waltham Geophotos (both); **p91** Tony Waltham Geophotos; **p92** Tony Waltham Geophotos (both); **p93** Tony Waltham Geophotos (all); **p95** Oxford Scientific Films (top), Tony Waltham Geophotos (bottom left and bottom right); **p96** GSF Picture Library; **p97** Panos/Jeremy Horner; **p103** Science Photo Library (left), Oxford Scientific Films (right); **p104** SPL/David Parker; **p105** SPL/Bernhard Edweiser, (top), Rex Features (bottom); **p112** The Stockmarket/Gabe Palmer; **p114** OUP (top left and bottom middle), Zefa (top middle), Environmental Images/T. Perry (top right), British Airways (bottom left), Barnaby's Picture Library (bottom right); **p115** OUP (all); **p116** Hutchison Picture Library; **p118** OUP; **p119** OUP; **p120** OUP (both); **p123** Science Photo Library/Crown Copyright/ Health & Safety Laboratory; **p124** OUP (top left and right), BMW (bottom); **p125** Synetix (top); **p127** Richard Anthony/Holt Studios; **p128** Mary Evans Picture Library; **p129** Cephas/M. Rock (top), OUP (bottom); **p130** Powerstock/ Zefa (top), Science Photo Library/B Murton/Southampton Oceanography Centre (bottom); **p132** OUP (left and middle), Andrew Lambert (right); **p133** Andrew Lambert (left and middle), Peter Gould (right); **p136** Science Photo Library/Charles D. Winters (left), OUP (right); **p140** ICI; **p141** Mary Evans Picture Library; **p142** Holt Studios/N. Cattlin (top left), NHPA/E. Soder (top middle), Corel Professional Photos (bottom left), Sc ience Photo Library (bottom middle), ICI (bottom right); **p143** ICI (both); **p144** Mary Evans Picture Library; **p145** Robert Harding Picture Library; **p146** Hulton Getty; **p148** Bridgeman Art Library (top), Sally &Richard Greenhill (bottom); **p150** Andrew Lambert; **p152** AGA (top), Science Photo Library/NASA (bottom); **p153** OUP; **p155** OUP (both); **p156** Science Photo Library; **p157** Coalite; **p158** OUP; **p159** Environmental Images/L. Garland; **p166** Corel Professional Photos; **p168** OUP (top left and right), De Beers (middle), Derby Museum & Art Gallery (bottom) ; **p170** Aerofilms; **p172** Barnaby's Picture Library; **p173** Andrew Lambert; **p174** Hulton Getty; **p176** Ancient Art & Architectural Collection/Sheridan (top), Mary Evans Picture Library (bottom); **p177** Science Photo Library (top), Corbis/Stockmarket (bottom); **p178** Science Photo Library; **p179** Science Photo Library/A. Barrington Brown; **p180** OUP (top left and right), Ace/J.D. Begg (bottom); **p182** OUP; **p184** OUP; **p186** GSF Picture Library (top left), OUP (top right); **p187** SPL/D Guyon (left), SPL (right); **p188** OUP (all); **p189** OUP (all); **p190** OUP (right); **p191** Frank Spooner Pictures/Gamma/Alain Morvan; **p192** Andrew Lambert (top), University of Liverpool/P Beahan (bottom left), GSF Picture Library (bottom right); **p193** Crafts Council; **p194** Andrew Lambert; **p195** Science Photo Library (both); **p196** John Birdsall Photography; **p197** OUP (all); **p199** Hulton Getty; **p200** OUP; **p201** Allsport/T. Warshaw; **p202** OUP; **p203** Robert Harding Picture Library; **p207** Severn Trent Water (top left), ICI (top right); **p214** Getty Images Stone; **p216** OUP (all); **p217** Powerstock; **p218** OUP; **p219** OUP; **p220** Allsport/J. McDonald (top left), OUP (top right), Cement & Concrete Association (bottom left), British Cement (bottom right); **p221** Science Photo Library/A & Hanns-Frieder Michler (top left), J Allan Cash (top right), OUP (bottom); **p222** Still Pictures/J.Schytte; **p223** Harvey Softeners; **p224** OUP; **p225** OUP (top), Ecowater (bottom).

The illustrations are by: Harriet Dell, Michael Eaton, Jeff Edwards, Nick Hawken, Illustra Graphics, Peter Joyce, Ben Manchipp, Mark Oliver, Tony Simpson, Julie Tolliday and Galina Zolfaghari, Art Construction, Chartwell Illustrators and Oxford Computer Illustration.

Cover image by Gary Thompson

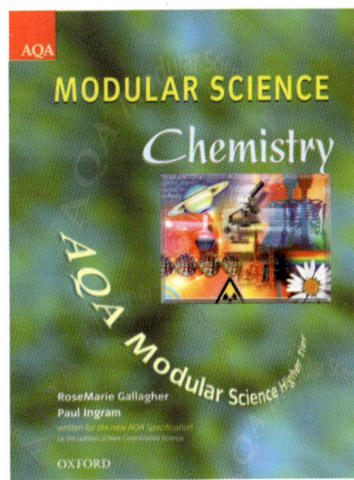

INTRODUCTION

Science is about asking questions. Chemistry is the science that asks questions about materials, the differences between them, how they react with one another, and how heat and other forms of energy affect them.

You will find this book useful if you are studying chemistry as part of the AQA Modular Science Single or Double Award GCSE science course.

Everything in this book has been organized to help you find out things quickly and easily. It is written in two-page units called spreads.

- **Use the contents page**
  If you are looking for information on a large topic, look it up in the contents page.

- **Use the index**
  If there is something small you want to check on, look up the most likely word in the index. The index gives the page number where you'll find information about that word.

- **Use the questions**
  Asking questions and answering them is a very good way of learning. There are questions at the end of every Module. At the end of the book there is a set of further exam-style questions and a selection of multiple-choice questions. Answers to numerical questions, and pointers to those requiring short answers, are provided.

- **Helping you revise**
  To help you revise, in addition to the questions, there are some revision notes, a glossary of important terms, and end-of-Module summaries.

Chemistry is an important and exciting subject. It is all around you: in fairgrounds, fields, farms, and factories. It is taking place deep in the Earth and far out in space. You'll find chemistry everywhere.

We hope that this book helps you with your studies, that you enjoy using it, and that at the end of your course, you agree with us that chemistry is exciting!

RoseMarie Gallagher
Paul Ingram

July 2001

Note: The *Further Topics* are included to satisfy the additional satutory requirements of the national curricula for students in Northern Ireland and Wales.

Contents

Routemap 6

Module 5
Metals 8

5.01	Metals and the periodic table	10
5.02	Group 1: the alkali metals	12
5.03	The transition metals	14
5.04	Metals and reactivity	16
5.05	The reactivity series	18
5.06	Metals in the Earth's crust	20
5.07	Extracting metals from their ores	22
5.08	Making use of metals	24
5.09	Extracting iron	26
5.10	Principles of electrolysis	28
5.11	Extracting aluminium	30
5.12	Purifying copper by electrolysis	32
5.13	Corrosion	34
5.14	Metals, civilization, and you	36
5.15	Acids and alkalis	38
5.16	The reactions of acids	40
5.17	A closer look at neutralization	42
5.18	Acids and alkalis in everyday life	44
	Questions for Module 5	46

Module 6
Earth materials 52

6.01	Limestone and its uses	54
6.02	To quarry, or not?	56
6.03	Crude oil	58
6.04	Separating oil into fractions	60
6.05	Cracking hydrocarbons	62
6.06	The alkanes and alkenes	64
6.07	Polymerization and plastics	66
6.08	Polythene – here to stay?	68
6.09	Oil and the environment	70
6.10	Global warming	72
6.11	The atmosphere past and present	74
6.12	The atmosphere in balance	76
6.13	What have we done to the ozone layer?	78
6.14	The oceans	80
6.15	The Earth's structure	82
6.16	From weathering to deposition	84
6.17	From sediment to sedimentary rock	86
6.18	Evidence from fossils	88
6.19	Igneous rock	90
6.20	Metamorphic rock	92
6.21	The rock cycle	94
6.22	The Earth's plates	96
6.23	A closer look at plate movements	98
6.24	Shaping the Earth's surface	100
6.25	Wegener: a scientist ahead of his time	102
6.26	Can we predict earthquakes and eruptions?	104
	Questions for Module 6	106

Module 7
Patterns of chemical change — 112

7.01	Rates of reaction	114
7.02	Measuring the rate of a reaction	116
7.03	Changing the rate of a reaction (I)	118
7.04	Changing the rate of a reaction (II)	120
7.05	Explaining rates	122
7.06	More about catalysts	124
7.07	Enzymes	126
7.08	Some traditional uses of enzymes	128
7.09	Some modern uses of enzymes	130
7.10	Exothermic and endothermic reactions	132
7.11	Explaining energy changes	134
7.12	Reversible reactions	136
7.13	Shifting the equilibrium	138
7.14	Making ammonia in industry	140
7.15	Fertilizers	142
7.16	The pros and cons of fertilizers	144
7.17	The masses of atoms	146
7.18	Percentage composition of a compound	148
7.19	The formula of a compound	150
7.20	Equations for chemical reactions	152
7.21	Calculations from equations	154
7.22	Calculating the volumes of gases	156
7.23	Calculations on electrolysis	158
	Questions for Module 7	160

Module 8
Structures and bonding — 166

8.01	Atoms, elements, and compounds	168
8.02	More about atoms	170
8.03	Isotopes and A_r	172
8.04	How electrons are arranged	174
8.05	How ideas of the atom developed	176
8.06	The atom: the inside story	178
8.07	Why compounds form	180
8.08	The ionic bond	182
8.09	Some other ions	184
8.10	Ionic compounds and their properties	186
8.11	The covalent bond	188
8.12	Covalent substances	190
8.13	Metals: more giant structures	192
8.14	How the periodic table developed	194
8.15	The patterns within Group 1	196
8.16	The non-metal groups in the periodic table	198
8.17	Reactions of the halogens	200
8.18	Compounds of the halogens	202
8.19	The electrolysis of sodium chloride	204
8.20	The chlor-alkali industry	206
	Questions for Module 8	208

Further topics — 214

FT 1	Revision: solids, liquids, and gases	216
FT 2	A closer look at gases	218
FT 3	Composite materials	220
FT 4	Soft and hard water	222
FT 5	Making hard water soft	224

Exam-style questions	226
Multiple-choice questions	232
Revision and exam guidance	236
Module summaries/checklists	238
Glossary	242
Appendix 1: Separating solids from mixtures	246
Appendix 2: Distillation and chromatography	248
Appendix 3: Working with gases in the lab	250
Appendix 4: Testing for ions in the lab	252
Appendix 5: Maths toolkit	254
Appendix 6: Periodic table and atomic masses	256
Appendix 7: Safety first	258
Numerical answers	259
Index	260

Routemap

Although you will probably follow the AQA specification module by module, this 'routemap' shows you different ways to work through the material. It is designed to help you understand the subject in small, manageable chunks by suggesting groupings of spreads relevant to particular topics and highlighting connections between them. It is especially useful when you are revising because it helps you to identify and revise logical sections of material at a time, and if you have missed any work, it can also help you to catch up.

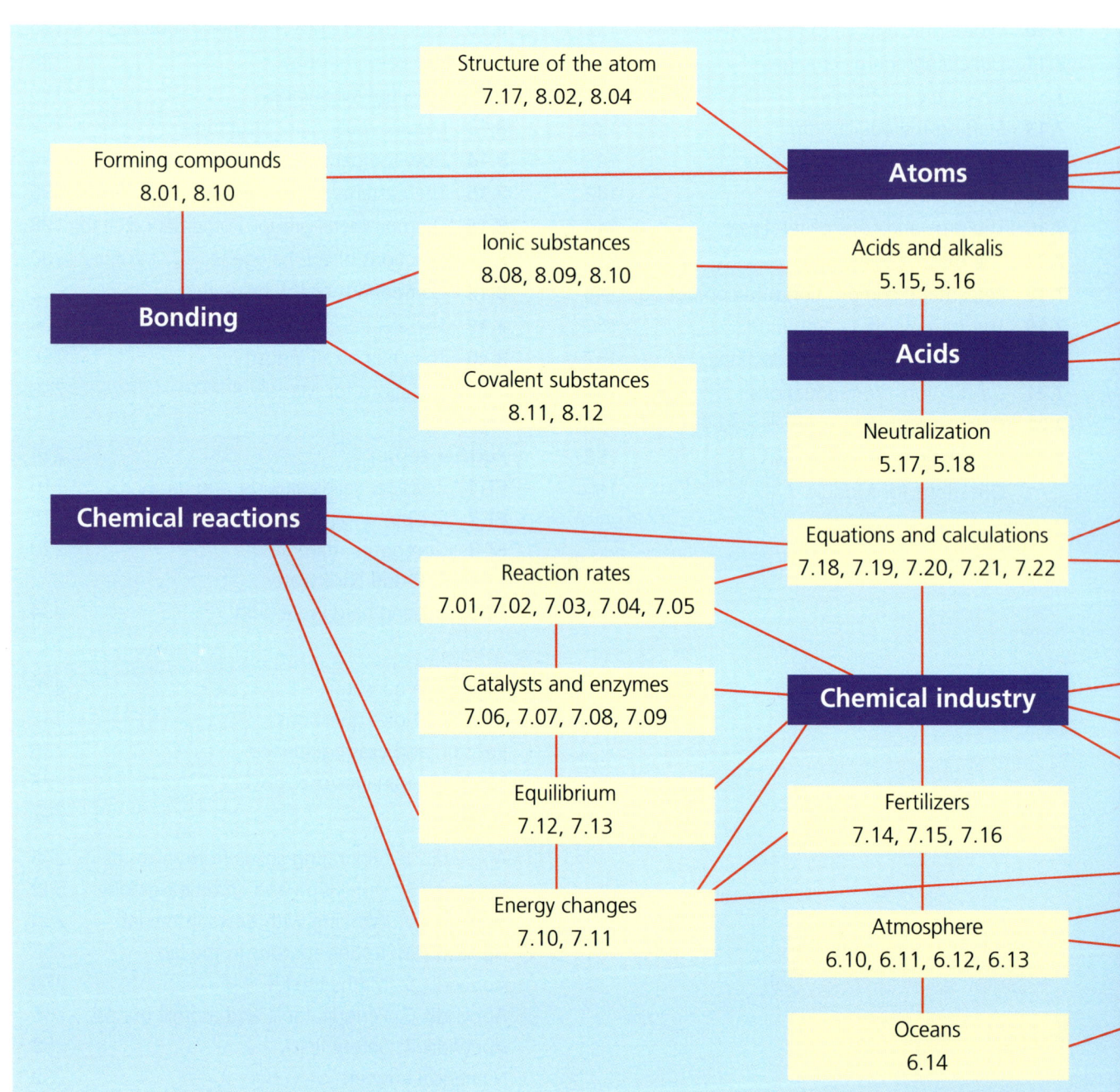

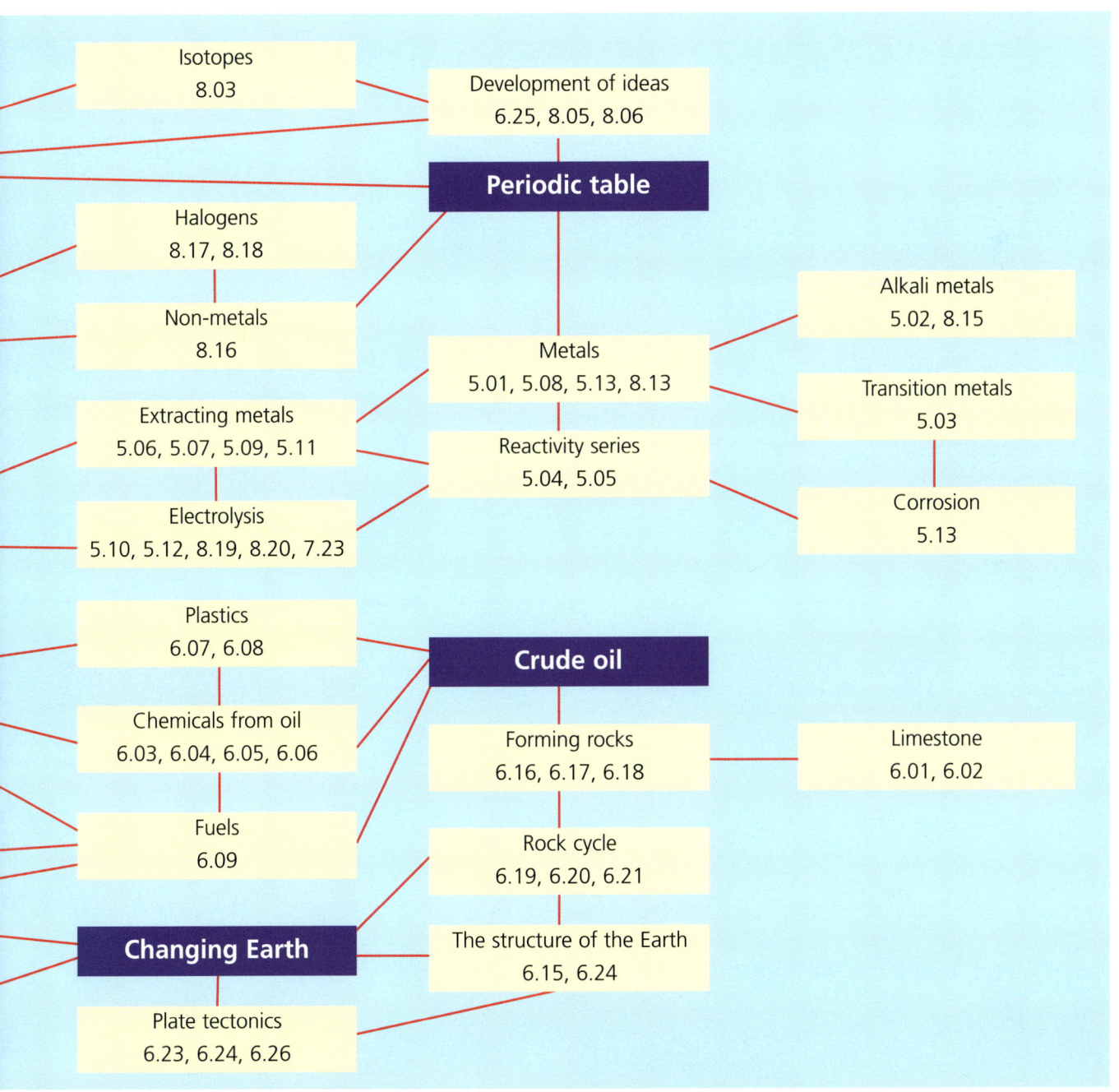

Module 5
Metals

The elements are the ingredients from which everything in the world is made. Over three-quarters of them are metals! The two most common metals in the Earth's crust are aluminium and iron, in that order.

We think of metals as strong, hard, and tough – great for building outdoor structures. But they are not all like that. Some are weak enough to pull apart and so reactive you have to keep them well away from air and water.

Iron is the world's most widely used metal, usually in the form of steel. Before it can be used, its ore has to be dug up and the iron extracted. Iron extraction was common as long as 3000 years ago. The first blast furnace was built in the 14th century and gave a tonne of iron a day. Modern blast furnaces produce over 13 000 tonnes a day, but the basic process hasn't really changed much.

In this module you will learn a great deal about metals: what properties they share, how they compare in reactivity, how they are extracted and used, and how to make their salts. ∎

5.01 Metals and the periodic table

Elements: a reminder

Everything in the world is made from **elements**. They combine with each other to give millions of compounds. Your body is built up from over 40 different elements – including enough iron to make a big nail!

Most of the elements are **metals**. There are over 80 of these. Chances are you've seen and touched a good many of them already. Think of the iron in gates and railings, aluminium cooking foil, the tin coating on 'tin' cans, silver and gold in jewellery, and magnesium in the lab.

So what exactly are metals?

Metals are really just the elements that share these properties:
- good conductors of heat and electricity.
- tend to have high melting points, so are solid at room temperature.
- tend to be hard, and malleable (they can be bent into different shapes), and ductile (they can be drawn out into wires).
- react to form compounds with that other group of elements, the non-metals. For example many metals react with oxygen.

It's elemental! The human body is built from 4 main elements – carbon, hydrogen, nitrogen, oxygen – and smaller amounts or just traces of 37 others. The main metal is calcium – 1.5% of body weight.

A big variety

But given that there are 80 metals, it is not surprising they vary a lot:

Sodium is so soft you can cut it with a knife, and it melts at only 98 °C. It floats on water – and reacts vigorously with it.

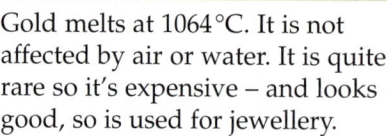

Gold melts at 1064 °C. It is not affected by air or water. It is quite rare so it's expensive – and looks good, so is used for jewellery.

We think of iron as a typical metal – malleable, strong, heavy. It melts at 1530 °C. But it will rust away if you don't keep it painted.

Metals and the periodic table

We humans have known about some metals, such as gold, for centuries. But nearly three-quarters of the metals we know today were discovered in just the last 200 years.

As more and more metals were found and studied, chemists were puzzled by one thing. Some seemed to resemble each other closely – they fell naturally into groups. Why was this? What made them similar? So chemists set off on the long search for the key to the puzzle. And the result is that famous summary of the elements: **the periodic table**.

10

The periodic table

[Periodic table shown with Groups 1-7 and 0, periods 1-7, and the lanthanide/actinide series below]

A quick reminder:

- The elements are arranged in order of increasing atomic number.
- The zig-zag line separates the metals from the non-metals in the table, with the metals on the left.
- The numbered columns are called **groups**. The number tells you how many outer electrons the atoms of that group have.
  (The 0 in Group 0 means a full shell.)
- The elements in a group behave in a similar way.
- The numbered rows are **periods**. This time the number tells you how many electron shells there are.

The metal groups and blocks

Note the two metal groups to the left. Group 1, the **alkali metals**, are called this because they react with water to give an alkaline solution. Chemists once used the word **earth** for an oxide that was not very soluble – so Group 2 was called the **alkaline earths**. (Their oxides are not very soluble in water, but give an alkaline solution.)

Look at the block of transition metals. These metals have a lot in common with each other, but are quite different from the Group 1 and 2 metals. For a start, they are much less reactive.

Some of the metals in the table are very common, and we use them a lot. For example aluminium makes up 8% of the Earth's crust, and iron 6%, and we use them everywhere. But some are very rare, and you will probably never see them, or even say their names!

One of the elements in the periodic table is named after this famous scientist. Which one?

Questions

1. You are given a sample of an element. Write down three clues that would tell you it was a metal.
2. Metals tend to be solid at room temperature. Name one that is not.
3. Think of one way in which gold behaves:
   a like iron   **b** differently from iron
4. Look at the metals in the periodic table.
   a Pick out the ones you have seen in real life, and write down their symbols.
   b Pick out three you think are named after:
   i places
   ii people

11

5.02 Group 1: the alkali metals

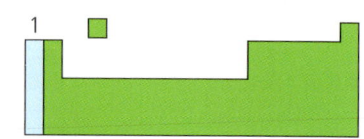

What are they?

The alkali metals are the six metals of Group 1 in the periodic table: lithium, sodium, potassium, rubidium, caesium, and francium. Only the first three are safe enough to keep in the school lab, so we concentrate on them in this unit:

Lithium   Li (2,1)   Sodium   Na (2,8,1)   Potassium   K (2,8,8,1)

Rubidium and caesium are dangerous. If they are added to water in a glass trough the violent reaction will shatter the glass. Francium is even more dangerous – and **radioactive**. Hardly any of it exists.

Their physical properties

The alkali metals are not typical metals.
- Like all metals, they *are* good conductors of heat and electricity.
- They are softer than most other metals. The samples shown above were cut with a knife.
- They are 'lighter' than most other metals – they have **low density**. The three metals above float in water (and immediately react with it).
- They have low melting and boiling points compared with most metals.

density of metal less than 1 g/cm³ so it floats (example: sodium)

water, density 1 g/cm³

density of metal greater than 1 g/cm³ so it sinks (example: iron)

Their chemical properties

The alkali metals are the most reactive of all metals. Here are some of their main reactions.

Reaction with water

They react violently with water, giving hydrogen and a hydroxide.

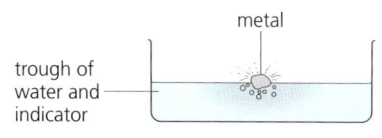

What you see
- lithium floats and fizzes
- sodium shoots across the water
- potassium melts with the heat of the reaction and the hydrogen catches fire

increasing reactivity

For sodium the reaction is:

sodium + water → sodium hydroxide + hydrogen

The hydrogen bubbles off. The hydroxide is alkaline so the indicator changes colour.

All the alkali metals react vigorously with water, releasing hydrogen gas and forming hydroxides. The hydroxides give alkaline solutions.

Reaction with non-metals

1 **With chlorine** Put heated alkali metals into chlorine gas – and they will burst into flame. They burn brightly, giving white solids called **chlorides**. The reaction with sodium is:

   sodium + chlorine → sodium chloride

Sodium chloride is also known as common salt (or table salt at home). It dissolves in water to give a colourless solution.

2 **With oxygen** They also burst into flame when you heat them and place them in oxygen, and burn even more fiercely, forming white solids called **oxides**. This time the flames have different colours:

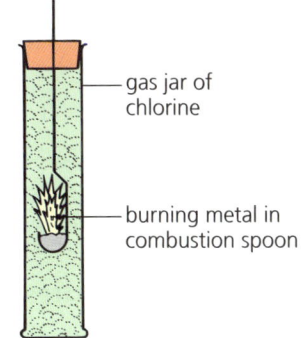

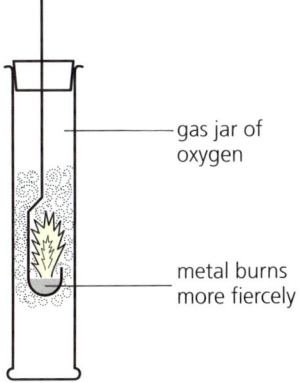

What you see
- a red flame for lithium
- a yellow flame for sodium
- a lilac flame for potassium

The reaction with sodium is:

   sodium + oxygen → sodium oxide

The oxide dissolves in water to give a colourless solution.

Ionic compounds

The compounds formed in these reactions are **ionic**. Sodium chloride is made of sodium ions (Na^+) and chloride ions (Cl^-). Sodium oxide is made of sodium ions and oxide ions (O^{2-}). So, during the reaction, sodium atoms have become sodium ions with a charge of 1+.

It's the same for the other alkali metals:
The alkali metals form ionic compounds in which the metal ion has a charge of 1+. The compounds are white solids that dissolve in water to form colourless solutions.

Because it reacts with oxygen, sodium is stored under oil.

Questions

1 Name the Group 1 elements and give their symbols.
2 What is the family name for the Group 1 elements? Why are they called that?
3 Which best describes the Group 1 metals:
   a soft or hard?
   b high density or low density?
   c high melting point or low melting point?
   d reactive or unreactive, with water?
4 a Write a word equation for the reaction of lithium with water.
   b The pH of the water changes. Why?
5 a What forms when potassium reacts with chlorine?
   b What colour is this compound?
   c What will you *see* when you dissolve it in water?
   d Do you think the solution will conduct electricity? Give a reason for your answer.

13

5.03 The transition metals

What are they?
The transition metals are the middle block of 30 elements in the periodic table. They include most of the metals you find in everyday use in the kitchen, in school, and around town.

Their physical properties
Here are three of the transition metals:

Iron: the most widely used; grey with a metallic lustre (shine)

Copper: reddish with a metallic lustre

Nickel: silvery with a metallic lustre

Here is some data for them, with sodium for comparison:

Metal	Symbol	Density / g/cm^3	Melts at / °C
Iron	Fe	7.9	1535
Copper	Cu	8.9	1083
Nickel	Ni	8.9	1455
Sodium	Na	0.97	98

The transition metals have these physical properties:

- **hard, tough and strong.** You could not cut them with a knife, as you can the Group 1 metals.
- **high melting points.** Look at the values in the table. But mercury is an exception. It is a liquid at room temperature. It melts at −39 °C.
- **malleable and ductile.** Look at the photo of a copper roof.
- **good conductors of heat and electricity.** Silver is the best electrical conductor of them all, and copper is next.
- **high density.** Unlike sodium, they sink in water, since their density is greater than the density of water (1 g/cm^3).

Some transition metals

iron	copper
nickel	zinc
silver	gold
platinum	mercury
chromium	cadmium

Transition metal compounds provide the colours in these pottery glazes.

Copper has often been used for roofing, partly because it weathers to a pleasant green colour over the years.

The colour is due to a very thin layer of copper(II) sulphate, formed by the action of sulphur dioxide and damp air. This layer prevents further corrosion, so copper roofs endure well.

Their chemical properties

- **They are much less reactive than the Group 1 metals.** For example:

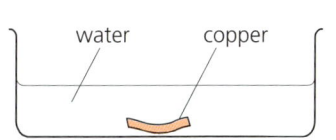

Copper does not react with water. (Sodium reacts vigorously.)

Copper does not burn in air. (Sodium burns vigorously.)

Salts of transition metals

- You can make salts of the transition metals by reacting their oxides or hydroxides with acids.
- The oxides and hydroxides of transition metals are **bases**.

Their low reactivity means they do not corrode very readily in air or water. But iron is an exception. It rusts very easily, and we spend a fortune every year on rust prevention.

- **Most of them form coloured compounds.** (Most compounds of Group 1 and 2 metals are white.)

Their uses

The properties of the transition metals lead to a wide range of uses.

- Their hardness and strength make them suitable for building structures. Iron is used by far the most for this. (It is abundant and is not too expensive.) It is usually used in the form of **steels**. These are **alloys** of iron – iron mixed with other substances to improve its properties.
- They are used for things that must let heat or a current pass through easily. Steel is used for radiators. Copper is used for electric cables.
- Many transition metals and their compounds acts as **catalysts** – they speed reactions up without being chemically changed themselves. Iron is a catalyst in the **Haber process** to make ammonia from nitrogen and hydrogen. Nickel is a catalyst for the **hydrogenation** of vegetable oils to margarine:

vegetable oils (liquid) + hydrogen $\xrightarrow{\text{nickel as catalyst}}$ margarine (soft solid)

The colours in stained glass also come from transition metal compounds.

Questions

1 Name five transition metals.
2 Which best describes the transition metals:
   a soft or hard? b high density or low density?
   c high melting point or low melting point?
   d reactive or unreactive with water?
3 What is unusual about mercury?
4 Most paints contain compounds of transition elements. Why do you think this is?
5 Suggest reasons why copper is used in hot water pipes while iron is not.

15

5.04 Metals and reactivity

Comparing reactivity

Group 1 metals are very reactive, and Group 2 metals a bit less so. But how do they compare with other metals? To find out you could test different metals with water, oxygen, and hydrochloric acid (where this is safe to use) to see *if* they react, and how fast.

The more reactive the metal, the faster and more violent its reactions will be.

Reactions with oxygen

This table shows what happened when a range of metals was heated in air.

Metal	Behaviour when heated in air	Order of reactivity	Product
Sodium	Catches fire with only a little heating. Burns with a bright yellow flame	most reactive	Sodium peroxide, Na_2O_2 a pale yellow powder
Magnesium	Catches fire easily. Burns with a blinding white flame		Magnesium oxide, MgO a white powder
Iron	Does not burn, but the hot metal glows and gives off yellow sparks		Iron oxide, Fe_3O_4 a black powder
Copper	Does not burn, but the hot metal becomes coated with a black substance		Copper oxide, CuO a black powder
Gold	No reaction, no matter how much the metal is heated	least reactive	————

Where the metal does react, it forms an **oxide**. Look at sodium, top of the reactivity league. Now look at gold.

Reactions with water

Compare these:

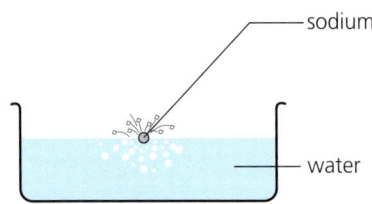

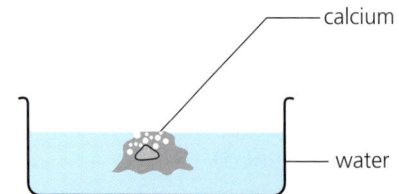

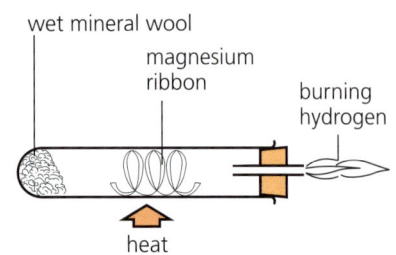

Sodium reacts violently with cold water, whizzing over the surface. Hydrogen gas and a clear solution of sodium hydroxide are formed.

The reaction between calcium and cold water is slower. Hydrogen bubbles off, and a cloudy solution of calcium hydroxide forms.

Magnesium reacts very slowly with cold water, but vigorously when heated in steam: it glows brightly. Hydrogen and solid magnesium oxide are formed.

Other metals were compared in their reaction with water. Look at the results.

Metal	Reaction with water	Order of reactivity	Product
Potassium	Very violent with cold water. Catches fire	most reactive ↑	Hydrogen and a solution of potassium hydroxide, KOH
Sodium	Violent with cold water		Hydrogen and a solution of sodium hydroxide, NaOH
Calcium	Less violent with cold water		Hydrogen and calcium hydroxide, $Ca(OH)_2$, which is only slightly soluble
Magnesium	Very slow with cold water, but vigorous with steam		Hydrogen and solid magnesium oxide, MgO
Zinc	Quite slow with steam		Hydrogen and solid zinc oxide, ZnO
Iron	Slow with steam		Hydrogen and solid iron oxide, Fe_3O_4
Copper Gold	No reaction	least reactive	————

Notice how the three most reactive of these metals produce **hydroxides** when they react. The others produce **oxides** – if they react at all.

Reactions with hydrochloric acid

This table shows the results of a test with hydrochloric acid.

Metal	Reaction with hydrochloric acid	Order of reactivity	Product
Magnesium	Vigorous	most reactive ↑	Hydrogen and a solution of magnesium chloride, $MgCl_2$
Zinc	Quite slow		Hydrogen and a solution of zinc chloride, $ZnCl_2$
Iron	Slow		Hydrogen and a solution of iron(II) chloride, $FeCl_2$
Lead	Slow, and only if the acid is concentrated		Hydrogen and a solution of lead(II) chloride, $PbCl_2$
Copper Gold	No reaction, even with concentrated acid	least reactive	

This time if a reaction occurs it produces a salt – a **chloride**.

Questions

1 Write word equations for the reactions of iron with oxygen, water, and hydrochloric acid.
2 Write word equations for the reactions of potassium with water, and magnesium with steam.
3 No attempt was made to test sodium with hydrochloric acid. Why not?
4 Now compare the three tables.
  a Which two metals come bottom of each list?
  b Which seems to be more reactive:
    i zinc or iron? ii magnesium or zinc?
5 From the tests, which appear to be least reactive, Group 1 metals, Group 2 metals, or transition metals?

5.05 The reactivity series

What is the reactivity series?

In the last spread we compared different metals to see how reactive they are. This table shows the final results. It is called the **reactivity series**.

Potassium, K	most reactive
Sodium, Na	
Calcium, Ca	
Magnesium, Mg	
Aluminium, Al	
Carbon	↑ increasing reactivity
Zinc, Zn	
Iron, Fe	
Lead, Pb	
Hydrogen	
Copper, Cu	
Silver, Ag	
Gold, Au	least reactive

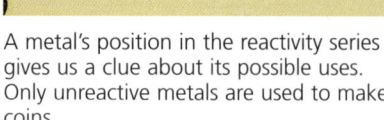

A metal's position in the reactivity series gives us a clue about its possible uses. Only unreactive metals are used to make coins.

Carbon and hydrogen are non-metals, but they have been included here for comparison. Note that the metals above hydrogen react with acids, displacing hydrogen. Metals below hydrogen *do not* react with acids.

Making use of the reactivity series

The reactivity series helps you predict what will happen in reactions. The more reactive a metal, the more it 'likes' to exist as compounds. Iron is above copper, which means it is more reactive than copper. So if those two compete to be compounds, iron will win as you'll see below.

1 Competing for oxygen

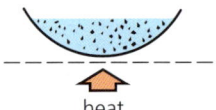
heat

This is a mixture of powdered iron and copper(II) oxide. On heating, the reaction starts.

The mixture glows, even after the Bunsen is removed. Iron(II) oxide and copper are formed.

Since the mixture keeps glowing without the Bunsen, the reaction must be giving out heat. It is **exothermic**. In the reaction, the iron competes with copper for oxygen – and wins:

iron + copper(II) oxide ⟶ iron(II) oxide + copper
Fe (s) + CuO (s) ⟶ FeO (s) + Cu (s)

When it takes away the oxygen from copper, iron acts as a **reducing agent**. Other metals can do the same.
When a metal is heated with the oxide of a less reactive metal, it takes the oxygen from it. The reaction is exothermic.

Using a competition reaction to repair railway lines. Aluminium and iron(III) oxide are being heated together, to give molten iron. This is run into gaps between rails. The process is called the Thermit process.

18

2 Competing to be in solution

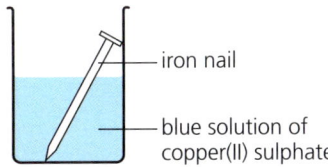
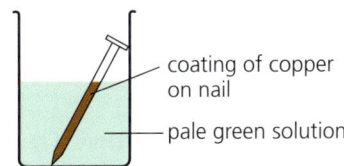

This time, an iron nail is placed in some blue copper(II) sulphate solution.

Soon a coat of copper appears on the nail. The solution turns pale green.

Here iron and copper compete to be the compound in solution. Once again iron wins. It drives out or **displaces** copper from the copper(II) sulphate solution. Green iron(II) sulphate forms:

iron + copper(II) sulphate ⟶ iron(II) sulphate + copper
$Fe(s)$ + $CuSO_4(aq)$ ⟶ $FeSO_4(aq)$ + $Cu(s)$
        blue   green

Other metals displace less reactive metals in the same way.
A metal will always displace a less reactive metal from solutions of its compounds. This is called a displacement reaction.

Now look at the way copper reacts with silver nitrate solution:

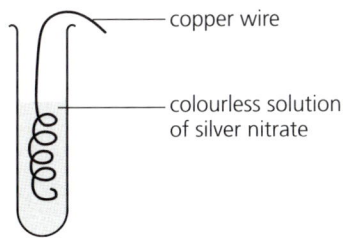
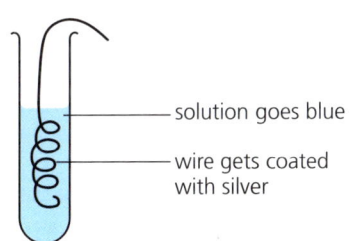

A coil of copper is placed in a solution of silver nitrate. The copper gradually dissolves . . .

. . . so the solution goes blue. At the same time, a coating of silver forms on the wire.

In this reaction copper displaces the silver from silver nitrate because copper is more reactive than silver:

copper + silver nitrate ⟶ copper(II) nitrate + silver
$Cu(s)$ + $2AgNO_3(aq)$ ⟶ $Cu(NO_3)_2(aq)$ + $2Ag(s)$
                               blue

Crystals of silver on a copper wire. The copper has displaced the silver from silver nitrate solution.

Questions

The table on the opposite page will help you answer some of these questions.
1 Describe how iron and lead react with hydrochloric acid. Write word equations for the reactions.
2 Write a rule for the reactions of a metal with:
   a oxides of other metals
   b solutions of compounds of other metals
3 Will copper react with lead(II) oxide? Explain why.
4 Will iron react with lead(II) oxide? If *yes*, write a word equation for the reaction.
5 What would you *see* when zinc is added to copper(II) sulphate solution? (Zinc sulphate is colourless.)
6 Explain how the Thermit process works.
7 Tin does not react with iron(II) oxide. But it reduces lead(II) oxide to lead. Arrange tin, iron, and lead in order of decreasing reactivity.

5.06 Metals in the Earth's crust

The composition of the Earth's crust

We get some metals from the sea, but most from the Earth's **crust**. The crust is the outer layer of the Earth. It is a huge mixture of many different compounds. It also contains *elements* such as sulphur, copper, silver, platinum, and gold. These occur uncombined, or **native**, because they are unreactive.

If you could dig up the Earth's crust and break down all the compounds, you would find it is almost half oxygen! Its composition is:

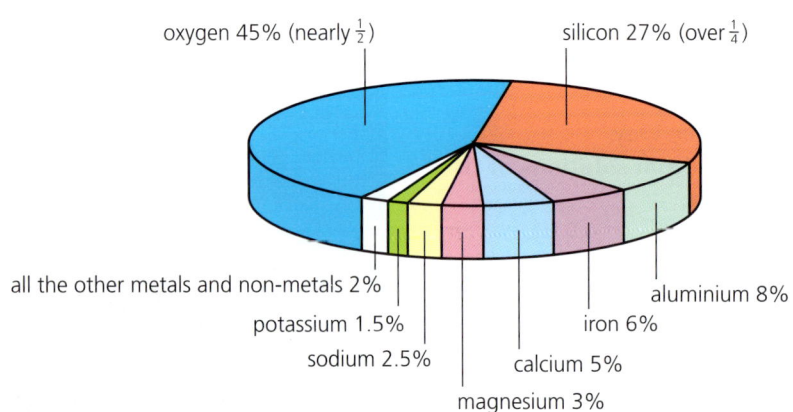

oxygen 45% (nearly ½)
silicon 27% (over ¼)
all the other metals and non-metals 2%
potassium 1.5%
sodium 2.5%
magnesium 3%
calcium 5%
iron 6%
aluminium 8%

We use about nine times more iron than all the other metals put together – for everything from bridges to needles.

Nearly three-quarters of the Earth's crust is made of just two non-metals, oxygen and silicon. These are found combined in compounds such as silicon dioxide (**silica** or **sand**). Oxygen also occurs in compounds such as aluminium oxide, iron oxide, and calcium carbonate.

The rest of the crust consists mainly of just six metals. Aluminium is the most abundant of these, and iron is next. None of the six is found uncombined, because they are all reactive metals, and reactive metals 'like' to form compounds.

Some metals are scarce

Look again at the pie chart above. It shows that there are just six plentiful metals. All the other metals *together* make up less than one-fiftieth of the Earth's crust, so they are not plentiful.

If a metal makes up less than one-thousandth of the Earth's crust, it is called **scarce**. Here are some of the scarce metals:

copper	zinc	lead	tin
mercury	silver	gold	platinum

The scarce metals are expensive. Even so, we are using them up quickly, and many people are worried that they will soon run out.

Light strong metals such as aluminium and titanium make space travel possible.

Metal ores

The rocks in the Earth's crust are a mixture of many different compounds. But some contain a large amount of a single metal compound, or a single metal, and it may be worth digging these up to extract the metal. Rocks from which metals are obtained are called **ores**. Below are some examples:

This is a chunk of **rock salt**, the main ore of sodium. It is mostly **sodium chloride**.

This is a piece of **bauxite**, the main ore of aluminium. It is mostly **aluminium oxide**.

Since gold is unreactive, it is found as a free element. This sample is almost pure gold.

To mine or not to mine?

Before starting to mine an ore, the mining company must decide whether it is economical. It must find the answers to questions like these:

1 How much ore is there?
2 How much metal will we get from it?
3 Are there any special problems about getting the ore out?
4 How much will it cost to mine the ore and extract the metal from it? (The cost will include roads, buildings, mining equipment, extraction plant, transport, fuel, chemicals, and wages.)
5 How much will we be able to sell the metal for?
6 So will we make a profit if we go ahead?

The answers to these questions will change from year to year. If the cost of fuel drops, and the selling price of a metal rises, even a low-quality or **low-grade** ore may become worth mining.

The mining company must also think about the local community, who may worry that the landscape will be ruined by dust, pits, and scars, and that the extraction plant will cause pollution. On the other hand, they may welcome the new jobs that mining will bring.

A landscape spoiled by mining. These days, mining companies are often forced to clear and restore the land when a mine is exhausted. That is another cost to consider.

Questions

1 Which is the main *element* in the Earth's crust?
2 Which is the most common metal in the Earth's crust? Which is the second?
3 Gold occurs *native* in the Earth's crust. Explain.
4 Is it true that the reactive metals are plentiful in the Earth's crust?
5 What is a *scarce* metal? Name four.
6 One metal is used more than all the others put together. Which one?
7 What is a metal ore?
8 Name the main ore of: **a** sodium **b** aluminium What is the main compound in each ore?

5.07 Extracting metals from their ores

Getting the metal out

An **ore** is rock from which a metal is obtained. After mining the ore the next step is to **extract** the metal from it. This can be quite easy or very difficult, depending on how **reactive** the metal is.

Native metals   A few metals – the most unreactive ones like silver, gold, platinum, and some copper – are found as **elements**. The metal is obtained by separating it from its impurities, just like removing stones from soil. The process does not involve any chemical reactions.

Other metals   The other metals exist as **compounds** in their ores. This means the metal has to be extracted using a chemical reaction.

Extraction by removing oxygen

Often the metal compound is an **oxide** or a compound that can easily be converted to an oxide. The metal is obtained by removing the oxygen from the oxide. This process is called **reduction**.

Reduction is the removal of oxygen. A substance that can remove oxygen is called a reducing agent.

$$\text{metal oxide} \xrightarrow[\text{(oxygen removed using a reducing agent)}]{\text{reduction}} \text{metal}$$
(ore)

Carbon and carbon monoxide are often used as reducing agents. As the box on the right shows, carbon is more reactive than zinc. So carbon and carbon monoxide can remove oxygen from zinc, iron, and lead oxides.

For example carbon monoxide is used to extract iron from iron ore:

iron(III) oxide + carbon monoxide → iron + carbon dioxide
$Fe_2O_3 (s)$ + $3CO (g)$ → $2Fe (l)$ + $3CO_2 (g)$

Extraction by electrolysis

Electrolysis means *splitting by electricity*. It is the most powerful way to extract a metal from its compounds. First the compound is melted or dissolved so that the ions are free to move. Then a current is applied to separate them. Since electricity is expensive, this method is used only for more reactive metals that are hard to extract in other ways, such as aluminium:

aluminium oxide → aluminium + oxygen
$2Al_2O_3 (l)$ → $4Al (l)$ + $3O_2 (g)$

Again oxygen is removed, which means reduction takes place. But how? First the aluminium oxide is separated into aluminium and oxide ions. Then the aluminium ions *gain electrons* to become aluminium atoms:

$Al^{3+} + 3e^- \rightarrow Al$

So this brings us to a wider definition of reduction:
Reduction is a gain of electrons.

Metals are a **non-renewable resource**. Once we dig the ore from the ground, no new ore forms. So it makes sense to **recycle** used metal – like the millions of empty food tins we throw out.

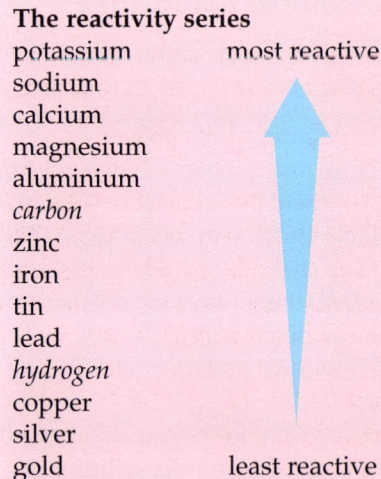

The reactivity series
potassium   most reactive
sodium
calcium
magnesium
aluminium
carbon
zinc
iron
tin
lead
hydrogen
copper
silver
gold   least reactive

Carbon and hydrogen are included for comparison. Both will reduce the oxides of metals below them in the list. For example, hydrogen will reduce copper(II) oxide, giving copper and water.

22

A summary of extraction methods

Metal			Method of extraction from ore		
Potassium Sodium Calcium Magnesium Aluminium	metals more reactive	ores more difficult to decompose	Electrolysis	method of extraction more powerful	method of extraction more expensive
Zinc Iron Lead			Heating with carbon or carbon monoxide		
Silver Gold			Occur naturally as elements so no chemical reaction needed		

So the method of extraction depends on the position of the metal in the reactivity series.

Reduction is the opposite of oxidation

Look again at the reduction of iron(III) oxide:

$$Fe_2O_3\ (s) + 3CO\ (g) \rightarrow 2Fe\ (l) + 3CO_2\ (g)$$

The iron(III) oxide lost oxygen. It was **reduced**.
But the carbon monoxide *gained* oxygen. It was **oxidized**.
Reduction is the opposite of oxidation. They *always* take place together.

Reduction and oxidation in electrolysis

As you saw on the last page, aluminium oxide is separated into aluminium and oxide ions, ready for electrolysis. The aluminium ions then gain electrons. They are **reduced**.

But oxidation and reduction *always* take place together. So let's look at what happens the oxide ions. They *lose* electrons:

$$O^{2-} - 2e^- \rightarrow O \quad \text{then} \quad O + O \rightarrow O_2$$

And this brings us to a wider definition for oxidation:
Oxidation is the loss of electrons.
The box on the right will help you to remember it.

Redox reactions

Since oxidation and reduction always take place together, the overall reaction is called a **redox** reaction. (**Red**uction plus **ox**idation.)

Oxidation and reduction – they always take place together!

iron(III) oxide + carbon monoxide

reduction   oxidation

iron + carbon dioxide

The reaction is a **redox** reaction.

LEARN AND REMEMBER!

Oxidation **I**s **L**oss of electrons

OILRIG

Reduction **I**s **G**ain of electrons

Questions

1 Why is no chemical reaction needed to get gold?
2 Lead is extracted by heating its oxide with carbon:
   lead oxide + carbon → lead + carbon monoxide
   a Why can carbon be used for this reaction?
   b One substance is *reduced*. Which one?
   c Which substance is the reducing agent?
   d Which substance is oxidized during the reaction?
3 Give *two* definitions for each of these:
   a reduction   b oxidation
4 The reaction in question 2 is a *redox* reaction. Why?
5 Sodium is extracted from rock salt (sodium chloride).
   a Electrolysis is needed for this. Explain why.
   b Write a word equation for the reaction.
   c Say which ions are oxidized, and which reduced.

5.08 Making use of metals

Pure metals and alloys

The way a metal is used depends on its **properties**:

Pure aluminium can be rolled into very thin sheets, which are quite strong but easily cut. So it is used for milk bottle tops and cooking foil.

Pure lead is soft, and bends easily without being heated. It also resists corrosion. So it is used to seal off brickwork around chimneys.

Pure copper is easily drawn into wires, and is an excellent conductor of electricity. So it is used for electrical wiring around the home.

Sometimes a metal is most useful when it is pure. For example copper is not nearly such a good conductor when it contains impurities. But many metals are most useful when they are *not* pure. Iron is almost never used pure:

Pure iron is no good for building things, because it is too soft and stretches easily, as you can see in the photo above. Besides, it rusts easily too.

But when a little carbon (0.5%) is mixed with it, the result is **mild steel**. This is hard and strong. It is used for buildings, bridges, ships, and car bodies.

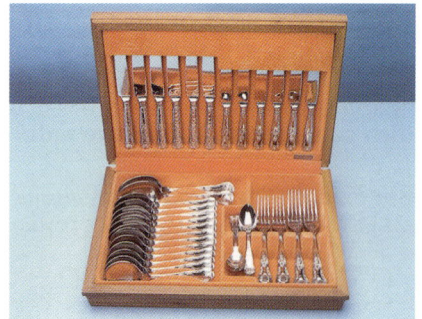

When nickel and chromium are mixed with iron, the result is **stainless steel**. This is hard and rustproof. It is used for car parts, kitchen sinks, and cutlery.

You can see that the properties of the iron have been changed by mixing other substances with it.

The properties of any metal can be changed by mixing other substances with it. The mixtures are called alloys.

The added substances are usually metals, but sometimes non-metals like carbon or silicon. An alloy is usually made by melting the main metal and then dissolving the other substances in it.

Turning a metal into an alloy increases its range of uses.

24

Uses of pure metals

This table summarizes some of the uses of pure metals:

Metal	Uses	Properties that make it suitable
Sodium	A coolant in nuclear reactors	Conducts heat well. Melts at only 98 °C, so the hot metal will flow along pipes.
	Extraction of titanium	Is more reactive than titanium and melts easily.
Aluminium	Overhead electricity cables (with a steel core for strength)	A good conductor of electricity (not as good as copper, but cheaper and much lighter); resists corrosion.
	CDs and CD-ROMs	Provides a cheap reflective coating.
Zinc	Coating iron, to give **galvanized** iron	Protects the iron from rusting.
Tin	Coating steel cans or 'tins'	Unreactive and non-toxic. Protects the steel from rusting.
Nickel	Electroplating steel	Resists corrosion, sticks well to steel, shiny and attractive to look at.
Titanium	Teeth implants and replacement hip joints	Light, strong, resists corrosion, non-toxic, and ductile so can be easily shaped.
	Tailpipes for aircraft	Light, strong, resists corrosion, ductile.

Uses of alloys

There are thousands of different alloys. Here are just a few!

Alloy	Made from	Special properties	Uses
Cupronickel	75% copper 25% nickel	Hard-wearing, attractive silver colour	'Silver' coins
Stainless steel	70% iron 20% chromium 10% nickel	Does not rust	Car parts, kitchen sinks, cutlery, tanks and pipes in chemical factories
Manganese steel (Hadfield steel)	85% iron 13.8% manganese 1.2% carbon	Very hard	Springs
A titanium alloy	92.5% titanium 5% aluminium 2.5% tin	High strength at high temperatures	Jet engine components
Brass	70% copper 30% zinc	Harder than copper, does not corrode	Musical instruments
Bronze	95% copper 5% tin	Harder than brass, does not corrode, chimes when struck	Statues, ornaments, church bells
Solder	70% tin 30% lead	Low melting point	Joining wires and pipes

Questions

1 Why is iron more useful when it is mixed with a little carbon?
2 What are *alloys*? How are they made?
3 Explain why tin is used to coat food cans.
4 Name an alloy that:
  a has a low melting point   b never rusts
5 Which metals are used to make:
  a stainless steel?   b brass?   c bronze?

5.09 Extracting iron

The blast furnace

This diagram shows the **blast furnace** used for extracting iron from its ore. It is an oven shaped like a chimney, at least 30 metres tall.

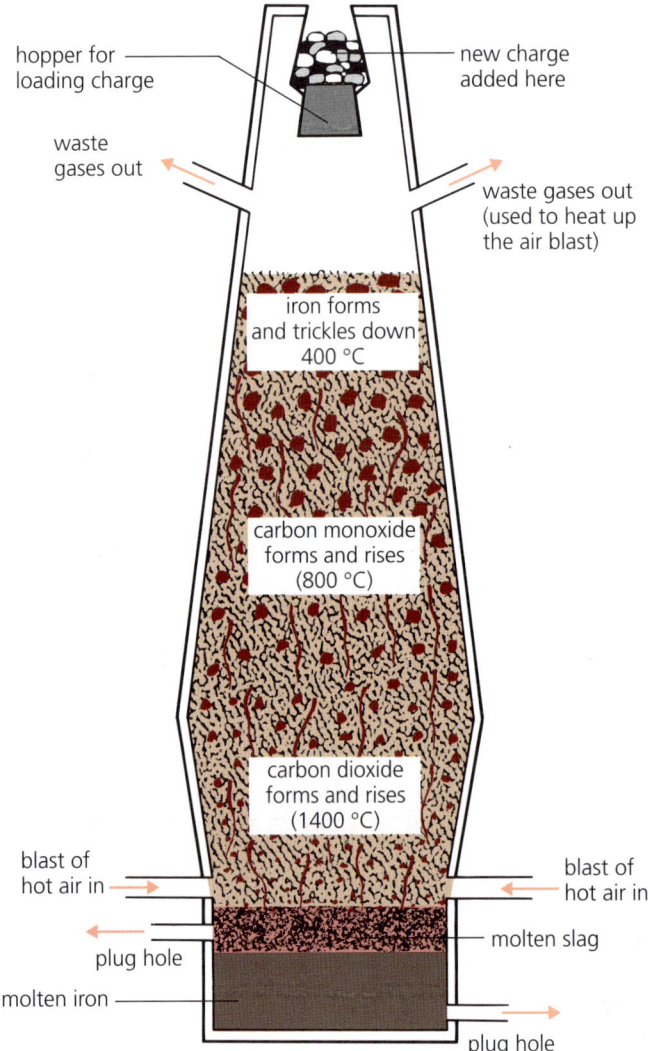

Blast furnaces run non-stop 24 hours a day.

A mixture called the **charge**, containing the iron ore, is added through the top of the furnace. Hot air is blasted in through the bottom. After a series of reactions, liquid iron collects at the bottom of the furnace.

What's in the charge?

The charge contains three things:

1 Iron ore. The chief ore of iron is called **haematite**. It is mainly iron(III) oxide, Fe_2O_3, mixed with sand.

2 Limestone. This is a common rock. It is mainly calcium carbonate, $CaCO_3$.

3 Coke. This is made from coal, and is almost pure carbon.

A stockpile of iron ore.

The reactions in the blast furnace

Reactions, products, and waste gases	Comments
Stage 1: The coke burns, giving off heat The blast of hot air starts the coke burning. It reacts with the oxygen in the air, giving carbon dioxide: carbon + oxygen → carbon dioxide $C(s) + O_2(g) \rightarrow CO_2(g)$	This is a **combustion** reaction (or burning). It is **exothermic** – it gives off heat which then heats the furnace. The blast of air provides the oxygen for the reaction. (Air is mainly nitrogen but this does not take part in the reaction.)
Stage 2: Carbon monoxide is made The carbon dioxide reacts with more coke, giving carbon monoxide: carbon + carbon dioxide → carbon monoxide $C(s) + CO_2(g) \rightarrow 2CO(g)$	In this reaction the carbon dioxide loses oxygen. It is **reduced**. The reaction is **endothermic** – it takes in heat. That's good because the next stage – the reduction with carbon monoxide – requires a lower temperature.
Stage 3: The iron(III) oxide is reduced Carbon monoxide reacts with the iron(III) oxide in the ore, giving liquid iron: iron(III) oxide + carbon monoxide → iron + carbon dioxide $Fe_2O_3(s) + 3CO(g) \rightarrow 2Fe(l) + 3CO_2(g)$ The iron trickles to the bottom of the furnace.	Carbon monoxide acts as a **reducing agent**. It removes oxygen from the iron(III) oxide, giving the metal. So the iron oxide is **reduced**. But at the same time the carbon monoxide gets **oxidized** (oxygen is added to it). So the overall reaction is a **redox** reaction.
The limestone reaction Limestone reacts with sand (silica) in the ore, to form **calcium silicate** or **slag**. limestone + silica → calcium silicate + carbon dioxide $CaCO_3(s) + SiO_2(s) \rightarrow CaSiO_3(s) + CO_2(g)$ The slag runs down the furnace and floats on the iron.	The purpose of this reaction is to remove impurities. Silica is acidic. The reaction of limestone with it is a **neutralization reaction**. The molten slag is drained off. When it solidifies it is sold, mostly for road building.
Waste gases These are **carbon dioxide** and **nitrogen**. They come out at the top of the furnace.	The carbon dioxide is from the reduction reaction. The nitrogen is from the air blast. It has not taken part in the reactions so has not been changed.

The molten iron obtained is impure. Some is allowed to harden into **cast iron** which is hard but brittle – so suitable only for things like iron railings. But most is turned into **steels** – alloys of iron containing controlled amounts of other substances to improve its properties. For example **mild steel** is 99.5% iron and 0.5% carbon, and strong enough for car bodies and buildings.

Questions

1 Name the raw materials for extracting iron.
2 Write an equation for the reaction that gives iron.
3 The calcium carbonate in the blast furnace helps to purify the iron. Explain how, with equations.
4 What is the 'blast' of the blast furnace?
5 Name the waste gases from the blast furnace.
6 The slag and waste gases are both useful. How?
7 What is: **a** cast iron?   **b** steel?

5.10 Principles of electrolysis

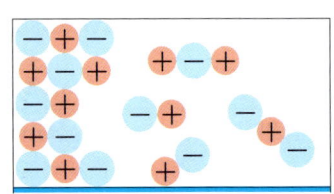

Electrolysis means *splitting by electricity*. Compounds are decomposed by an electric current. This can only happen if the compound is:

- made of ions
- melted or dissolved in water so that the ions are free to move.

When an ionic solid melts its ions become free to move.

The electrolysis of molten lead bromide
This diagram shows the apparatus.
The graphite rods are called **electrodes**.
They will carry current into and out of the molten lead bromide when the switch is closed.

Lead bromide is the **electrolyte** – the liquid you electrolyse.

When the switch is closed, bromine vapour starts bubbling off from around the positive electrode. After some time a bead of molten lead forms below the negative electrode.

So the electrical energy from the battery has caused a **chemical change**:

lead bromide ⟶ lead + bromine
$PbBr_2 \, (l)$ ⟶ $Pb \, (l)$ + $Br_2 \, (g)$

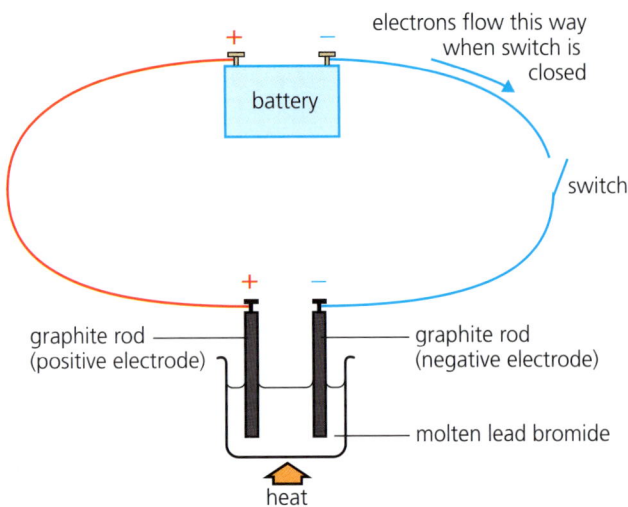

Why the molten lead bromide decomposes
When lead bromide melts, its lead ions and bromide ions become free to move. When the switch is closed, the electrodes become charged, and the ions are immediately attracted to them:

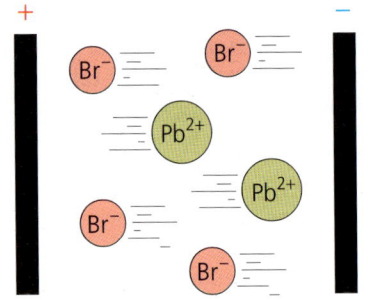 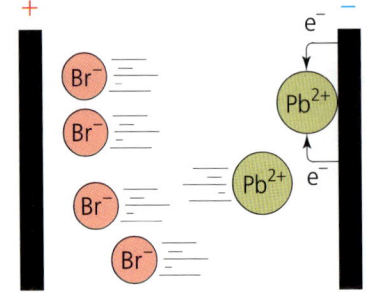 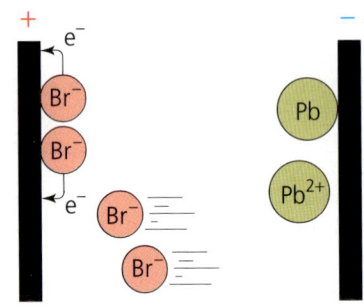

Opposite charges attract. So the positive lead ions (Pb^{2+}) move to the negative electrode. The negative bromide ions (Br^-) move to the positive electrode.	**At the negative electrode:** the lead ions each receive two electrons and become lead atoms: $Pb^{2+} + 2e^- \rightarrow Pb$ Lead collects on the electrode and eventually drops off it.	**At the positive electrode:** bromide ions each give up an electron and become bromine atoms: $Br^- - e^- \rightarrow Br$ These pair up to form molecules: $Br + Br \rightarrow Br_2$ The bromine bubbles off as a gas.
Free ions move.	Ions gain electrons: **reduction**.	Ions lose electrons: **oxidation**.
Remember OILRIG: Oxidation Is Loss, Reduction Is Gain	**Overall, electrolysis is a redox reaction.** Reduction at the negative electrode, oxidation at the positive.	

The electrolysis of other molten compounds

Look at the examples in this table:

Electrolyte	The decomposition	What forms at the negative electrode?	What forms at the positive electrode?
Sodium chloride NaCl	sodium chloride → sodium + chlorine 2NaCl (l) → 2Na (l) + Cl$_2$ (g)	Sodium Na$^+$ + e$^-$ → Na	Chlorine Cl$^-$ − e$^-$ → Cl Cl + Cl → Cl$_2$
Potassium iodide KI	potassium iodide → potassium + iodine 2KI (l) → 2K (l) + I$_2$ (g)	Potassium K$^+$ + e$^-$ → K	Iodine I$^-$ − e$^-$ → I I + I → I$_2$
Copper(II) bromide CuBr$_2$	copper(II) bromide → copper + bromine CuBr$_2$ (l) → Cu (l) + Br$_2$ (g)	Copper Cu^{2+} + 2e$^-$ → Cu	Bromine Br$^-$ − e$^-$ → Br Br + Br → Br$_2$

So when a molten ionic compound is electrolysed, a metal forms at the negative electrode and a non-metal at the positive electrode.

The electrolysis of solutions

When an ionic compound dissolves in water, its ions are free to move. So the *solution* can be electrolysed. But the products may be different than for the *molten* compound because water produces some ions:

some water molecules → hydrogen ions + hydroxide ions
H$_2$O (l) → H$^+$ (aq) + OH$^-$ (aq)

The H$^+$ ions go to the negative electrode with the metal ions.
The OH$^-$ ions go to the positive electrode with other negative ions.
And at the electrodes there's a competition! Look at these results:

Compound dissolved in water	What forms at the negative electrode?	What forms at the positive electrode?
Potassium iodide	Hydrogen gas	Iodine
Sodium chloride	Hydrogen gas	Chlorine
Magnesium sulphate	Hydrogen gas	Oxygen
Zinc chloride	Zinc + hydrogen gas	Chlorine
Lead nitrate	Lead	Oxygen
Copper(II) bromide	Copper	Bromine

So when you electrolyse aqueous solutions of compounds:
- hydrogen gas is released at the negative electrode in preference to a reactive metal
- a halogen gas is released at the positive electrode if the compound is a halide. If not, you get oxygen.

Example: electrolysis of sodium chloride solution

Four types of ion present: Na$^+$ and Cl$^-$ from the salt, and H$^+$ and OH$^-$ from water.

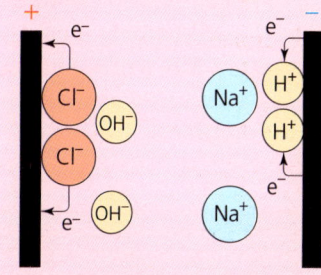

At the electrodes:
H$^+$ ions accept electrons more readily than Na$^+$ ions do.
Cl$^-$ ions give up electrons more readily than OH$^-$ ions do.
So **hydrogen** and **chlorine** form.

The Na$^+$ and OH$^-$ ions remain in solution – giving sodium hydroxide solution!

Questions

1. What type of compound undergoes electrolysis?
2. For the electrolysis of molten lead bromide, draw diagrams to show:
   a. how the ions move when the switch is closed
   b. what happens at the positive electrode
   c. what happens at the negative electrode
3. Name the products at the electrodes when these molten compounds are electrolysed: a magnesium chloride b sodium iodide d calcium bromide
4. When an aqueous solution of potassium chloride is electrolysed, hydrogen gas is produced. Explain why, with the help of a diagram.

5.11 Extracting aluminium

From rocks to rockets

Aluminium is the most abundant metal in the Earth's crust. Its main ore is **bauxite**, which is aluminium oxide mixed with impurities like sand and iron oxide. The impurities make it reddish brown.

These are the steps in obtaining aluminium:

1 First, geologists test rocks to find out how much bauxite there is, and whether it is worth mining. If the tests are satisfactory, mining begins.

2 Bauxite usually lies near the surface, so it is easy to dig up. This is a bauxite mine in Jamaica. Everything gets coated with red-brown bauxite dust.

3 From the mine, the ore is taken to a bauxite plant, where it is treated to remove the impurities. The result is white **aluminium oxide**, or **alumina**.

4 The alumina is taken to another plant for electrolysis. Much of the Jamaican alumina is shipped to plants like this one, in Canada or the USA . . .

5 . . . where electricity is cheaper. There it is electrolysed to give aluminium. The metal is made into sheets and blocks, and sold to other industries.

6 It is used to make beer cans, cooking foil, saucepans, racing bikes, TV aerials, electricity cables, ships, aeroplanes, and even space rockets.

A closer look at the electrolysis

In step **5** above you saw that aluminium is obtained from alumina by electrolysis. Two tonnes of alumina will give one tonne of aluminium.

The electrolysis is carried out in a huge steel tank. The tank is lined with carbon which acts as the negative electrode. Huge blocks of carbon hang in the middle of the tank and act as the positive electrode.

Pure alumina melts at 2045 °C. It would be expensive, and dangerous, to keep the tank at that temperature. Instead, the alumina is dissolved in molten **cryolite** for the electrolysis. (Cryolite is sodium aluminium fluoride, and it has a much lower melting point.)

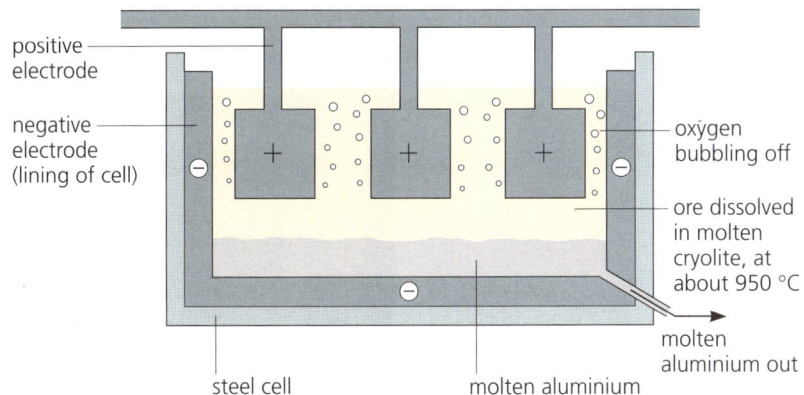

Aluminium being dissolved in molten cryolite, for electrolysis. The carbon electrodes are hung from the yellow beams.

When the alumina dissolves, its aluminium ions and oxide ions become free to move.

At the negative electrode   The aluminium ions gain electrons. This is reduction:

$Al^{3+} + 3e^- \longrightarrow Al$

The aluminium atoms collect together and drop to the bottom of the cell as molten metal. This is run off at intervals.

At the positive electrode   The oxygen ions lose electrons. This is oxidation:

$2O^{2-} - 4e^- \longrightarrow O_2$

Oxygen gas bubbles off. It reacts with the carbon electrodes to form carbon dioxide:

$C\ (s) + O_2\ (g) \longrightarrow CO_2\ (g)$

So the electrodes get eaten away and need replacing from time to time.

Some properties of aluminium

1 Aluminium is a bluish-silver, shiny metal.
2 Unlike most metals, it has a low density – it is 'light'.
3 It is a good conductor of heat and electricity.
4 It is malleable and ductile.
5 It is non-toxic.
6 It is not very strong when pure, but it can be made stronger by mixing it with other metals to form alloys.

This underground train is made of aluminium strengthened with small amounts of other metals.

Questions

1 Copy and complete: The chief ore of aluminium is called ………. . It is first purified to give ………. ……….. or ………. , which has the formula ………. . Then this is ………. to give aluminium.
2 Draw the cell for the electrolysis of alumina.
3 Why do the aluminium ions move to the negative electrode? What happens to them there?
4 Why must the electrodes be replaced now and then?
5 List six uses of aluminium. For each, say what properties of the metal make it suitable.

5.12 Purifying copper by electrolysis

Copper(II) sulphate and carbon electrodes

Look what happens when a solution of copper(II) sulphate is electrolysed using carbon electrodes:

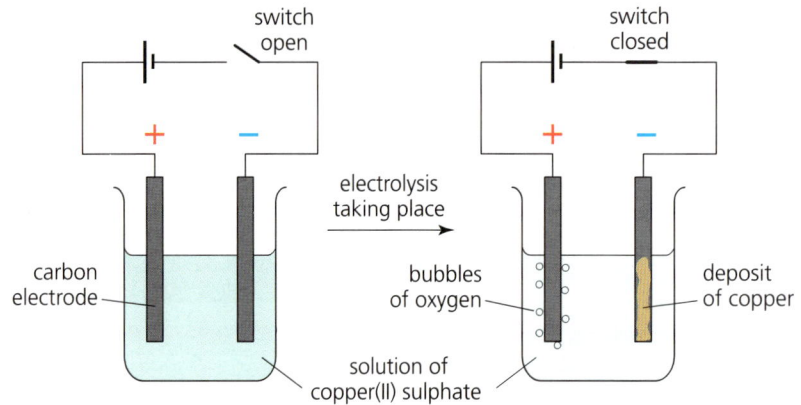

Very pure copper is used for fast computer chips.

Copper is obtained at the negative electrode and oxygen gas bubbles off at the positive electrode. (This is what you would expect.) The blue colour of the starting solution was due to copper ions. It fades as the copper is deposited.

Copper(II) sulphate and copper electrodes

If the carbon electrodes are replaced with *copper* electrodes, something unusual happens:

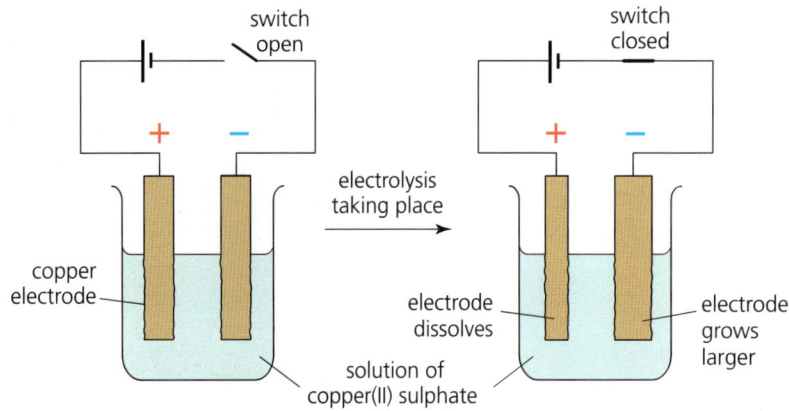

The electrolysis of copper(II) sulphate solution using copper electrodes. The red wire goes to the positive electrode. Which electrode will grow heavier?

Copper is again obtained at the negative electrode. But no bubbles of oxygen form at the positive electrode. Instead the positive electrode gets smaller. It is *dissolving*.

This time the colour of the solution does not change. This is because as copper ions become atoms at the negative electrode, the same number of copper atoms become ions at the positive electrode. So the overall number of copper ions in the solution remains unchanged.

This electrolysis is very useful because it can be used to purify copper. The purer copper is, the better it is at conducting electricity so the better it is for electrical wires and connections.

Purifying copper by electrolysis

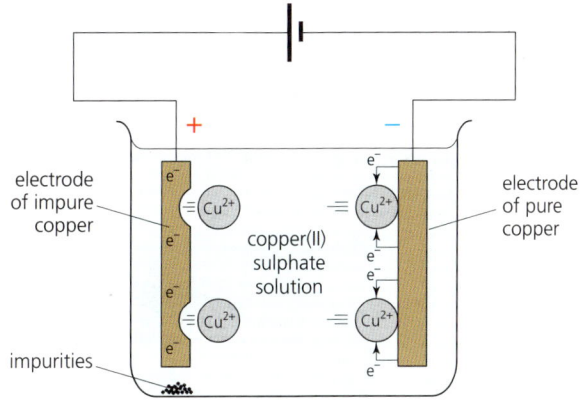

At the positive electrode (impure copper)	At the negative electrode (pure copper)
Copper atoms **lose electrons** to become ions: $Cu - 2e^- \rightarrow Cu^{2+}$ This is **oxidation**. Only the copper ions *dissolve*. The impurities just fall to the bottom. The ~~cathode~~ anode gets thinner.	Copper ions **gain electrons** to become atoms: $Cu^{2+} + 2e^- \rightarrow Cu$ This is **reduction**. The copper atoms cling to the pure copper cathode. So the cathode grows thicker.

Electrolysis can produce copper that is over 99.999% pure. In industry the impurities that fall to the bottom of the tank are not just thrown away. They may contain valuable metals such as gold, silver, platinum, and selenium. These are recovered and sold.

Electroplating

In the electrolysis above, the negative copper electrode was coated with more copper. You can also use electrolysis to coat a negative electrode with a *different* metal. This is called **electroplating**.

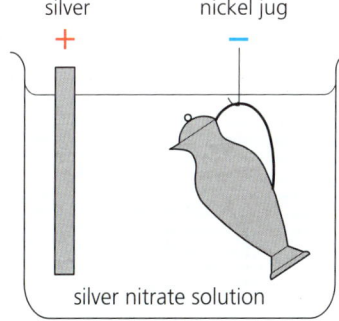

Suppose you want to electroplate a nickel jug with silver. You could set up an electrolysis with the jug as the negative electrode and silver foil as the positive electrode, in a solution of silver nitrate.

At the positive electrode The silver dissolves, forming silver ions:

$Ag - e^- \longrightarrow Ag^+$

At the negative electrode Silver ions receive electrons and form a layer of silver on the jug. When the layer is thick enough the jug is removed:

$Ag^+ + e^- \longrightarrow Ag$

Electroplating is used to plate car bumpers with chromium to make them look good and to protect them from rust. Steel is plated with tin to make 'tins' for food.

Rules for electroplating

To plate an object with metal X:

negative electrode = object
positive electrode = strip of X
solution = compound of X

Questions

1 What effect does replacing carbon electrodes with copper electrodes have in the electrolysis of copper(II) sulphate solution?
2 a Explain how copper is purified by electrolysis.
  b What happens to the impurities?
3 Why is it necessary to obtain very pure copper?
4 a Draw a labelled diagram to show how you would plate an iron nail with nickel.
  b Write equations to show the changes that would take place at the positive and negative electrodes.

33

5.13 Corrosion

When a metal is attacked by air, water, or other substances in its surroundings, the metal is said to **corrode**.

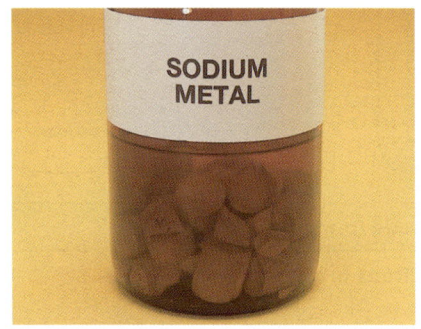

Damp air quickly attacks sodium, turning it into a pool of sodium hydroxide solution. (That is why sodium is stored under oil!)

Iron also corrodes in damp air. So do most steels. The result is **rust** and the process is called **rusting**. It happens especially quickly near sea water.

But gold is unreactive and never corrodes. This gold mask of King Tutankhamun was buried in his tomb for over 3000 years, and still looks as good as new.

In general, the more reactive a metal is, the more readily it corrodes.

The corrosion of iron

The corrosion of iron is called **rusting**. The iron is **oxidized**. Rusting needs both air and water, as these tests show.

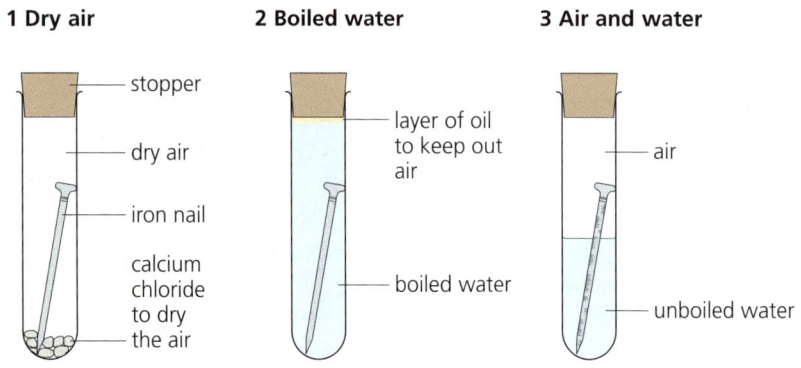

Nails 1 and 2 do not rust. Nail 3 does rust.

How to prevent rusting

Iron and most steels (alloys of iron) rust. Below are ways to prevent them rusting. Most involve *coating the metal with something to keep out air and water:*

1 Paint. Steel bridges and railings are usually painted. Paints that contain lead or zinc are mostly used, because these are especially good at preventing rust. For example, 'red lead' paints contain an oxide of lead, Pb_3O_4.

2 Grease. Tools, and machine parts, are coated with grease or oil.

Steel girders for buildings are first sprayed with special paint.

3 **Plastic.** Steel is coated with plastic for use in garden chairs, bicycle baskets, and dish racks. Plastic is cheap and can look attractive.
4 **Galvanizing.** Iron for sheds and dustbins is usually coated with zinc. It is called **galvanized iron**.
5 **Tin plating.** Baked beans come in 'tins' which are made from steel coated on both sides with a fine layer of tin. Tin is used because it is unreactive, and non-toxic. It is deposited on the steel by electrolysis, in a process called **tin plating**.
6 **Chromium plating.** Chromium is used to coat steel with a shiny protective layer, for example on car bumpers. Like tin, the chromium is deposited by electrolysis.
7 **Sacrificial protection.** Magnesium is more reactive than iron. So when a bar of magnesium is attached to the side of a steel ship, it corrodes instead of the steel. This is called **sacrificial protection**, because the magnesium is sacrificed to protect the steel.

Steel cans for food are plated with tin.

Now look at this experiment:

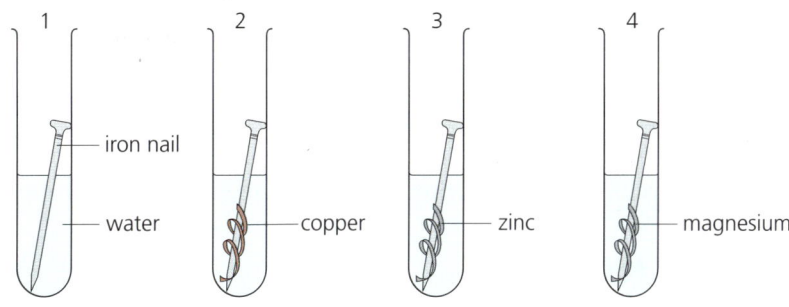

Nails 1 and 2 rust. Nails 3 and 4 do not rust. So zinc can also be used for sacrificial protection. But copper will not work because it is less reactive than iron.

Does aluminium corrode?

Aluminium is more reactive than iron, so you might expect it to corrode faster in damp air. In fact, clean aluminium starts corroding immediately, but the reaction quickly stops.

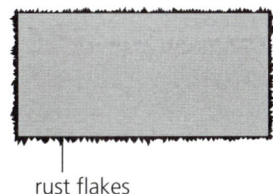

rust flakes

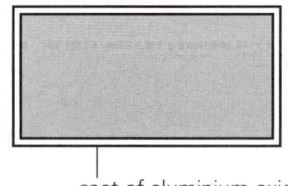

coat of aluminium oxide

When iron corrodes, rust forms in tiny flakes. Damp air can get past the flakes, to attack the metal. In time, it rusts right through.

But when aluminium reacts with air, an even coat of aluminium oxide forms. This seals the metal surface, preventing further corrosion.

A thin coat of oxide protects aluminium. You can make it thicker by electrolysis, giving **anodized** aluminium. This is used for window frames.

Questions

1 What is *corrosion*?
2 Which two substances cause iron to rust?
3 Name an alloy of iron that doesn't rust. (Kitchen sinks!)
4 Iron that is tin-plated does not rust. Why?
5 a What does the *sacrificial protection* of iron mean?
  b Both magnesium and zinc can be used for it. Why?
  c But copper will not work. Explain why.
6 Why won't aluminium corrode right through?

5.14 Metals, civilization, and you

No you without metals

Without metals you probably would not exist. You certainly would not be reading this book. The world would not have over 6 billion people.

Our human history has been shaped largely by the history of metals and their discovery. That's why two of our eras are known as the Bronze Age and Iron Age.

Metals and civilization

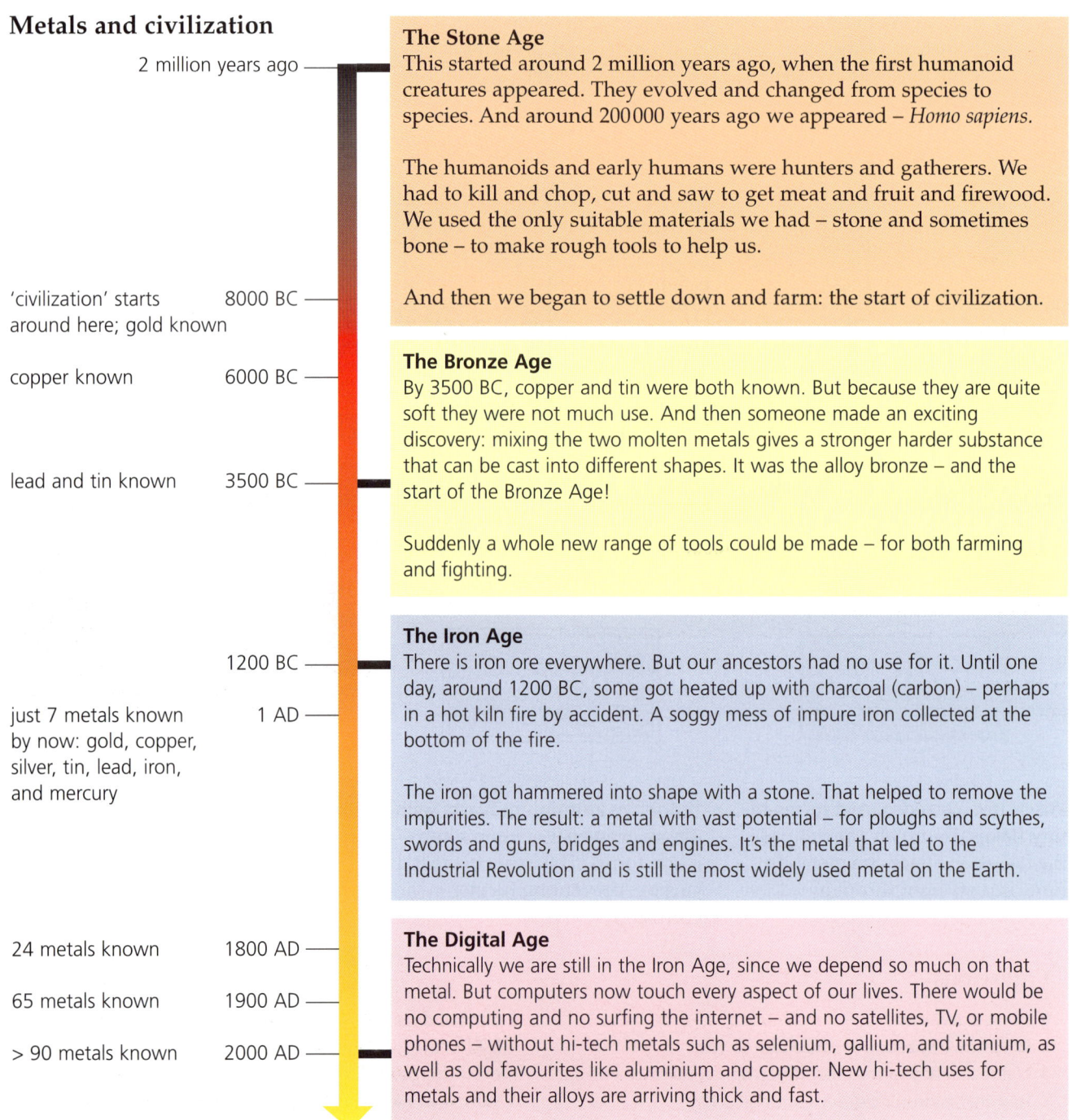

2 million years ago

The Stone Age
This started around 2 million years ago, when the first humanoid creatures appeared. They evolved and changed from species to species. And around 200 000 years ago we appeared – *Homo sapiens*.

The humanoids and early humans were hunters and gatherers. We had to kill and chop, cut and saw to get meat and fruit and firewood. We used the only suitable materials we had – stone and sometimes bone – to make rough tools to help us.

'civilization' starts around here; gold known — 8000 BC

And then we began to settle down and farm: the start of civilization.

copper known — 6000 BC

The Bronze Age
By 3500 BC, copper and tin were both known. But because they are quite soft they were not much use. And then someone made an exciting discovery: mixing the two molten metals gives a stronger harder substance that can be cast into different shapes. It was the alloy bronze – and the start of the Bronze Age!

lead and tin known — 3500 BC

Suddenly a whole new range of tools could be made – for both farming and fighting.

1200 BC

The Iron Age
There is iron ore everywhere. But our ancestors had no use for it. Until one day, around 1200 BC, some got heated up with charcoal (carbon) – perhaps in a hot kiln fire by accident. A soggy mess of impure iron collected at the bottom of the fire.

just 7 metals known by now: gold, copper, silver, tin, lead, iron, and mercury — 1 AD

The iron got hammered into shape with a stone. That helped to remove the impurities. The result: a metal with vast potential – for ploughs and scythes, swords and guns, bridges and engines. It's the metal that led to the Industrial Revolution and is still the most widely used metal on the Earth.

24 metals known — 1800 AD

The Digital Age
Technically we are still in the Iron Age, since we depend so much on that metal. But computers now touch every aspect of our lives. There would be no computing and no surfing the internet – and no satellites, TV, or mobile phones – without hi-tech metals such as selenium, gallium, and titanium, as well as old favourites like aluminium and copper. New hi-tech uses for metals and their alloys are arriving thick and fast.

65 metals known — 1900 AD

> 90 metals known — 2000 AD

Where next?

This bronze helmet, over 2000 years old, was worn by a Celtic warrior. It was found in the River Thames. Perhaps he was killed by the Romans?

The lead ore galena (lead sulphide) was used as eye make-up in ancient Egypt. Lead was probably produced by heating galena in an open fire.

Some early milestones

Who knows how gold was first discovered. Perhaps one of your ancestors saw a shiny lump in a stream. A few thumps with a stone and it was flattened. A few twists and there was a bracelet.

The first metals found were the unreactive ones that exist as elements. But these were too soft to be useful for anything other than ornament. Tin, lead, and iron occur as ores so reduction was needed for them.

It could have happened by accident – a lump of ore dropped into a fire where wood had been turned to charcoal (carbon). Or pottery kilns could have played a part. Pottery was first made around 5000 BC. The kilns were heated by closed-in fires, much hotter than open fires. The melted metals could have been discovered in the kiln fires – or even on the surface of pots, from the clay.

The invention of electrolysis

By 1800, only carbon and hydrogen had been used to reduce metal ores. So only the less reactive metals had been discovered.

But in 1800 two English scientists, Anthony Nicholson and William Carlisle, carried out the first ever electrolysis (of water). It caught the attention of Humphry Davy. He tried electrolysing molten compounds, with spectacular results. In 1807 he discovered potassium and sodium, and the next year magnesium, calcium, strontium, and barium.

Humphry Davy died at 52 of an illness probably caused by the harmful vapours produced in his electrolysis experiments.

As you know, reactive metals 'prefer' to exist as compounds. They will react with compounds of less reactive metals, driving those metals out. Once sodium and other reactive metals had been obtained they could be heated up with all kinds of compounds – and this led in turn to the discovery of many other metals.

Questions

1 Gold was one of the first metals to be discovered. Suggest a reason.
2 Lead was discovered quite early on. Suggest a reason for this.
3 Sodium and potassium were not discovered until just over two hundred years ago. Why not?
4 Titanium is a 'hi-tech' metal. It is made by heating titanium chloride with molten magnesium. It is hard, strong, and light. A thin coat of oxide forms on its surface and makes it unreactive.
  a Which is more reactive, titanium or magnesium?
  b Titanium is used for replacement hip joints. Why?

5.15 Acids and alkalis

Acids

One important group of chemicals is called **acids**.

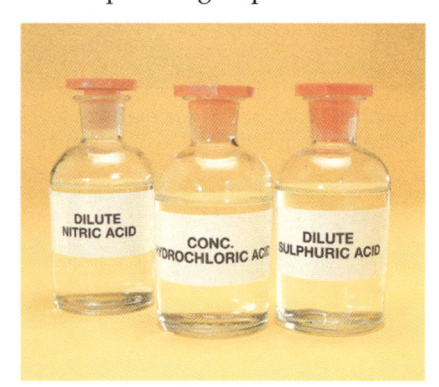

Here are some acids. You have probably seen them in the lab. They are all liquids. In fact they are **solutions** of pure compounds in water.

They must be handled carefully, especially the concentrated acids, for they are **corrosive**. They can eat away metals, skin, and cloth.

But some are not so corrosive even when concentrated. They are **weak** acids. Ethanoic acid is the natural weak acid found in vinegar.

You can tell if something is an acid, by its effect on **litmus**. Litmus is a purple dye. It can be used as a solution, or on paper:

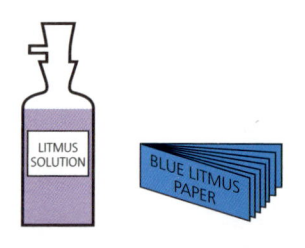

Litmus solution is purple. Litmus paper for testing acids is blue.

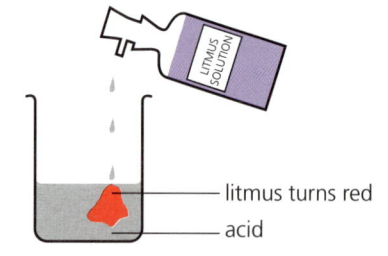

Acids will turn litmus solution red.

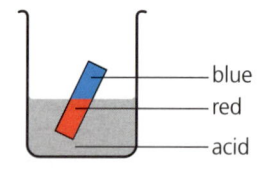

They will also turn blue litmus paper red.

Some common acids   The main ones are:

hydrochloric acid	HCl (aq)
sulphuric acid	H_2SO_4 (aq)
nitric acid	HNO_3 (aq)
ethanoic acid	CH_3COOH (aq)

There are many other weak acids which, like ethanoic acid, occur naturally. **Citric acid** is found in oranges and lemons, and **tannic acid** in tea. Ant and nettle stings contain **methanoic acid**.

Strong acids
Hydrochloric acid
Sulphuric acid
Nitric acid

Weak acids
Ethanoic acid
Citric acid
Carbonic acid

Alkalis

There is another group of chemicals that also affect litmus, but in a different way to acids. They are the **alkalis**.

Alkalis turn litmus solution blue, and red litmus paper blue.

Like acids, they must be handled carefully, because they can burn skin.

38

Some common alkalis Most alkalis are solids. But they are usually used in the laboratory as aqueous solutions. The main ones are:

sodium hydroxide	NaOH (*aq*)
potassium hydroxide	KOH (*aq*)
calcium hydroxide	Ca(OH)$_2$ (*aq*)
ammonia solution	NH$_3$ (*aq*)

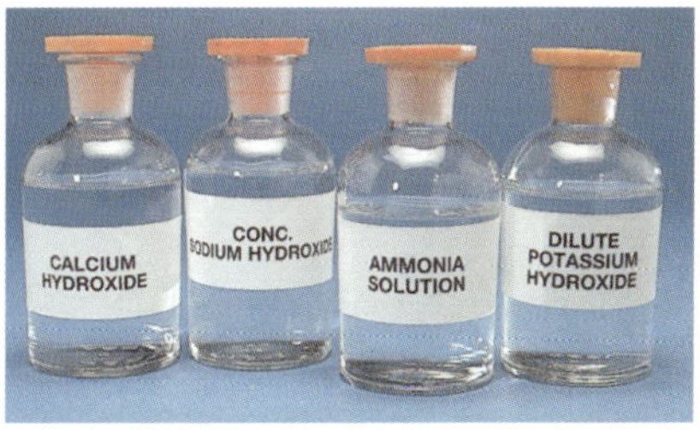

Common laboratory alkalis.

Strong alkalis
Sodium hydroxide
Potassium hydroxide
Calcium hydroxide

Weak alkali
Ammonia

Neutral substances

Many substances do not affect the colour of litmus solution, so they are not acids or alkalis. They are **neutral**. Sodium chloride and sugar solutions are both neutral.

The pH scale

Some acids are weaker than others. It is the same with alkalis. The strength of an acid or alkali is shown using the **pH scale** which goes from 0 to 14:

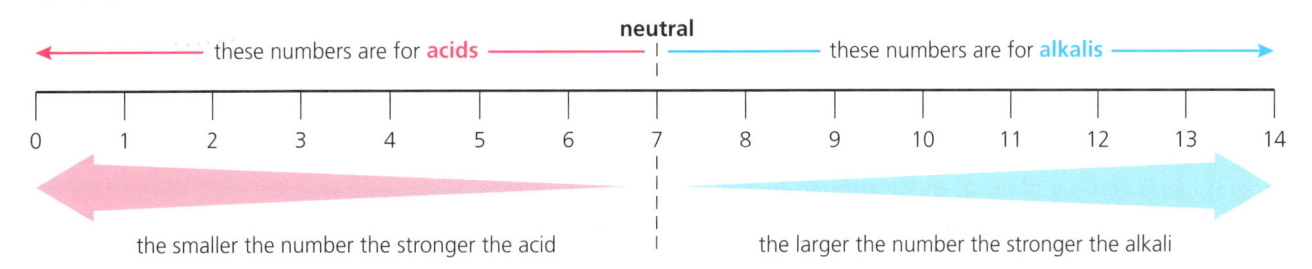

← these numbers are for **acids** — neutral — these numbers are for **alkalis** →

0 1 2 3 4 5 6 7 8 9 10 11 12 13 14

the smaller the number the stronger the acid   the larger the number the stronger the alkali

On this scale:

An acidic solution has a pH number less than 7.
An alkaline solution has a pH number greater than 7.
A neutral solution has a pH number of exactly 7.

You can find the pH of any solution by using **universal indicator**. Universal indicator is a mixture of dyes. Like litmus, it can be used as a solution, or as universal indicator paper. It goes a different colour at different pH values, as shown in this diagram:

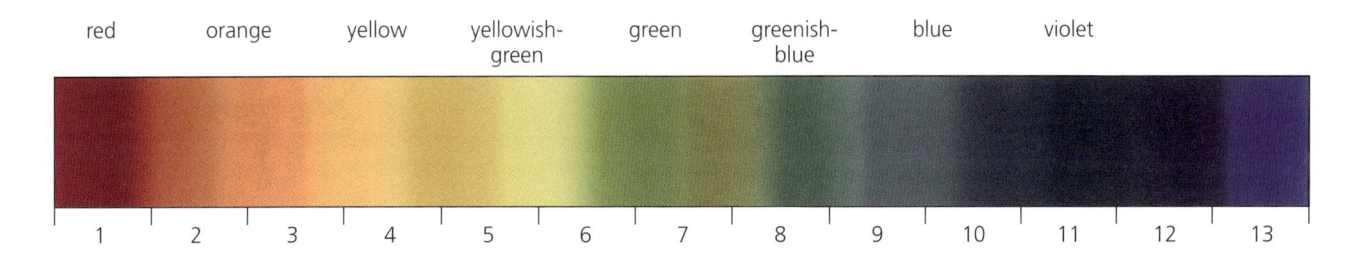

red	orange	yellow	yellowish-green	green	greenish-blue	blue	violet
1	2 3	4	5	6 7	8 9	10 11	12 13

Questions

1 What does *corrosive* mean?
2 How would you test a substance to see if it is an acid?
3 Write down the formula for:
   sulphuric acid   nitric acid
   calcium hydroxide   ammonia solution
4 What effect do alkalis have on litmus solution?
5 Say whether a solution is acidic, alkaline, or neutral, if its pH number is: 9 4 7 1 10 3
6 What colour would universal indicator show, in an aqueous solution of sugar? Why?

39

5.16 The reactions of acids

With alkalis

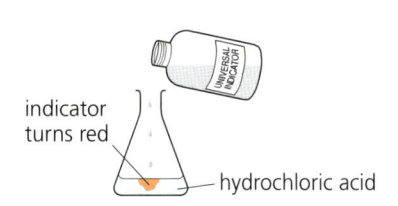

The flask contains hydrochloric acid. A drop of universal indicator is added. It turns red, as you'd expect.

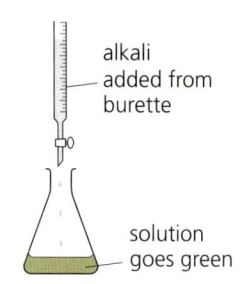

As sodium hydroxide is dripped in, the solution goes green. It is now neutral, like water! The alkali has **neutralize**d the acid.

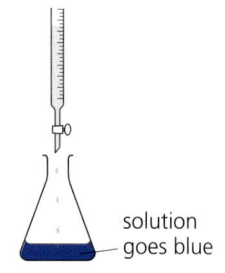

If you add more sodium hydroxide, the solution goes blue. That means it is now alkaline. There is no acid left in the beaker.

So a **neutralization reaction** has take place. The equation for it is:

hydrochloric acid + sodium hydroxide → sodium chloride + water
HCl (aq)   + $NaOH$ (aq)   → $NaCl$ (aq)   + H_2O (l)

The metal has taken the place of the hydrogen in the acid, giving sodium chloride, a **salt**. All acids are neutralized by alkalis in the same way:

acid + alkali → salt + water

Here are two more examples:

sulphuric acid + magnesium hydroxide → magnesium sulphate + water
H_2SO_4 (aq)   + $Mg(OH)_2$ (aq)   → $MgSO_4$ (aq)   + $2H_2O$ (l)

nitric acid + potassium hydroxide → potassium nitrate + water
HNO_3 (aq) + KOH (aq)   → KNO_3 (aq)   + H_2O (l)

Naming salts
It depends on the acid you start with:

nitric → **nitrates**
hydrochloric → **chlorides**
sulphuric → **sulphates**

With metals

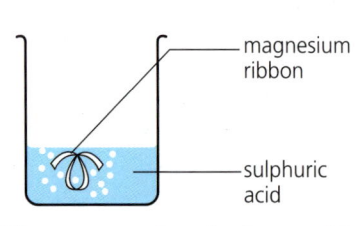

When magnesium is dropped into dilute sulphuric acid, hydrogen quickly bubbles off.

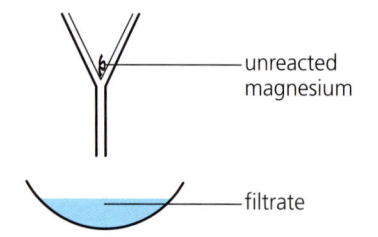

The bubbles stop when the reaction is over. The unreacted magnesium is then removed by filtering.

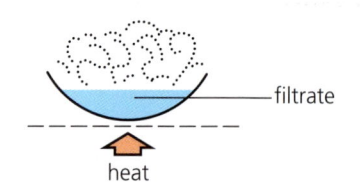

The filtrate is heated to evaporate the water. A white solid is left behind. This solid is **magnesium sulphate**.

The equation for the reaction is:

magnesium + sulphuric acid → magnesium sulphate + hydrogen
Mg (s)   + H_2SO_4 (aq)   → $MgSO_4$ (aq)   + H_2 (g)

Once again a metal has replaced the hydrogen in the acid. The salt magnesium sulphate has been formed.

With carbonates

When calcium carbonate is added to dilute hydrochloric acid, as shown on the right, carbon dioxide quickly bubbles off. The reaction is:

calcium carbonate + hydrochloric acid → calcium chloride + water + carbon dioxide

$CaCO_3$ (s) + 2HCl (aq) → $CaCl_2$ (aq) + H_2O (l) + CO_2 (g)

Once again a salt, calcium chloride, has been formed.

With bases

The oxides and hydroxides of many metals, including the transition metals, are insoluble in water. But they can neutralize acids to form salts, so they are **bases**.

Copper(II) oxide is a base. Look at this reaction:

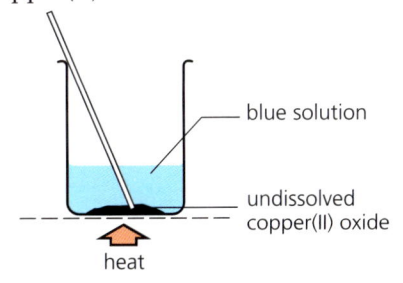

Calcium carbonate reacting with dilute hydrochloric acid.

Some copper(II) oxide is added to dilute sulphuric acid. On warming, it dissolves and the solution turns blue. More is added until no more dissolves.

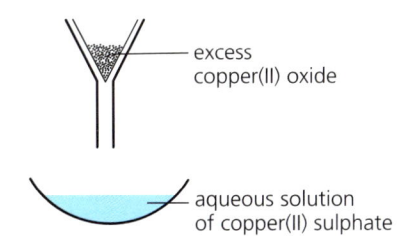

That means that all the acid has been used up. The excess solid is removed by filtering. This leaves a blue solution of copper(II) sulphate in water.

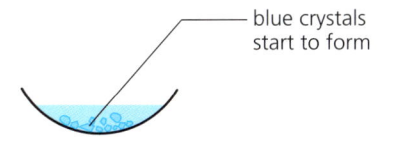

The solution is heated to evaporate some of the water. Then it is left to cool. Blue crystals of copper(II) sulphate start to form.

The equation for the reaction is:

copper(II) oxide + sulphuric acid → copper(II) sulphate + water
CuO (s) + H_2SO_4 (aq) → $CuSO_4$ (aq) + H_2O (l)

A salt, copper sulphate, has been formed.

A summary of acid reactions

acid + metal hydroxide → metal salt + water
acid + metal → metal salt + hydrogen
acid + metal carbonate → metal salt + water + carbon dioxide
acid + metal oxide → metal salt + water

Each reaction produces a salt. So you could use one of them (as long as it is safe) to produce a salt in the lab. The salt you get depends on the metal (or metal compound) and the acid chosen.

What's a base?
A base is anything that can neutralize an acid. All these are bases:
- alkalis
- insoluble metal oxides and hydroxides
- metal carbonates and hydrogencarbonates
- ammonia solution.

Questions

1 Write a word equation for the reaction of dilute sulphuric acid with:
   a zinc   b sodium hydroxide   c sodium carbonate
2 Never react sodium with acid in the lab. Why not?
3 Name the acid that forms salts called nitrates.
4 Which salt is forming in the photo at the top of this page? How would you obtain pure crystals of it?
5 What is a base?
6 Zinc oxide is a base. Suggest a way to make zinc nitrate from it. Write a word equation for the reaction.

41

5.17 A closer look at neutralization

What makes an acid acidic?
All acids react in a similar way. There must be something in them that makes them act alike. That 'something' is hydrogen ions.
Acids contain hydrogen ions.

For example in hydrochloric acid:

$HCl\ (aq) \rightarrow H^+\ (aq) + Cl^-\ (aq)$

Acidic solutions contain the hydrogen ion: H^+

What makes an alkali alkaline?
Alkalis dissolve in water to give alkaline solutions. All these react in a similar way too, so they must have something in common.
Alkaline solutions contain hydroxide ions.

In sodium hydroxide solution, they are produced like this:

$NaOH\ (aq) \rightarrow Na^+\ (aq) + OH^-\ (aq)$

Alkaline solutions contain the hydroxide ion: OH^-

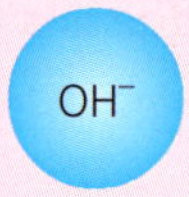

The neutralization reaction
Let's look more closely at what goes on in neutralization.

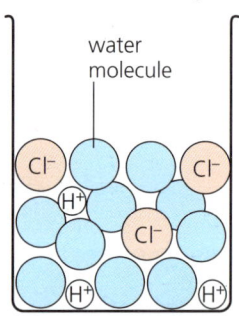

This is a solution of hydrochloric acid. It contains H^+ ions and Cl^- ions. It will turn litmus red.

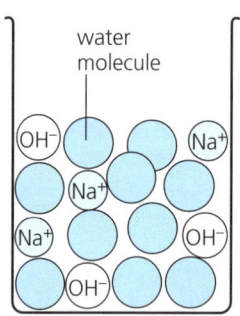

This is a solution of sodium hydroxide. It contains Na^+ ions and OH^- ions. It will turn litmus blue.

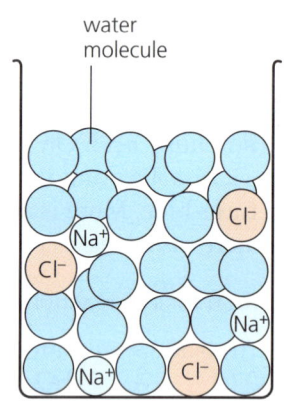

When the two are mixed, the H^+ and OH^- ions join to form **water molecules**. The result is a neutral solution of sodium chloride with no effect on litmus.

The equation for this reaction is:

$HCl\ (aq) + NaOH\ (aq) \longrightarrow NaCl\ (aq) + H_2O\ (l)$

You could also write it showing only the ions that combine:

$H^+\ (aq) + OH^-\ (aq) \longrightarrow H_2O\ (l)$

The alkali has **neutralized** the acid by removing its H^+ ions, and turning them into water.

During a neutralization reaction, the H^+ ions of the acid are turned into water.

A neutralization reaction is exothermic – it gives out heat. So the temperature of the solution rises a little.
You could obtain solid sodium chloride by heating the last solution above. The water evaporates leaving the solid behind.

Neutralization with ammonia

A solution of ammonia in water is alkaline. So it must contain hydroxide ions. They form when the gas dissolves in water, like this:

$NH_3\ (g) + H_2O\ (l) \rightarrow NH_4^+\ (aq) + OH^-\ (aq)$

The hydroxide ions then react with the hydrogen ions of acids.

Making sure neutralization is complete

Suppose you want to make the salt sodium chloride by the reaction on the opposite page. How can you tell when neutralization is complete? Use an indicator! **Phenolphthalein** is a good choice. It is pink in alkaline solutions but colourless in neutral solutions. To make the salt:

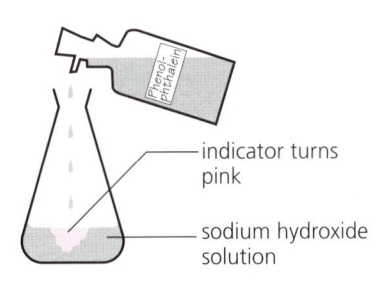

1 25 cm³ of sodium hydroxide solution is measured into a flask, using a pipette. Then two drops of phenolphthalein are added.

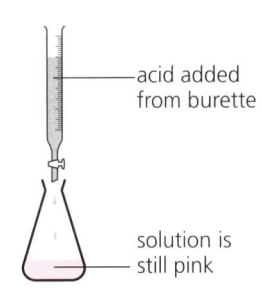

2 The acid is added from a burette, a little at a time. The flask is swirled in a controlled way, to allow the acid and alkali to mix.

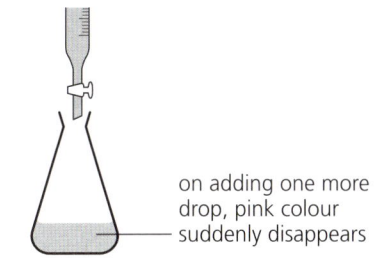

3 When all the alkali has been used up, the indicator suddenly turns colourless, showing that the solution is neutral.

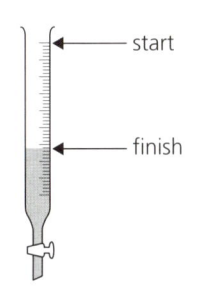

4 You can tell how much acid was added, using the scale on the burette. So now you know how much acid is needed to neutralize 25 cm³ of alkali.

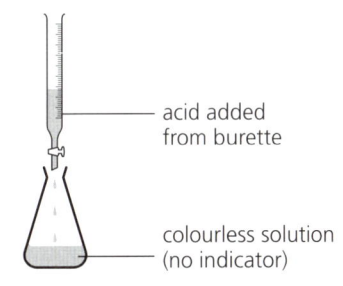

5 The reaction is carried out again, but this time there is no need for an indicator. 25 cm³ of alkali is put in the flask, and the correct amount of acid added.

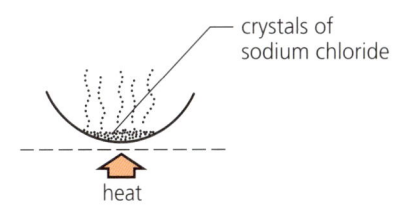

6 The solution from the flask is heated, to let the water evaporate. You will find dry crystals of sodium chloride are left behind.

In step 5 the reaction has to be carried out again, *without* indicator, because indicator would make the salt impure.

A similar method can be used for making potassium salts from potassium hydroxide, and ammonium salts from ammonia solution.

Questions

1 a Which ion is common to all acids?
 b Which ion is common to all alkalis?
 c Write an ionic equation showing what happens to these ions during a neutralization reaction.
2 The formula for ammonia is NH₃. Explain why its solution in water is alkaline.
3 Universal indicator is added to hydrochloric acid. Then enough sodium hydroxide solution is added to neutralize the acid. What happens to the:
 a H⁺ ions? b temperature? c indicator?
4 Write instructions for making *pure* potassium sulphate starting with potassium hydroxide.

5.18 Acids and alkalis in everyday life

increasing alkalinity

| pH | 8.8 | 9.5 | 10.0 | 11.0 | 11.0 | 11.9 | 12.5 | 13.0 |

Acids and alkalis in the kitchen

Many kitchen cleaners, like the ones in this photo, contain sodium hydroxide or ammonia, because alkalis are good at attacking grease. They feel soapy to the touch because they attack the grease on your skin. So handle them with care!

Some cleaners contain strong acids to dissolve the scale (calcium carbonate) left by hard water, and to remove stains. *Never* mix different cleaners because they can react together in a dangerous way.

Neutralization to the rescue

Several problems in life are caused by too much acidity – or too little. They can be solved by neutralization reactions. Here are some examples.

Insect stings   When a bee stings, it injects an acidic liquid into the skin. The sting can be neutralized by rubbing on **calamine lotion**, which contains zinc carbonate, or **baking soda**, which is sodium hydrogencarbonate.

Wasps stings are alkaline, and can be neutralized with vinegar. Why?

Indigestion   It may surprise you to know that you carry hydrochloric acid around in your stomach. It is a very dilute solution, and you need it for digesting food. But too much of it leads to **indigestion**, which can be very painful. To cure indigestion, you must neutralize the excess acid with a drink of sodium hydrogencarbonate solution (baking soda), or an indigestion tablet.

Soil treatment   Most plants grow best when the pH of the soil is close to 7. If the soil is too acidic, or too alkaline, the plants grow badly or not at all.

Chemicals can be added to soil to adjust its pH. Most often, soil is too acidic, so it is treated with **quicklime** (calcium oxide), **slaked lime** (calcium hydroxide), or **chalk** (calcium carbonate).

Factory waste   Liquid waste from factories often contains acid. If it reaches a river, the acid will kill fish and other river life. This can be prevented by adding slaked lime to the waste, to neutralize it.

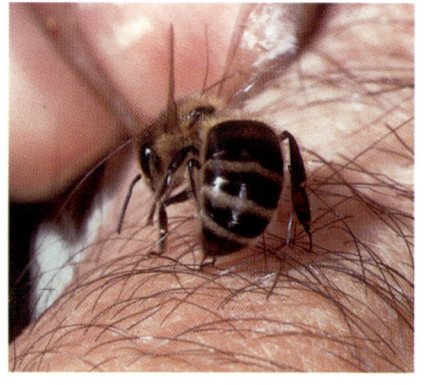

Bee stings are acidic.

Slaked lime being spread on fields where the soil is too acidic. Like limestone and quicklime, it is used because it is cheap. It is only slightly soluble, so rain won't wash it away. (Note that quicklime turns to slaked lime in rain.)

Burning fuels produces acidic gases.

Stonework worn away by acid rain.

Attacked by acid rain

We burn oil, coal, and gas in factories and homes to provide heat, and in power stations to give electricity. We burn petrol to make cars move. All these reactions release waste gases into the air, including sulphur dioxide and oxides of nitrogen.

Rainfall is naturally slightly acidic, since carbon dioxide from the air dissolves in it giving **carbonic acid**, a weak acid:

carbon dioxide + water → carbonic acid
CO_2 (g) + H_2O (l) → H_2CO_3 (aq)

But sulphur dioxide reacts with air and water to become sulphuric acid, and the oxides of nitrogen become nitric acid. These strong acids make the rain much more acidic.

Acid rain reacts with any vulnerable surfaces it falls on. Many buildings are made of limestone (calcium carbonate) so it eats away at their stonework. Concrete and cement contain calcium carbonate so they suffer too. Bridges, iron railings, and car bodies all come under attack. Acid rain also kills trees and other plants. It dissolves aluminium from minerals in rocks and soil, and carries it into rivers where it poisons fish.

Many factories and power stations try to solve the problem by spraying waste gases with jets of wet **slaked lime** (calcium hydroxide) before they leave the chimneys, to neutralize them. One product is calcium sulphate, a hard material which can be sold for road building.

These trees were killed by sulphur dioxide, which dissolved in rain to give acid rain.

Questions

1 Some kitchen cleaners are shown on the opposite page. Which one do you think is the most powerful? Why? What precaution would you take when using it? Why?

2 Nettle stings contain methanoic acid. Suggest something you could use to neutralize them.

3 Indigestion tablets contain a base. Why?

4 Slaked lime is made from limestone, a common rock. It is not very soluble. Give four reasons to explain why slaked lime is used on soil that is too acidic.

5 a Rain is naturally acidic. Why?
 b Rain in industrial areas may be more acidic. Why?
 c What damage does acid rain do?

Questions for Module 5

1. This extract from the periodic table shows the symbols for the first 20 elements.

H																	He
Li	Be											B	C	N	O	F	Ne
Na	Mg											Al	Si	P	S	Cl	Ar
K	Ca																

   Look at the row of elements from lithium (Li) to neon (Ne).
   a What is this row of the periodic table called?
   b On what basis are the elements arranged in it?
   c Going from left to right in this row, do the elements change from metals to non-metals, or vice versa?
   d For this row, give the symbol for:
      i the alkali metal
      ii the alkaline earth metal
   Look at the column of elements from beryllium (Be) to calcium (Ca).
   e What is this column of the periodic table called?
   f What can you say about the chemical reactions of these three elements?

2. In both appearance and reactions the transition metals are very different to the alkali metals. But they also share some properties.
   a Where are the transition metals found, in the periodic table?
   b From the list of properties below, select those which:
      i belong to the transition metals only
      ii belong to the alkali metals only
      iii belong to both groups
      iv belong to neither
      A good conductors of electricity
      B burn fiercely in oxygen
      C dissolve very easily in water
      D form coloured compounds
      E may have a density of less than 1 g/cm^3
      F cannot be cut with a knife
      G their salts can be made by reacting their oxides or hydroxides with acids
      H usually have very high melting points
      I react violently with water
      J conduct heat well
      K form hydroxides that dissolve in water to give solutions which turn litmus paper blue
      L are hard and strong enough to be used for building structures
      M do not usually form ions
      N can easily be shaped and bent by hammering

3. Read the following passage about the physical properties of metals.
   Elements are divided into metals and non-metals. All metals are <u>electrical conductors</u>. Many of them have a high <u>density</u> and they are usually <u>ductile</u> and <u>malleable</u>. All these properties influence the way the metals are used. Some metals are <u>sonorous</u> and this leads to special uses for them.
   a Explain the meaning of the words underlined.
   b Copper is ductile. How is this property useful in everyday life?
   c Aluminium is hammered and bent to make large structures for use in ships and aeroplanes. What property is important in the shaping of this metal?
   d Name one metal that has a *low* density.
   e Some metals are cast into bells. What property must the chosen metals have?
   f Copy and add the correct word: *Metals are good conductors of _____ and electricity.*
   g Name one other physical property of metals and give two examples of how this property is useful.

4. Choose one metal to fit each description below. (You must choose a different one each time.) Write down the name of the metal. Then write a balanced equation for the reaction that takes place.
   a a metal which burns in oxygen
   b a metal which reacts with oxygen without burning
   c a metal which reacts gently with dilute hydrochloric acid
   d a metal which floats on water and reacts vigorously with it
   e a metal which reacts quickly with steam but not with water

5. This is a list of metals in order of reactivity:
   sodium (*most reactive*)
   calcium
   magnesium
   zinc
   iron
   lead
   copper
   silver
   a Which metal is stored in oil?
   b i Which metals react with *cold* water?
      ii Which gas is given off in this reaction?

c i Choose one metal that will not react with cold water but will react with steam.
   ii Name two products of this reaction.
d i Name one metal that reacts slowly with dilute hydrochloric acid.
   ii Name the two products of this reaction.
e Which of the metals will *not* react with oxygen when heated?
f Which metal forms an oxide when heated in air but does not react with dilute acid?

6 When magnesium powder is added to copper(II) sulphate solution, solid copper forms. This change occurs because magnesium is more reactive than copper.

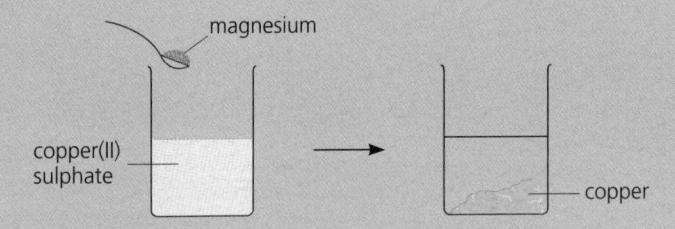

a Write a word equation for the reaction.
b Use the reactivity series of metals to decide whether these will react together:
   i iron + copper(II) sulphate solution
   ii silver + calcium nitrate solution
   iii zinc + lead(II) nitrate solution
c For those which react:
   i describe what you would *see*
   ii write the word equation
d What name is given to reactions of this type?

7 Nickel is below iron but above lead in the reactivity series. Use the series to predict the reaction of nickel:
a on heating in air
b with cold water
c with concentrated hydrochloric acid
d on heating with copper(II) oxide
e when it is placed in a solution of lead nitrate

8 Look again at the list of metals in question 5. Because zinc is more reactive than iron, it will remove the oxygen from iron(III) oxide, on heating.
a Write a word equation for the reaction.
b Decide whether these chemicals will react together, when heated:
   i magnesium + lead(II) oxide
   ii copper + lead(II) oxide
   iii magnesium + copper(II) oxide

c For those which react:
   i describe what you could *see*
   ii write a word equation
d What name is given to reactions of this type?

9 Use the reactivity series to explain why:
a calcium is not used for household articles
b copper is used for water pipes
c powdered magnesium can be used in fireworks
d gold is an excellent material for filling teeth

10 Write word equations for the reactions that occur when:
a magnesium reacts with steam
b zinc reacts with dilute sulphuric acid
c aluminium is heated with iron(III) oxide
d copper wire is placed in silver nitrate solution

11 Only a few elements are found uncombined in the Earth's crust. Gold is one example. The rest occur as compounds, and have to be extracted from their ores. This is usually carried out by heating with carbon, or by electrolysis.
Some information about the extraction of three different metals is shown below.

Metal	Formula of main ore	Method of extraction
Iron	Fe_2O_3	Heating with carbon
Aluminium	Al_2O_3	Electrolysis
Sodium	NaCl	Electrolysis

a Give the chemical name of each ore.
b Arrange the three metals in order of reactivity.
c How are the more reactive metals extracted from their ores?
d i How is the least reactive metal extracted from its ore?
   ii Why is this a reduction reaction?
   iii Why can't this method be used for the more reactive metals?
e Which of the methods would you use to extract:
   i potassium?
   ii lead?
   iii magnesium?
f Gold is a metal found native in the Earth's crust. Explain what *native* means.
g Where should gold go, in your list for **b**?
h Name another metal which occurs native.

12 Explain why the following metals are suitable for the given uses. (There should be more than one reason in each case.)
   a aluminium for window frames
   b iron for bridges
   c copper for electrical wiring
   d lead for roofing
   e zinc for coating steel

13 Many metals are more useful when mixed with other elements than when they are pure.
   a What name is given to the mixtures?
   b i What metals are found in these mixtures?
         brass   solder   stainless steel
      ii Describe the useful *properties* of the mixtures.
      iii Give one use for each of these mixtures.
   c Name another mixture of metals and the useful properties it has.

14 *aluminium   gold   iron   tin   magnesium
   mild steel   calcium   stainless steel*
   a In the above list of metals and alloys, only four are resistant to corrosion.
      i Which are they?
      ii Explain why each is resistant to corrosion.
   b Which of the other metals or alloys will corrode most quickly? Explain your answer.

15 a Draw a diagram of the blast furnace. Show clearly on your diagram:
      i where air is 'blasted' into the furnace
      ii where the molten iron is removed
      iii where the second liquid is removed
   b i Name the three raw materials added at the top of the furnace.
      ii What is the purpose of each material?
   c i What is the name of the second liquid that is removed from the bottom of the furnace?
      ii When it solidifies, does it have any uses? If so, name one.
   d i Name a waste gas that comes out at the top of the furnace.
      ii Does this gas have a use? If so, what?
   e Write an equation for the chemical reaction which produces the iron.

16 The electrolysis of lead bromide can be investigated using the following apparatus.

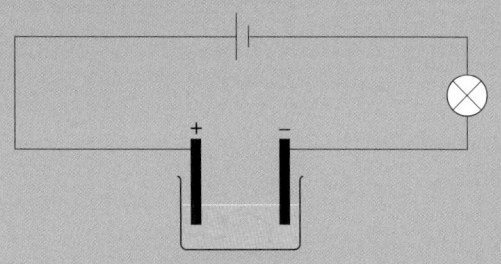

   a What must be done to the lead bromide before the bulb will light?
   b What would be *seen* at the positive electrode during the experiment?
   c Name the substance in **b**.
   d What is formed at the negative electrode?
   e Write an equation for the reaction at each electrode.
   f i At which electrode do positive ions gain electrons?
      ii What name is given to this process?
   g What happens to the other electrode?

17 The diagram below shows stages in obtaining copper from a low-grade ore. This ore contains copper(II) sulphide, CuS. It may also contain small amounts of silver, gold, platinum, iron, zinc, cadmium, and arsenic.
   a What is an *ore*?
   b What is a *low-grade* ore?
   c i How much waste rock is removed per tonne in step 1?
      ii What percentage of the ore is finally extracted as pure copper metal?
   d Why might it be economic to extract copper from a low-grade ore like this?
   e i Which type of chemical reaction occurs in Step 2?
      ii With what could the copper ore be reacted, to obtain the metal?
      iii The copper in C is 99% pure. Suggest some impurities it may contain.
   f i What process is carried out at step 3 to purify the metal?
      ii What will the main cost in this process be?
      iii As well as pure copper, this process may produce other valuable substances. Explain.
   g List all the environmental problems that may arise in going from A to D.

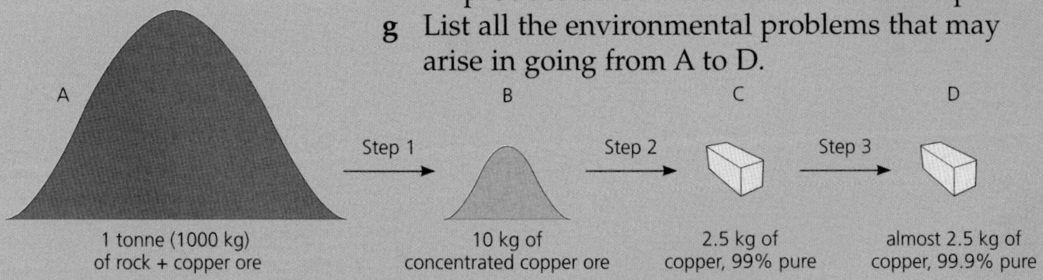

A — 1 tonne (1000 kg) of rock + copper ore
Step 1
B — 10 kg of concentrated copper ore
Step 2
C — 2.5 kg of copper, 99% pure
Step 3
D — almost 2.5 kg of copper, 99.9% pure

18 The diagram shows apparatus that can be used to purify impure copper by electrolysis.

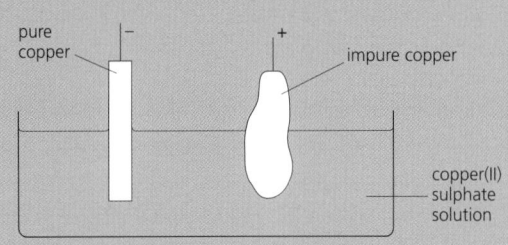

a Describe what you would see during the electrolysis.
b What name is given to the copper(II) sulphate solution in this apparatus?
c Write a half equation for the reaction at each electrode.
d Copper is used for making high-speed computer chips and water pipes. What physical and chemical properties of copper make it suitable for each of these uses?
e Copper can be beaten into sheets to make roofs. What word describes this property of copper?

19 Aluminium is the most abundant metal in the Earth's crust, making up about 8%. Iron is the next most abundant metal, about 5% of the Earth's crust. Iron and aluminium are extracted from their ores in large amounts. The table below summarizes the two extraction processes.

	Iron	Aluminium
Chief ore	Haematite	Bauxite
Formula of relevant compound in ore	Fe_2O_3	Al_2O_3
Energy source used	Combustion of coke in air	Electricity
Other substances used in the extraction	Limestone	Carbon (graphite) Cryolite
Temperature at hottest part of reactor /°C	1900	1000
How the metal separates from the reaction mixture	Melts and collects at the bottom	Melts and collects at the bottom
Other products	Carbon dioxide Sulphur dioxide Slag	Carbon dioxide

a In each extraction, is the metal oxide oxidized or reduced when it is converted to the metal?
b Explain briefly how each of the following substances is used.
  i limestone in the extraction of iron
  ii carbon in the extraction of aluminium
  iii cryolite in the extraction of aluminium
c Describe any two similarities in the extraction processes for aluminium and iron.
d Give a use for the slag produced as a by-product in the extraction of iron.
e Aluminium costs more than three times as much per tonne as iron. Suggest two reasons why aluminium is more expensive than iron even though it is more abundant.
f Both steel and aluminium are recycled in the UK. Why is it important to recycle these metals?

20 Match each solution from list A with the correct formula from list B.

List A (solutions)	List B (formulae)
Sodium hydroxide	H_2SO_4 (aq)
Hydrochloric acid	HNO_3 (aq)
Ammonia	HCl (aq)
Calcium hydroxide	CH_3COOH (aq)
Sulphuric acid	NH_3 (aq)
Nitric acid	$Ca(OH)_2$ (aq)
Ethanoic acid	NaOH (aq)

21 Five solutions A to E were tested with universal indicator solution, to find their pH. The results are shown below.

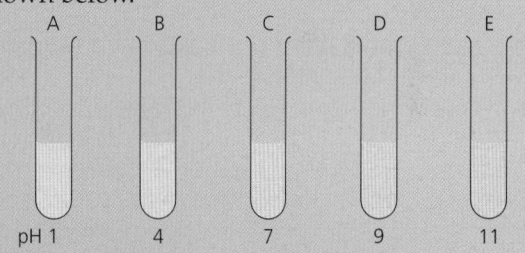

pH 1   4   7   9   11
    A   B   C   D   E

a What colour would each solution be?
b Which solution is:
  i neutral?   ii strongly acidic?
  iii weakly acidic?   iv strongly alkaline?
c The five solutions were known to be sodium chloride, sulphuric acid, ammonia solution, sodium hydroxide, and ethanoic acid. Now identify each of the solutions A to E.

22 Rewrite the following, choosing the correct word from each pair in brackets.
Acids are compounds which dissolve in water giving (hydrogen/hydroxide) ions. Sulphuric acid is one example. It is a (strong/weak) acid, which can be neutralized by (acids/alkalis) to form salts called (nitrates/sulphates).   *(continued over)*

49

Many (metals/non-metals) react with acids to give a gas called (hydrogen/carbon dioxide). Acids also react with (chlorides/carbonates) to give (chlorine/carbon dioxide).
Solutions of acids are (good/poor) conductors of electricity. They also affect indicators. For example, phenolphthalein turns (pink/colourless) in acids, while litmus turns (red/blue).
Alkalis are compounds that dissolve in water giving (hydrogen/hydroxide) ions. They have a pH number (greater/less) than 7.

23 A and B are white powders. A is insoluble in water but B is soluble and its solution has a pH of 3. A mixture of A and B bubbles or effervesces in water. A gas is given off and a clear solution forms.
  a One of the white powders is an acid. Is it A or B?
  b The other white powder is a carbonate. What gas is given off in the reaction?
  c Although A is insoluble in water, a clear solution forms when the mixture of A and B is added to water. Explain why.

24 The chemical name for aspirin is **2-ethanoyloxybenzoic acid**.
This acid is soluble in hot water.
  a How would you expect an aqueous solution of aspirin to affect litmus paper?
  b Do you think it is a strong acid or a weak one? Explain why you think so.
  c What would you expect to see when baking soda is added to an aqueous solution of aspirin? (Baking soda is sodium hydrogencarbonate. It reacts like carbonates do, with acids.)

25 Do these properties belong to acids, alkalis, or both?
  a soluble in water
  b pH values greater than 7
  c soapy to touch
  d dangerous to handle
  e usually react with metals forming the gas hydrogen
  f may be strong or weak
  g neutralize bases
  h form ions in water
  i turn universal indicator red
  j form salts with certain other chemicals
  k found in the stomach to aid digestion
  l contain the hydroxide ion

26 This is a brief description of a neutralization reaction.
'25 cm³ of potassium hydroxide solution were placed in a flask and a few drops of phenolphthalein were added. Dilute hydrochloric acid was added until the indicator changed colour. It was found that 21 cm³ of acid were used.'
  a What piece of apparatus should be used to measure 25 cm³ of sodium hydroxide solution accurately?
  b What piece of apparatus should be used to measure accurately the volume of acid added?
  c What colour was the solution in the flask before the acid was added?
  d What colour did it turn when the alkali had been neutralized?
  e Name the salt formed in this neutralization.
  f Write a word equation for the reaction.
  g How would you obtain *pure* crystals of the salt?

27 These diagrams show the stages in the preparation of copper(II) ethanoate, a salt of ethanoic acid.

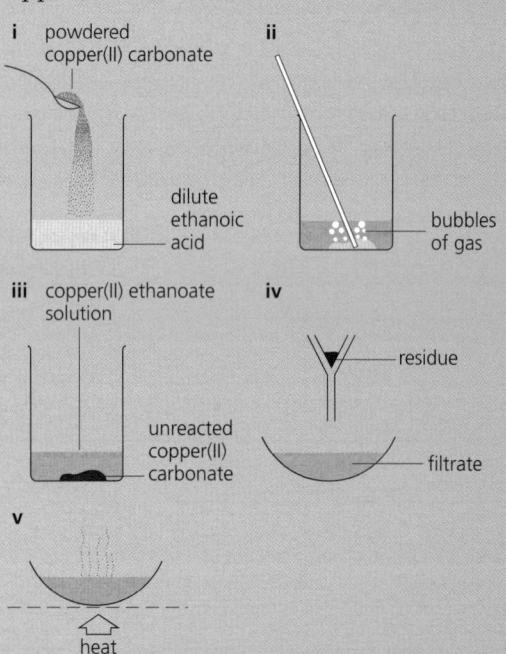

i powdered copper(II) carbonate / dilute ethanoic acid
ii bubbles of gas
iii copper(II) ethanoate solution / unreacted copper(II) carbonate
iv residue / filtrate
v heat

  a Which gas is given off in stage **ii**?
  b Write a word equation for the reaction.
  c How can you tell when the reaction is complete?
  d Which reactant is completely used up in the reaction? Explain your answer.
  e Why is copper(II) carbonate powder used, rather than lumps?

f Name the residue in stage **iv**.
g Write a list of instructions for carrying out this preparation in the laboratory.
h Suggest another copper compound that could be used instead of copper(II) carbonate, to make copper(II) ethanoate.

28 a Name the salt obtained from the following chemical reactions.
   i   calcium hydroxide and nitric acid
   ii   zinc and hydrochloric acid
   iii potassium hydroxide and sulphuric acid
   iv   sodium carbonate and nitric acid
   v   magnesium and sulphuric acid
   vi   sodium hydroxide and nitric acid
   vii copper oxide and sulphuric acid
   viii calcium carbonate and hydrochloric acid
b Which of the reactions would also produce a gas?
c Which reactions also have water as one of the products?

29 The graph shows how the pH of the solution changes as sodium hydroxide is added slowly to hydrochloric acid.

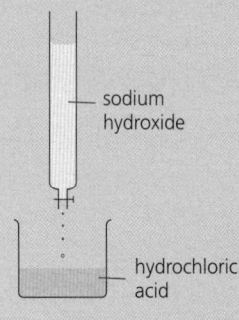

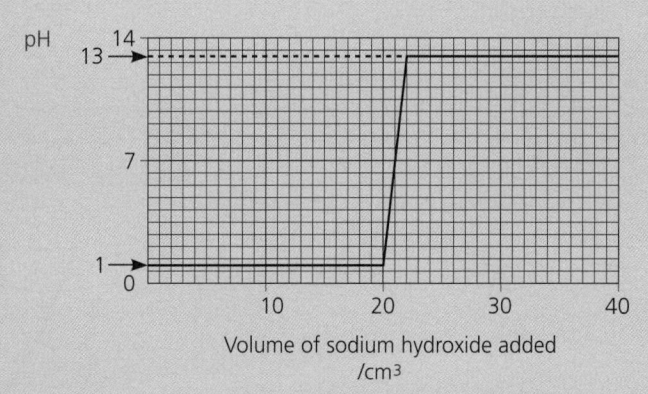

a Describe how the pH changes as the alkali is added.
b What is present in the beaker when the pH is 7?
c What volume of sodium hydroxide neutralizes all the hydrochloric acid in the beaker?
d What is present in the beaker when the pH is 13?
e Sketch the graph that would be obtained if the acid had been added to the sodium hydroxide rather than the other way around.

30 Magnesium sulphate is the chemical name for Epsom salts. They can be made in the laboratory from magnesium oxide in a neutralization reaction.
a What type of substance is magnesium oxide?
b Which acid should be used to make the salt?
c Write a word equation for the reaction.
d Describe in detail how a student could obtain pure Epsom salt cystals in the laboratory.

31 Using the apparatus shown in question 29, a solution of nitric acid is added to a solution of potassium hydroxide in the beaker.
a i   Write a word equation for the chemical reaction that occurs in the beaker.
   ii Which of these terms correctly describes the reaction?
       decomposition, displacement, neutralization, reduction, oxidation
   iii Is the reaction exothermic, or endothermic?
   iv Will the temperature in the beaker rise or fall?
b i   Name the apparatus that contains the nitric acid.
   ii Why is it used in this type of experiment?
c i   Name an indicator (not universal indicator) that could be used to show when the reaction was complete.
   ii What colour change will you see with this indicator?

32 Sulphur dioxide from power stations is a major cause of acid rain. The waste gases from the power station can be sprayed with a slurry of calcium hydroxide, to deal with this problem.
a Explain why sulphur dioxide causes acid rain.
b Explain why acid rain attacks metal structures such as iron railings.
c Acid rain can also damage limestone buildings. Why?
d What type of chemical reaction is brought about by the calcium hydroxide?
e One product of this reaction is a salt. Name it.
f Name a weak acid that is always found in rain.
g Suggest a way to test rain in different areas, to see how acidic it is.

51

Module 6
Earth materials

Planet Earth – your home. Born 4.6 billion years ago from a mass of particles swirling around the Sun. Surrounded by a thin blanket of gas, the atmosphere. Almost covered by a thin film of water, the oceans. Spinning non-stop on its imaginary axis, and circling endlessly around the Sun.

From space, the Earth looks beautiful, but cold, solid, and unchanging. Zoom in closer and you find a different story. A planet teeming with hundreds of thousands of different species. Where over 6 billion of us humans live. Where the temperature varies from desert heat to Arctic cold. Where every living thing depends on, and interacts in some way with, the atmosphere, rocks, and oceans. And where 'solid' is only skin deep. For much of this planet is soft inside, even liquid in places. And the surface is continually on the move, driven by powerful and dangerous forces.

In this module you will learn more about this home of yours: the atmosphere, the oceans, the Earth's interior, and how the crust, where you live, is in continual change. You will also learn about two important raw materials that result from those changes: crude oil and limestone. ∎

6.01 Limestone and its uses

Limestone: a sedimentary rock

Many of the creatures that live in the sea have shells and skeletons made of calcium carbonate. Over millions of years their dead remains have fallen to the ocean floor. Slowly, over millions of years, pressure has turned the layers of shell and bone fragments into limestone rock.

Over millions more years, powerful forces have raised some sea beds upwards, draining and folding them to form mountains. That explains why limestone is found inland, miles from the sea!

In this way, bits of dead sea creatures have turned into one of our major raw materials. Every year over 60 million tonnes of limestone are blasted from UK quarries. This is what we use it for:

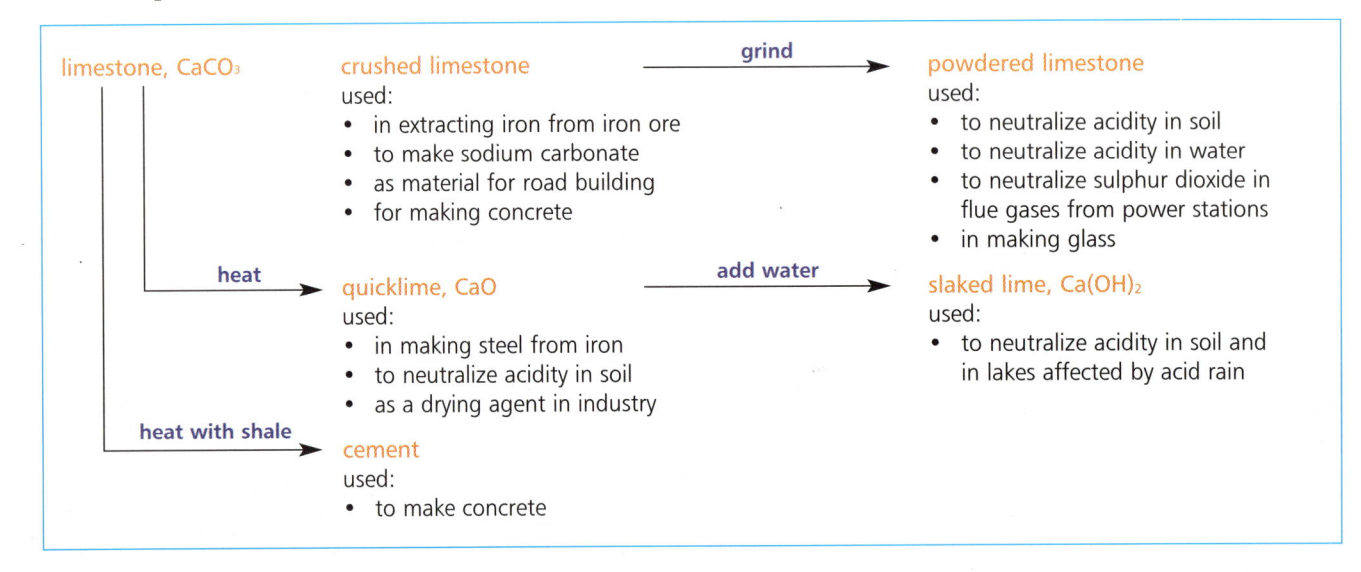

limestone, $CaCO_3$

→ **crushed limestone**
used:
- in extracting iron from iron ore
- to make sodium carbonate
- as material for road building
- for making concrete

→ grind → **powdered limestone**
used:
- to neutralize acidity in soil
- to neutralize acidity in water
- to neutralize sulphur dioxide in flue gases from power stations
- in making glass

heat → **quicklime, CaO**
used:
- in making steel from iron
- to neutralize acidity in soil
- as a drying agent in industry

→ add water → **slaked lime, $Ca(OH)_2$**
used:
- to neutralize acidity in soil and in lakes affected by acid rain

heat with shale → **cement**
used:
- to make concrete

The substances made from limestone

Quicklime When limestone is heated, it breaks down:

$$CaCO_3\,(s) \longrightarrow CaO\,(s) + CO_2\,(g)$$
calcium carbonate   calcium oxide   carbon dioxide
(limestone)   (quicklime)

This is an example of **thermal decomposition**. (*Thermal* tells you it is brought about by heat.) Other carbonates break down in the same way.

This shows a rotary kiln for making quicklime. Limestone is fed in one end, and quicklime comes out the other. Note its uses, in the table above.

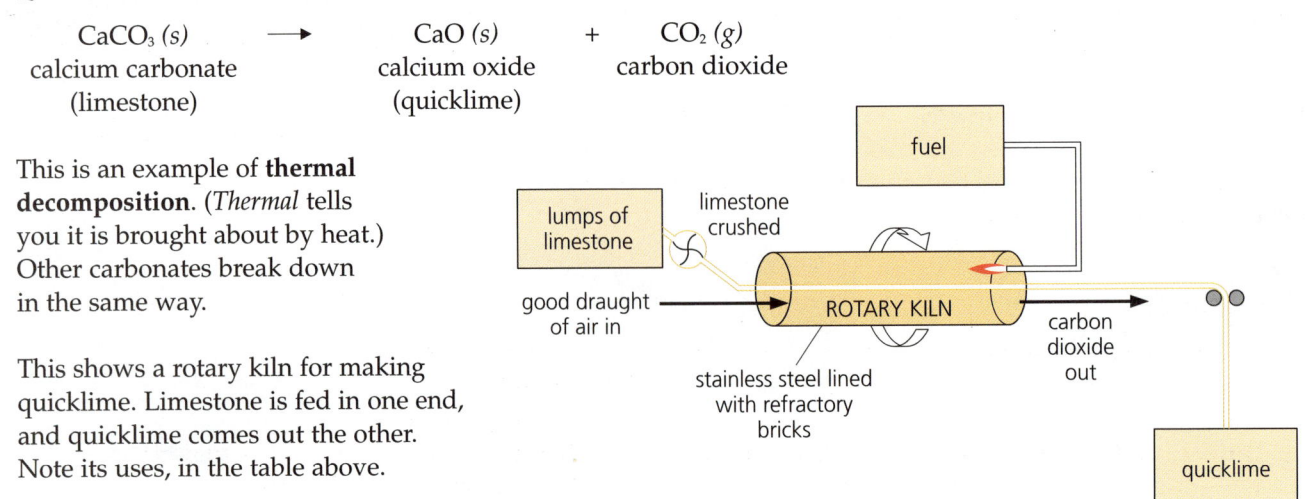

A rotary kiln for making quicklime.

54

Slaked lime When water is added to quicklime the reaction is exothermic. The mixture hisses and steams. Conditions are carefully controlled so that the slaked lime forms as a fine powder:

$$CaO\,(s) + H_2O\,(l) \longrightarrow Ca(OH)_2\,(s)$$
calcium oxide → calcium hydroxide
(quicklime) (slaked lime)

It is used to neutralize acidity in soil and in lakes.

Cement This grey powder is a mixture of calcium silicates and aluminates. When it is mixed with water it dries into a hard solid. It is made by mixing limestone with shale, heating the mixture very strongly, adding gypsum (calcium sulphate), and grinding up the final lumps of solid:

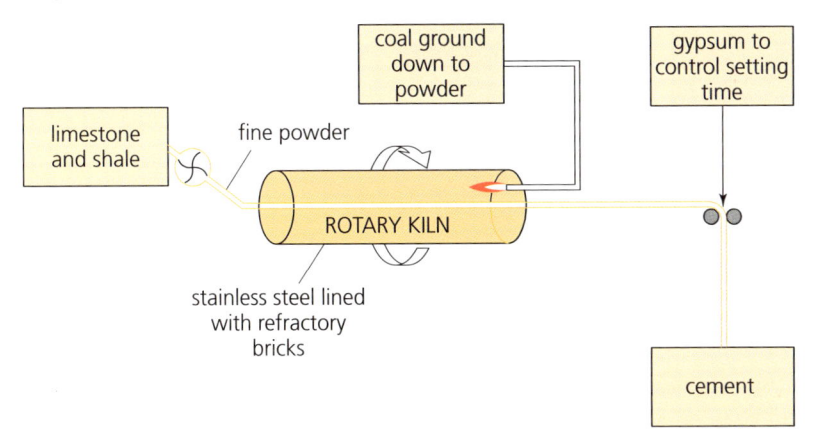

A rotary kiln for making cement.

Concrete For this, cement is mixed with **aggregate** (a mixture of stone chips and sand) and water added. As the concrete 'sets', the compounds in the cement bond to form a hard mass of **hydrated** crystals – that is, crystals with water molecules bonded inside them.

The setting process is exothermic. If it's too fast, the heat inside a big concrete structure (such as a dam) could be enough to crack the concrete. It's also a nuisance if the concrete you mix on building sites hardens before you're ready to use it. The purpose of adding gypsum to cement is to slow the setting process down.

Glass It is made by heating limestone, sand (silicon dioxide), sodium carbonate, and recycled glass in a furnace to around 1300 °C. In the furnace the carbonates break down to give a molten mixture of oxides. On cooling this hardens to form glass.

> **Limewater**
> **Limewater** is a solution of calcium hydroxide in water (it is only slightly soluble). Limewater is used to test for **carbon dioxide**. It turns milky when the gas bubbles through it because a precipitate of calcium carbonate forms. This then dissolves again.

You can pour and shovel wet concrete, which is convenient. But don't walk on it until it's set.

Questions

1 How is limestone formed?
2 How is quicklime made? Write a balanced equation for the reaction.
3 How is slaked lime made? Write a balanced equation.
4 Give two uses for: **a** crushed limestone **b** quicklime
5 **a** What is *cement*?
   b How is it made?
   c Why is gypsum added?
6 What ingredients are mixed to make concrete?
7 List the ingredients for making glass.

55

6.02 To quarry, or not?

The quarrying dilemma

Limestone is an essential material. Concrete, cement, and glass, steel for cars, buildings, and bridges, even our roadways all depend on limestone. No wonder the UK uses 60 million tonnes of it a year.

The problem is – this has to be quarried from somewhere. Which means noise, dust, and large gashes in the landscape – and limestone landscapes are often scenic spots.

It is a special problem in the Peak District, where the Eldon Hill quarry (above) is located. The Peak District is one of our National Parks, whose aims include 'to conserve and enhance natural beauty'. Quarrying doesn't really help!

Quarrying in the Peak District

This map shows the geology of the Peak District. There has been limestone quarrying here since Roman times.

After World War II, large quantities of building materials were needed. The limestone in the Peak District is high quality, and was in great demand. So licences were granted to remove large amounts.

The quantity rose rapidly – from 1.5 million tonnes in 1951 to a peak of 8.2 million tonnes in 1991.

Limestone is a bulky material. Try to imagine several million tonnes of it! As well as noise, dust, and holes in the landscape, it means a stream of trucks in an area with narrow country roads. But it also means jobs for local people.

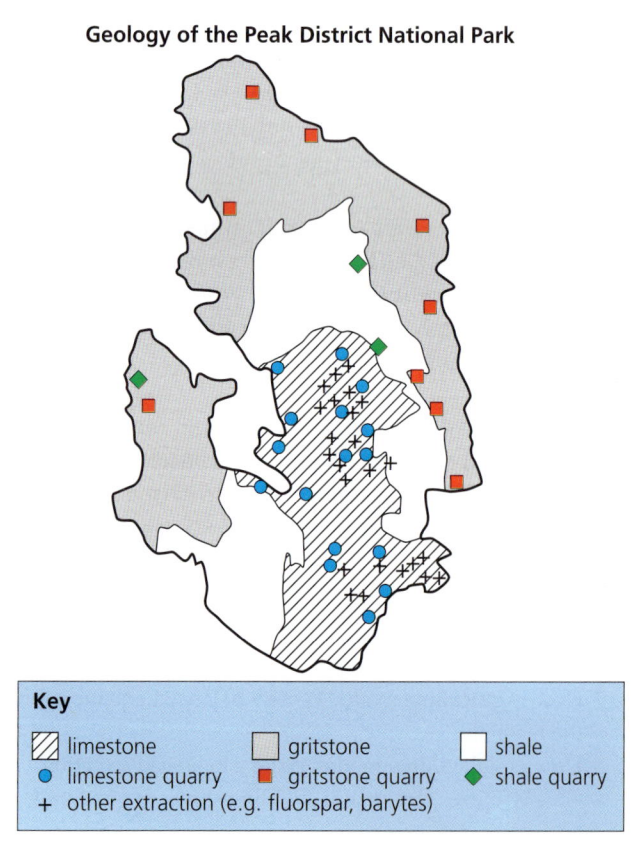

Geology of the Peak District National Park

Key
- limestone
- gritstone
- shale
- ● limestone quarry
- ■ gritstone quarry
- ◆ shale quarry
- + other extraction (e.g. fluorspar, barytes)

What is the Peak District limestone used for?

Over half the Peak District limestone goes for aggregate – chips for building and making roads. Look at this table:

Uses of Peak District limestone (millions of tonnes)

	1986	1987	1988	1989	1990	1991	1992	1993	1994	1995
Aggregate	3.1	3.5	4.4	4.1	4.5	4.7	4.1	3.7	3.4	3.4
Other uses	1.9	2.8	3.1	3.3	4.0	3.5	2.5	2.4	2.6	2.6
Total	5.0	6.3	7.5	7.4	8.5	8.2	6.6	6.1	6.0	6.0

Keeping it under control

Limestone is not the only material taken from the Peak District. Some **shale** and **gritstone** are quarried. **Fluorspar** (calcium fluoride) and barytes (barium sulphate) are mined. But the National Park Authority has to protect the area. In deciding whether to grant licences, it asks questions like these. (The answers for limestone are in blue.)

1 Is this product really needed, or could another be used instead?
Limestone is essential for some uses, like iron and steel making. But other stone could be used for building roads.
2 Is there another source for this material?
There are other limestone areas in the UK, which are not National Parks.
3 What will the effect on traffic be? Could the material go by rail?
Some quarries already move limestone by rail. Some plan to.
4 How will the local residents, landscape, and environment be affected?
A quarry still means noise and dust and holes in the landscape – but these can be screened from the public. It means a certain amount of traffic. AND jobs for local people.

The heavy trucks are just one of the problems caused by quarrying in a National Park.

The National Park Authority and many local people feel too much limestone is being removed. So when a licence runs out it may not be renewed. In 1997 the Eldon Hill quarry was refused an extension.

Just moving the problem elsewhere?

If less limestone is quarried in the Peak District, more must be quarried elsewhere. Or must it? This is part of a wider issue. Do we need more new roads? Do we need *all* this aggregate? Is this really progress? Now people are starting to think harder about **sustainable development**. That means only building and making and using things that do no harm in the long run to ourselves, the environment, or other living things.

Questions

1 How many limestone quarries operate in the Peak District?
2 Write down any ways in which limestone quarrying benefits the Peak District.
3 Now write down any problems it brings.
4 Over half the limestone from the Peak District is used for *aggregate*. What does that mean?
5 What does *sustainable development* mean?

6.03 Crude oil

It's a fossil fuel
Crude oil is a **fossil fuel**. That means it is the remains of plants and animals that lived millions of years ago – some of them long before the dinosaurs roamed the Earth. Coal and natural gas are also fossil fuels. Oil and natural gas are often found together.

How did crude oil form?
This is what scientists think happened:

Tiny marine plants and animals die and fall to the ocean floor. They are buried under sediment (sand and mud). The layer grows deeper and deeper and in time compacts into sedimentary rock.

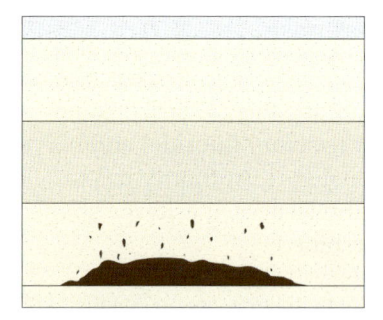

The remains can't rot away since there is no oxygen. But movements in the Earth's crust increase the heat and pressure on them, causing other chemical changes. They slowly turn into **oil** and **gas**.

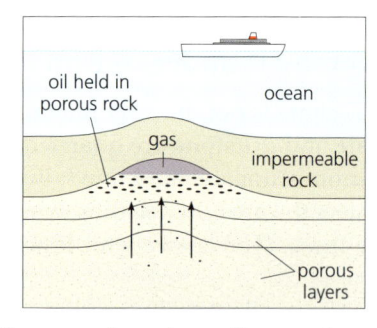

These are less dense than water. They move upwards through permeable rock until they can get no further. They are in an **oil trap**. The rock they collect in is called **reservoir rock**.

And that's where the oil and gas remain until an oil company arrives. (Coal is formed in a similar way, on land, from buried vegetation.)

Getting the crude oil out
Oil companies survey for oil by bouncing shock waves off the sea bed. They feed the data into a computer and build up a 3-D image of the rock structure. If this looks promising they drill out rock samples and test them for the presence of oil. This is not such an easy job – oil is often 3 or 4 km below the surface.

If the news is good a production platform is set up. A hole is drilled into the reservoir rock and lined with steel. The oil is pumped up through it. Pipes carry the oil to a tanker terminal or all the way ashore to an oil refinery.

Once the production platform is operating it will run 24 hours a day and for as long as it's profitable – which could be 20 years or more. Production is measured in barrels of oil per day. (A barrel of oil is about 160 litres.)

Crude oil varies in appearance and quality from oil field to oil field. Some is runny and almost colourless. Some is dark and heavy.

An oil platform in the North Sea. There are over 120 oil fields in the North Sea in production or being developed.

What's crude oil made of?

Crude oil is a smelly mixture of hundreds of different compounds. They are **organic compounds** which means they originated in living things. Most are **hydrocarbons** which means they contain *only* carbon and hydrogen. Here are three hydrocarbon molecules:

This molecule has a chain of 5 carbon atoms.

Here a chain of 6 carbon atoms has formed a ring.

Here 6 carbon atoms form a **branched** chain.

Note that the carbon atoms form the spine of each molecule, and the hydrogen atoms are bonded to them. Some of the molecules in crude oil contain up to 50 carbon atoms.

Liquid gold

The compounds obtained from oil have thousands of different uses.

Some of the compounds are used as fuels for cars, trucks, planes, and ships.

Some are the starting point for detergents, drugs, dyes, paints, and cosmetics.

Some are the starting point for nylon, PVC, and other plastics. Most plastics start off as oil.

Chances are that many of the things you use every day started off as oil – your toothbrush, deodorant, shampoo, shower gel, and alarm clock just for a start. Just think about that tomorrow morning!

A non-renewable resource

Crude oil is still forming, slowly, somewhere under the oceans. But we are using oil up much faster than it can form, so it is considered a **non-renewable** resource.

It is hard to estimate when oil will run out. We have not found it all. We are being encouraged to use less, in the fight against global warming. But at the present rate of use, and known reserves, some experts estimate that we have enough for about 40 more years.

Scientists and oil
Many kinds of scientist work in the oil industry. For example:
- geologists and seismologists explore for oil (seismologists study shockwaves
- computer scientists handle data
- analytical chemists test samples
- research chemists develop new uses for compounds from oil.

Questions
1 Explain why oil is called a fossil fuel.
2 Under what conditions do dead marine plants and animals turn into oil and gas?
3 What is reservoir rock?
4 What is a hydrocarbon?
5 What does crude oil contain?
6 Explain why oil is such a valuable resource.
7 Oil is called a *non-renewable* resource. Explain why.

6.04 Separating oil into fractions

Refining oil

Crude oil contains *hundreds* of different hydrocarbons. A big mixture like that is not much use. So the first step is to separate it into smaller groups of compounds with molecules of similar size. This is called **refining** the oil. It is carried out by **fractional distillation**.

Fractional distillation of crude oil in the lab

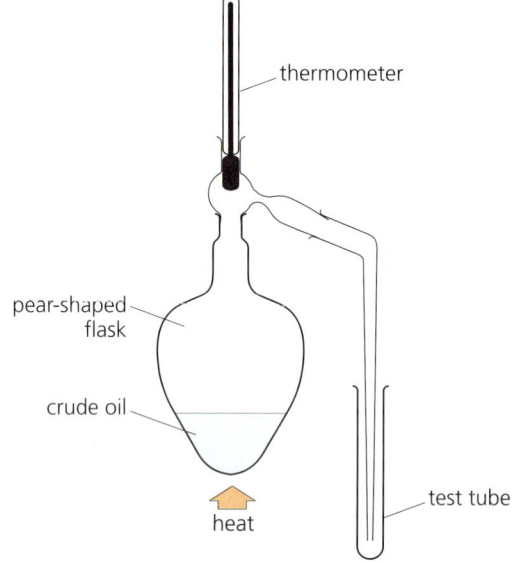

1 As you heat the oil, the compounds **evaporate**. The ones with smaller lighter molecules go first, since it takes less energy to free these from the liquid.

2 As the hot vapours rise, so does the thermometer reading. The vapours condense in the cool test tube.

3 When the thermometer reading reaches 100 °C, replace the test tube with an empty one. The liquid in the first test tube is your first **fraction** from the distillation.

4 Collect three further fractions in the same way, replacing the test tube at 150 °C, 200 °C, and 300 °C.

Comparing the fractions

Compare the fractions – how runny they are, how easily they ignite, and so on. If you try igniting them on a watch glass you will find:

fraction 1 
It catches fire easily. The flame burns high over the liquid, showing that the liquid is **volatile**: it evaporates easily.

fraction 2 
It catches fire quite easily. The flame burns closer to the liquid – so this fraction is less volatile than fraction 1.

fraction 3 
It seems to be less volatile than fraction 2. It does not catch fire so readily or burn so easily – it is not so **flammable**.

fraction 4 
This one does not ignite easily, and you need to use a wick to keep it burning. It is the least flammable of the four.

This table summarizes all the trends you will notice:

Fraction	Boiling point range	How easily does it flow?	How volatile is it?	How easily does it burn?	Molecule sizes
1	Up to 100 °C	Very runny	Volatile	Very easily	small
2	100 °C to 150 °C	Runny	Less volatile	Easily	
3	150 °C to 200 °C	Not very runny	Even less volatile	With some difficulty	
4	200 °C to 300 °C	Rather sticky or **viscous**	Least volatile	Only with a wick	large

What the results indicate

The results indicate that the larger the molecules in a hydrocarbon:
- the higher its boiling point is
- the less volatile it is
- the less easily it flows (or the more viscous it is)
- the less easily it ignites (or the less flammable it is)

Fractional distillation in the oil refinery

The same idea is used to separate the crude oil into fractions in the oil refinery. The fractional distillation is carried out in a tall tower which is kept very hot at the base, and cooler towards the top. Crude oil is pumped in at the base, and the compounds start to boil off. Those with the smallest molecules have the lowest boiling points. So they boil off first and rise to the top of the tower. Others rise only part of the way, depending on their boiling points, and then condense.

An oil tanker delivering oil to a jetty near a refinery.

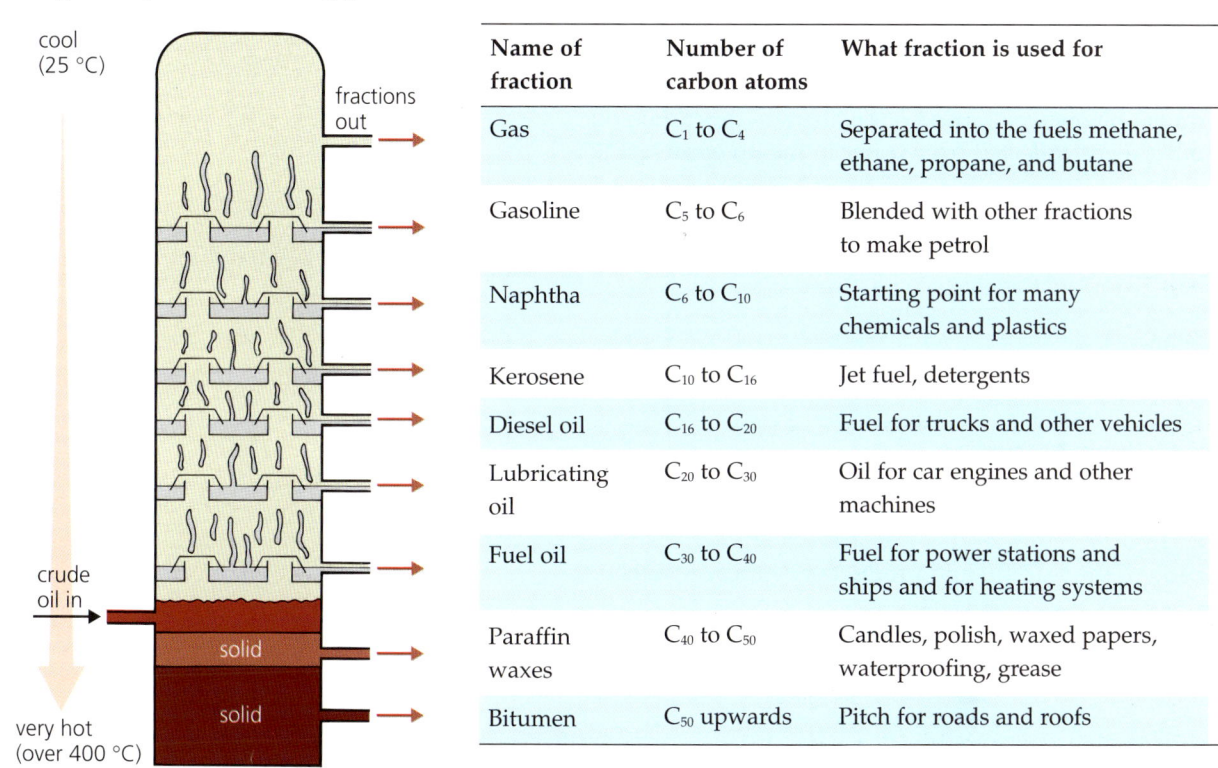

Name of fraction	Number of carbon atoms	What fraction is used for
Gas	C_1 to C_4	Separated into the fuels methane, ethane, propane, and butane
Gasoline	C_5 to C_6	Blended with other fractions to make petrol
Naphtha	C_6 to C_{10}	Starting point for many chemicals and plastics
Kerosene	C_{10} to C_{16}	Jet fuel, detergents
Diesel oil	C_{16} to C_{20}	Fuel for trucks and other vehicles
Lubricating oil	C_{20} to C_{30}	Oil for car engines and other machines
Fuel oil	C_{30} to C_{40}	Fuel for power stations and ships and for heating systems
Paraffin waxes	C_{40} to C_{50}	Candles, polish, waxed papers, waterproofing, grease
Bitumen	C_{50} upwards	Pitch for roads and roofs

boiling points and viscosity increase

As molecules get larger, the fractions get less runny, or more **viscous**: from gas at the top of the tower to solid at the bottom. They also get less **flammable**. Grease burns less easily than petrol, for example. So the lower fractions are not used as fuels.

Questions

1. Which two opposite processes take place during fractional distillation?
2. A group of compounds collected together during fractional distillation is called a ____?
3. What does it mean? **a** volatile **b** viscous
4. List four ways in which the properties of different fractions differ.
5. Name the oil fraction that: **a** is used for petrol **b** has the smallest molecules **c** is the most viscous **d** has molecules with 20 to 30 carbon atoms

6.05 Cracking hydrocarbons

After fractional distillation...

Crude oil is separated into **fractions** by fractional distillation. But that's not the end of the story. The fractions all need further treatment before they can be used.

1 They contain impurities. These are mainly sulphur compounds. If left in fuel, they will burn to form poisonous sulphur dioxide gas.

2 Some fractions are separated into single compounds or smaller groups of compounds. For example the gas fraction is separated into methane, ethane, propane, and butane.

3 Part of a fraction may be **cracked**. Cracking breaks down molecules into smaller ones.

This plant removes sulphur impurities.

Cracking a hydrocarbon in the lab

The experiment below was carried out using a hydrocarbon oil. The product is a gas collected over water in the inverted test tube:

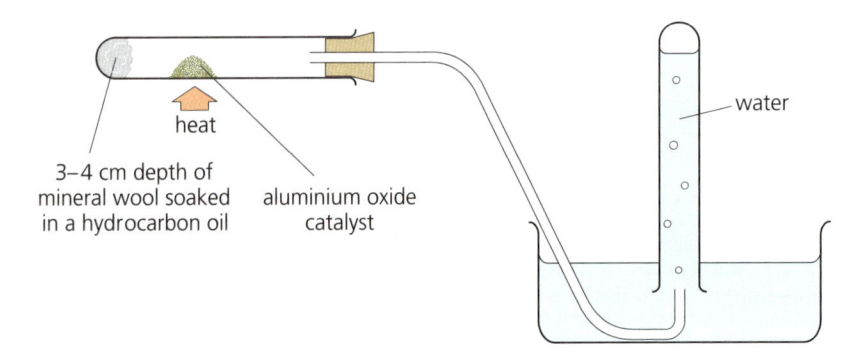

A comparison of the reactant and product gave these results:

	The reactant	The product
Appearance	Thick colourless liquid	Colourless gas
Smell	No smell	Pungent smell
Flammability	Difficult to burn	Burns readily
Reactions	Few chemical reactions	Many chemical reactions

The results indicate that heating has caused the hydrocarbon oil to break down. A **thermal decomposition** has taken place. Note these things:

- The reactant had a high boiling point and was not flammable, which tells us it had large molecules with long chains of carbon atoms.
- The product has a low boiling point and is very volatile – so it must have small molecules with short carbon chains.
- The product must also be a hydrocarbon, since nothing new was added during the reaction.
- So the molecules of the hydrocarbon oil have been cracked.
- The product is more reactive – so could be more useful in industry.

Cracking in the oil refinery

In an oil refinery cracking is carried out in much the same way. The long-chain hydrocarbon is heated to vaporize it, and the vapour is usually passed over a hot catalyst. Thermal decomposition takes place.

Cracking is one of the most important reactions in the oil industry:

1 It lets you turn long-chain molecules into more useful shorter ones. For example you may have plenty of naphtha, but not enough gasoline for making petrol. So you could crack some naphtha to get molecules the right size for petrol.
2 It *always* produces short-chain compounds that have a double bond between carbon atoms. The double bond makes these compounds **reactive**. So they can be used to make plastics and other substances.

When a catalyst is used, cracking is often called **catalytic cracking**.

This is the cracking plant at Esso's Fawley refinery.

Example: cracking the naphtha fraction

This is the kind of reaction that takes place when you crack naphtha:

decane, $C_{10}H_{22}$ from naphtha fraction

↓ 540°C, catalyst

pentane, C_5H_{12} suitable for petrol   +   propene, C_3H_6 +   ethene, C_2H_4

The decane has been broken down into smaller molecules. Look at the propene and ethene molecules. They have carbon–carbon double bonds. They are called **alkenes**.

Cracking ethane…

ethane →(steam, >800 °C) ethene + H_2

Ethane is already a small molecule – so even small molecules can be cracked. The hydrogen produced may be used to make ammonia.

Questions

1 What happens during cracking?
2 Cracking is a thermal decomposition. Explain why.
3 Describe the usual conditions needed for cracking a hydrocarbon.
4 What is always produced in a cracking reaction?
5 Explain why cracking is so important.
6 a A straight-chain hydrocarbon has the formula C_5H_{12}. Draw a diagram of one of its molecules.
  b Now show what might happen when the compound is cracked.

6.06 The alkanes and alkenes

Hydrocarbon families

Hydrocarbons are compounds that contain only carbon and hydrogen. Here we look at two families of hydrocarbons: the **alkanes** and **alkenes**.

The alkanes

This table shows the first four members of the alkane family:

Name	Methane	Ethane	Propane	Butane
Formula	CH_4	C_2H_6	C_3H_8	C_4H_{10}
Structure of molecule	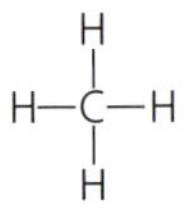 H—C—H with H above and below	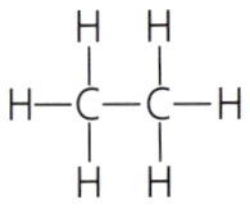 H—C—C—H with H's	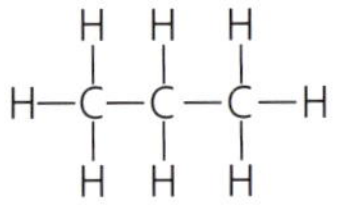 H—C—C—C—H with H's	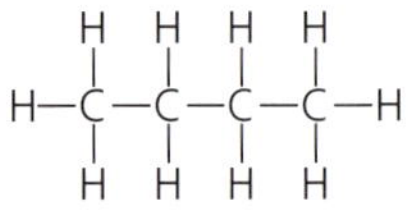 H—C—C—C—C—H with H's
Number of carbon atoms in chain	1	2	3	4
Boiling point	−164 °C	−87 °C	−42 °C	−0.5 °C

boiling point increases with chain length

One more carbon atom is added each time. It could reach thousands! The general formula for the family is C_nH_{2n+2} where n is a number. Can you see how this formula fits the alkanes in the table?

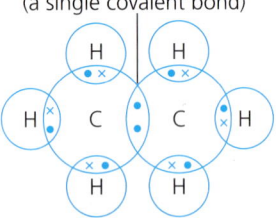

Points to note about the alkanes

1 They are found in oil and natural gas. Natural gas is mostly methane, with small amounts of ethane, propane, and butane. Oil contains alkanes with up to 50 carbon atoms.

2 The longer the chain, the higher the boiling point. So the first four alkanes are gases at room temperature, the next twelve are liquids, and the rest are solids.

3 In an alkane molecule, each carbon atom forms four single covalent bonds. So alkanes are said to be **saturated**. The bonding in ethane is shown above right.

4 Alkanes are unreactive. They are not affected by acids or alkalis.

5 But they burn well in a good supply of oxygen, forming carbon dioxide and water vapour, and giving out plenty of heat. For example, propane burns like this:

$C_3H_8 (g) + 5O_2 (g) \longrightarrow 3CO_2 (g) + 4H_2O (g) + \text{heat}$

So alkanes are used as fuels. Natural gas (North Sea gas) is used for cooking and heating in homes. Propane and butane are used as camping gas, and Calor gas is mainly butane. Petrol contains butane and liquid alkanes, as well as many other hydrocarbons.

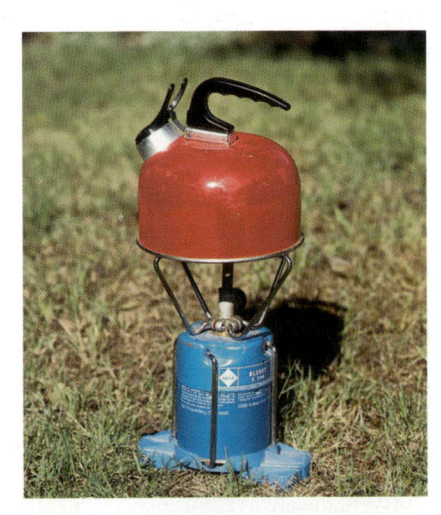

The bonding in ethane.

Camping gas is butane.

The alkenes

Here are the first two members of the alkene family:

Name	Ethene	Propene
Formula	C_2H_4	C_3H_6
Structure of molecule	![ethene structure] H₂C=CH₂	![propene structure] H₂C=CH–CH₃

The general formula for the alkene family is C_nH_{2n}. Can you see why?

Points to note about the alkenes

1. They are made from alkanes by **cracking**.
2. An alkene molecule contains a **double covalent bond** between carbon atoms. The bonding in ethene is shown on the right.
3. Alkenes are much more reactive than alkanes. This is because the double bond can break to form single bonds and add on other atoms. For example, ethene reacts with hydrogen like this:

   $$CH_2=CH_2 \text{ (g)} + H_2 \text{ (g)} \longrightarrow CH_3-CH_3 \text{ (g)}$$
   ethene → ethane

   This is called an **addition reaction**.
4. In the same way alkene molecules can also add on to each other, to make long-chain molecules called **polymers**.
5. Because the double bond allows them to add on more atoms, alkenes are said to be **unsaturated**.

The bonding in ethene.
two shared pairs of electrons (a double covalent bond)

A test for unsaturation

You can test whether a hydrocarbon is an alkane or alkene using **bromine water**. It is an orange solution of bromine in water. It turns colourless in the presence of an alkene because the bromine adds on to the alkene, giving a colourless compound:

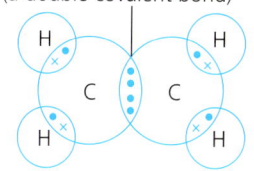

$$CH_2=CH_2 \text{ (g)} + Br_2 \text{ (aq)} \longrightarrow CH_2Br-CH_2Br \text{ (l)}$$
ethene + (orange) → 1,2-dibromoethane (colourless)

Unsaturated hydrocarbons don't burn cleanly. Crude oil contains many unsaturated hydrocarbons.

Questions

1. Will alkanes conduct electricity? Explain.
2. Draw a diagram to show the *bonding* in methane.
3. What are the differences between an *ethane* molecule and an *ethene* molecule?
4. Alkenes are *unsaturated*. What does that mean? What effect does it have?
5. Describe how bromine water can be used to distinguish between alkanes and alkenes.

65

6.07 Polymerization and plastics

The polymerization of ethene

Cracking produces **unsaturated** hydrocarbons with carbon–carbon double bonds. These compounds are very useful because the double bond makes them reactive.

For example, look what happens when ethene (a gas) is heated at very high pressure:

ethene molecules (monomers)

↓ polymerization

part of a polythene molecule (a polymer)

The double bonds break, and the ethene molecules join up to make very long molecules with thousands of carbon atoms. The result is **polythene**, the plastic used for plastic bags.

Polythene is a **polymer**. The small starting molecules are called **monomers**. The reaction is called an **addition polymerization** because the monomers just add on to each other. No other substance is formed.

One of the many uses for polythene.

Other unsaturated compounds will polymerize too!

Below are three more examples. Note the shorthand equations.

The monomer	Part of the polymer molecule	The equation for the reaction
chloroethene (vinyl chloride)	poly(chloroethene) or polyvinyl chloride (PVC)	n stands for a large number!
propene	poly(propene)	
phenylethene (styrene)	poly(phenylethene) or poly(styrene)	

Some uses of these polymers

Polymer	Uses
Poly(ethene) or polythene	Plastic bags and bottles, clingfilm
Poly(chloroethene) or PVC	Drain pipes, wellingtons, covering for electrical cables
Poly(propene)	Crates, ropes
Poly(styrene)	Fast-food cartons, plastic cups, packaging materials

Plastics are polymers

The materials we call **plastics** are in fact polymers. (We say plastic bags, not polythene bags.) These are the general properties of plastics:

1 They do not conduct electricity and are poor conductors of heat.
2 They are unreactive. Most are not affected by air, water, or chemicals. (It makes them useful – but hard to get rid of too!)
3 They are strong. This is because their long molecules are attracted to each other. So plastics are hard to tear or pull apart.

a pair of polymer molecules

as molecules get longer....

force of attraction between them increases

Melt or char?

One important property of a plastic is how it behaves on heating. Some melt when you test them with a hot nail. Others char. Why?

In some plastics the long polymer chains are attracted to each other but not linked together. Like this:	In others, the chains are cross-linked by strong covalent bonds, like this:
This plastic is flexible, and will soften and melt when you heat it, because the chains can bend and slide over each other. It is called a **thermosoftening** plastic because heat softens it. On cooling it hardens into its new shape. (But you can melt it again.)	This plastic is rigid and will break rather than bend. It won't melt or soften on heating – it chars instead. This is because the cross-links stop the chains bending or sliding. It is called a **thermosetting** plastic, because it sets into shape while it is being made.
Examples: polythene, PVC	Examples: Bakelite, melamine

The hot nail test: melt or char?

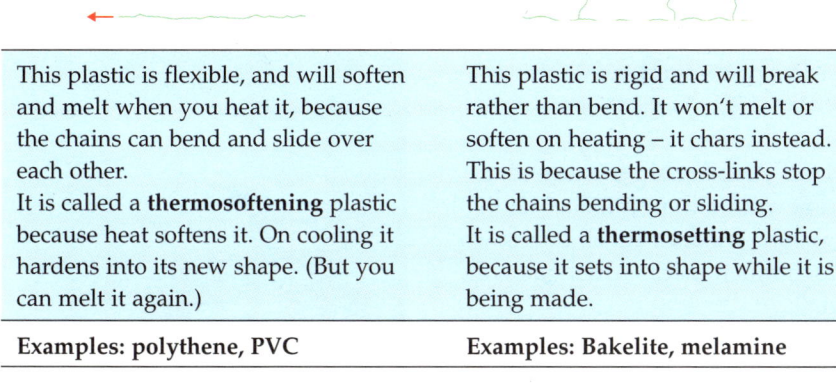

So polythene is no good for light bulb sockets – but Bakelite is just fine!

They were put in an oven for just a few seconds. Which one is made from a thermosoftening plastic?

Questions

1 What is: **a** a monomer? **b** a polymer?
2 What is *addition polymerization*?
3 Draw a diagram to show what happens when chloroethene molecules polymerize.
4 Write the equation for the polymerization of ethene in a short form (using *n* to stand for a large number).
5 **a** What is a thermosetting plastic?
   b What is a thermosoftening plastic?
   c Draw diagrams to show their structures.
6 Give three reasons why:
   a Bakelite is used for the fittings that hold light bulbs
   b polythene is used for plastic bags

67

6.08 Polythene – here to stay?

A happy accident

Often chemists make their best discoveries by accident. That's how it was with polythene.

It is 1935. At ICI, two chemists are carrying out an experiment to study the effects of very high pressure on reactions. They are heating ethene in a reaction vessel. Suddenly the pressure drops. The equipment has sprung a leak – but they don't know that. Puzzled, they pump more ethene in. Some oxygen gets in too, by accident. A little later they open the reaction vessel and find a strange white waxy solid. It's polythene.

No leak, no oxygen. No oxygen, no polythene. The oxygen had acted as a catalyst for the polymerization reaction.

They can't think of a use for the white substance. But someone else hears about it. His company needs a material to cover underwater telephone cables. Could polythene do the job? Yes: it is unreactive and an electrical insulator. So ICI open a small factory to make polythene – just as World War II is starting. Suddenly, urgently, British fighter planes need a lightweight insulating material to allow them to carry radar. Polythene is the answer. It plays a vital part in the war.

And another happy accident

It is 1953. The war is over. Polythene is becoming popular as wrapping film, but it's too light and flexible to make anything more sturdy. In his lab, a German scientist is trying to make a sample of the polymer. But he can't get the monomers to join. This is weird! What's going on? In the reaction vessel he finds traces of a nickel compound from an earlier experiment. He hadn't cleaned it properly. The compound is inhibiting the polymerization.

So he decides to try out other compounds too. And finds a mixture that acts as a great catalyst – and gives a brand new kind of polythene! High density and a high melting point. Soon it will be used for lunch boxes, buckets, milk jugs, drainage pipes, and much much more. Polythene will become the most used plastic in the world.

Their polythene
At the high temperature and very high pressure they used, the two chemists Eric Fawcett and Reginald Gibson got polythene with branched chains:

The branches stop the chains getting very close together. So this polythene is **low density**.

His polythene
The German chemist Karl Zeigler used a catalyst, medium pressure and low temperature – and got straight chains:

The chains can get close together, so this polythene is **high density**. Today polythene is made in a range of densities.

High density polythene has all sorts of everyday uses.

And now the bad news...

Polythene is so unreactive that we can't get rid of it. It is not affected by air and water. It is not **biodegradable** – bacteria can't break it down. The first polythene ever used could still be around, unless it got burned! And it's the same with other plastics. Around a million tonnes of plastic waste lies buried in the UK's landfill sites. Much of it could still be there, intact, 50 years from now.

Some plastics are recycled. But there are so many different types that this is not easy. And they are often turned into things like bin liners, shoe soles and fleeces that will just get thrown away next time round.

A happy ending?

In 1999 a British company launched a range of SPI-TEK™ polythene packaging. This contains an additive that promotes the breakdown of polythene to carbon dioxide and water. Heat, sunlight, or just tearing the plastic will start the process off, and bacteria take over at a later stage. How long the breakdown takes depends on how much additive is present. This can be varied for different purposes – for example to make rubbish sacks that will break down within weeks.

Several supermarkets sell SPI-TEK™ bin liners, and fruit packed in SPI-TEK™ packaging. It is too early to say whether the product is a success. But there could at last be a happy ending.

Polythene and other plastics are not biodegradable (unlike natural materials). So they hang around in landfill sites for years.

The leaves will break down within months. The SPI-TEK™ sack will break down even sooner.

Questions

1 What two properties of polythene made it useful during World War II?
2 Two 'mistakes' led to the discovery of two different forms of polythene. What were these mistakes?
3 a Draw labelled diagrams to show the structures of the two different forms of polythene.
  b Which type do you think is used for:
    i carrier bags? ii plastic milk cartons?
4 Why is polythene difficult to dispose of?
5 Describe some negative effects of plastics on the environment. (The upper photo will help.)
6 The SPI-TEK™ polythene will break down or *degrade*.
  a How has this been achieved?
  b What does it break down to form?
  c Why is it a good idea to make bin liners from this form of polythene?

6.09 Oil and the environment

Oil – the world's number one fuel

Oil is the world's main source of energy, as this pie chart shows. Around *3500 million tonnes* of petroleum fuel products are produced each year from oil, and used in cars, trucks, planes, and power stations.

Oil keeps us on the move, keeps us warm, and provides some of our electricity – in addition to being the starting point for many useful products. But there is a down side: its impact on the environment.

Where the world gets its energy
(pie chart showing: oil, coal, gas, hydro, nuclear, other)

The impact of oil on the environment

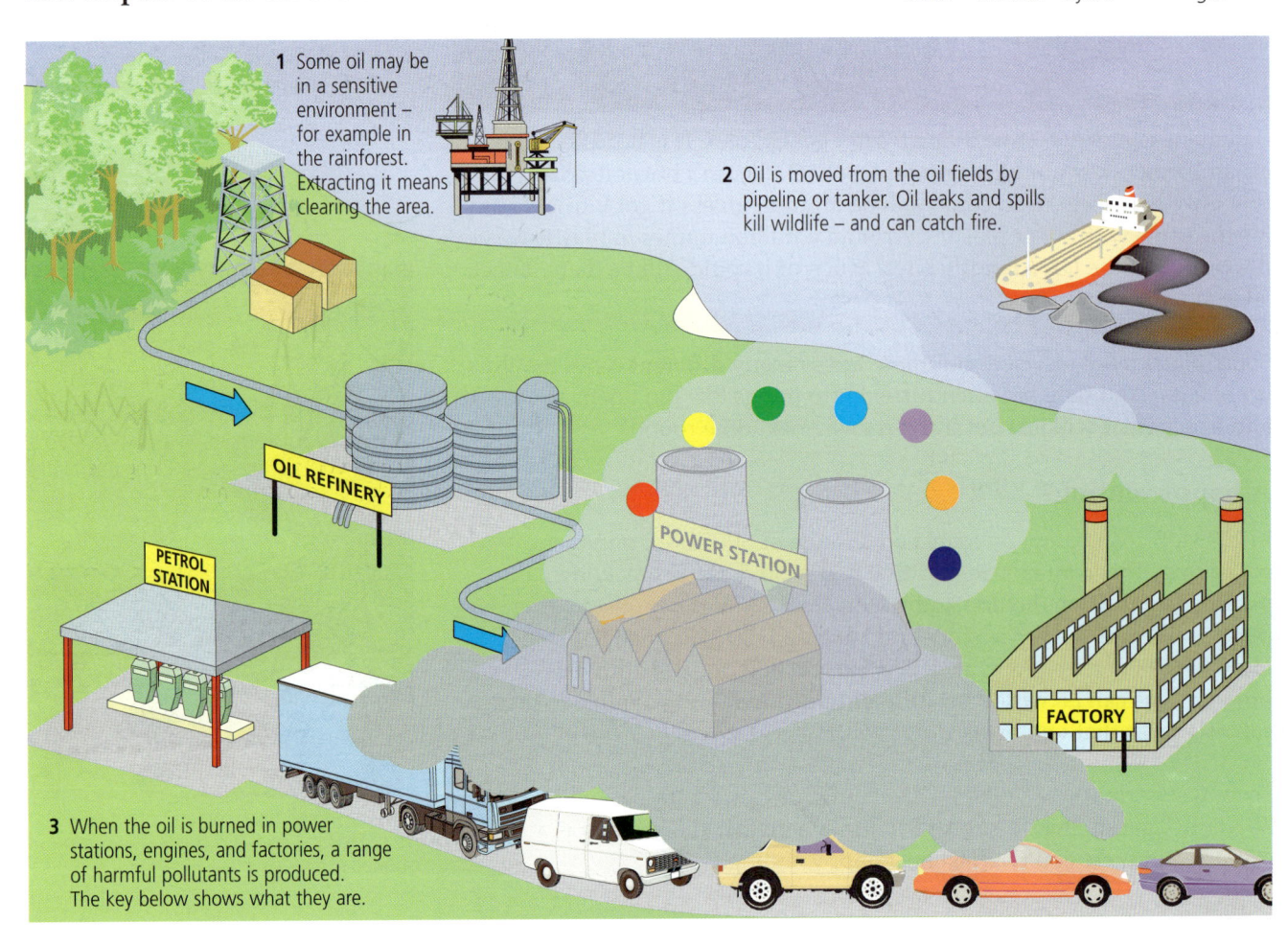

1 Some oil may be in a sensitive environment – for example in the rainforest. Extracting it means clearing the area.

2 Oil is moved from the oil fields by pipeline or tanker. Oil leaks and spills kill wildlife – and can catch fire.

3 When the oil is burned in power stations, engines, and factories, a range of harmful pollutants is produced. The key below shows what they are.

Key

- ● carbon dioxide. It is a key factor in global warming.
- ● sulphur dioxide ⎫
- ● nitrogen oxides ⎬ These gases can damage your lungs. They mix with rain (already mildly acidic thanks to carbon dioxide) to form **acid rain**. This kills trees and fish, attacks metal structures, and eats into brickwork.
- ● carbon monoxide, a poisonous gas
- ● soot which makes things grimy
- ● water vapour
- ● other substances such as unburnt gases (for example benzene vapour from petrol), and tiny particles of lead and other solids. All these can damage lungs and may cause cancer.

More about some of those pollutants

Sulphur dioxide   Crude oil can contain up to 6% sulphur, which is why sulphur dioxide is produced. Since this gas is so harmful, many oil-burning power stations are being forced to install equipment to remove it from exhaust gases. For example they can spray it with a slurry of slaked lime (calcium hydroxide) to neutralize it.

Nitrogen oxides   Here engines and furnaces are the problem, rather than oil or petrol. Air gets so hot in them that nitrogen, a very unreactive gas at room temperature, reacts with oxygen.

Carbon monoxide   Hydrocarbons burn in plenty of air to give carbon dioxide and water. For example pentane (in petrol) burns like this:

$$C_5H_{12}\ (l) + 8O_2\ (g) \longrightarrow 5CO_2\ (g) + 6H_2O\ (g)$$
pentane + oxygen ⟶ carbon dioxide + water vapour

But in car engines there is only a limited supply of air. So **incomplete combustion** takes place. This is a typical reaction:

$$2C_5H_{12}\ (l) + 11O_2\ (g) \longrightarrow 10CO\ (g) + 12H_2O\ (g)$$
pentane + oxygen ⟶ carbon monoxide + water vapour

(Some pentane will burn even less completely, giving carbon as **soot**.)

Carbon monoxide combines with the **haemoglobin** in blood, preventing it from carrying oxygen. The gas causes headaches and drowsiness in low concentrations. At a high enough concentration you suffocate.

In homes, carbon monoxide is produced by incomplete combustion of gas at faulty burners in water heaters and gas fires. It has no colour or smell, so people don't notice. It kill around 170 people a year in the UK.

Oiled feathers lose their ability to insulate, and the bird also swallows oil in its attempts to clean itself.

When a boiler is serviced the engineer checks for carbon monoxide.

Oil is not the only culprit

Like oil, natural gas and coal are also fossil fuels. Gas and coal are burned in power stations and furnaces, and gas in home central heating systems.

The table on the right compares the quantities of pollutants they produce for a given quantity of heat energy.

Comparing pollution from the fossil fuels
(kg of pollutant per billion British Thermal Units of energy)

Pollutant	Oil	Gas	Coal
Carbon dioxide	74 500	53 200	95 000
Nitrogen oxides	200	40	210
Sulphur dioxide	510	0.3	1180
Particulates (tiny particles of soot and other substances)	38	3	1250

Questions

1 **a** Which gas produced by burning fossil fuels is also a gas we breathe out?
    b So why is this gas considered harmful?
2 Name *three* products from burning oil that help to make rain acidic.
3 What does *incomplete combustion* mean?
4 Explain why carbon monoxide can kill.
5 People with gas fires are advised to get these checked regularly. Suggest a reason.
6 Car exhausts give out a certain amount of soot. Explain why.
7 All the fossil fuels produce pollutants when they burn. From the table above, which fuel appears to be:
    a the least harmful?   **b** the worst culprit?

6.10 Global warming

Two rising trends

Carbon dioxide in the atmosphere

Average global temperature

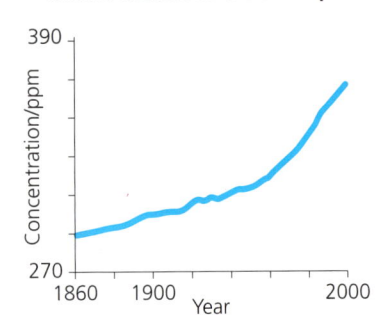

The concentration of carbon dioxide in the atmosphere is rising, because of the amount of fossil fuels we are burning. It has been rising steadily since the Industrial Revolution.

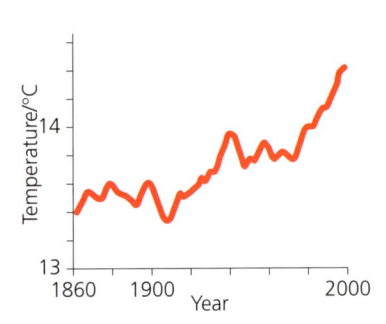

The average global temperature shows the same overall trend. **Global warming** is taking place, and most scientists agree that carbon dioxide is the chief culprit.

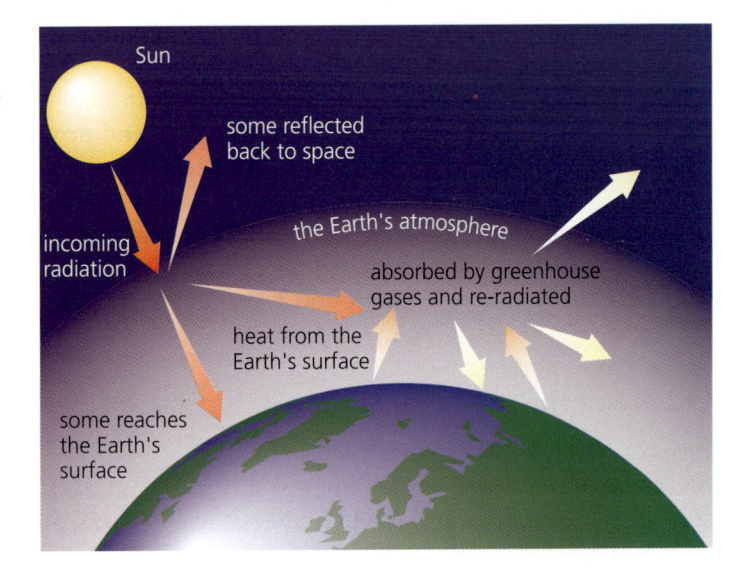

Carbon dioxide: a greenhouse gas

Carbon dioxide is a **greenhouse gas**: it helps to trap heat around the Earth.

It absorbs incoming radiation from the Sun, as well as heat being reflected away from the Earth, and radiates it again in all directions – some back towards the Earth.

It is not the only greenhouse gas: water vapour, methane, and dinitrogen oxide have the same effect. But it is the one we are pumping into the air in the most rapidly rising quantities.

What are the consequences?

These are some of the predictions:
- As temperatures rise, the polar icecaps will melt. It is happening already: the Arctic ice floes are thinning.
- So sea levels will rise. Low lying areas of the world – including parts of the UK – will flood. Some Pacific islands will be drowned.
- Climate patterns will change. Some regions will suffer drought and crops will fail. People will starve. There will be more refugees.
- Other regions will have more frequent storms and flooding.
- The UK could get milder in winter – or much colder, depending on the Gulf Stream. This warm current keeps us warm in winter. If it is pushed away by melting ice, prepare for freezing winters.
- If Europe gets *warmer* in winter, ski resorts may shut down.
- Storms and floods will cause £ billions of damage all over the world.
- The poorest countries will be least able to cope, so will suffer most. (This is ironic, since they burn least fossil fuel.)

In the winter of 2000/2001 many places in Britain were flooded. Are we already suffering the consequences of global warming?

As the icecaps melt, huge icebergs break off and float away. The habitat of polar bears and other Arctic animals is getting smaller.

So – what are we doing about it?

Scientists have been drawing attention to the link between global warming and fossil fuels for decades. They have monitored carbon dioxide levels continually since 1957. (To measure it for previous years, samples were taken from air trapped in Arctic ice.)

A string of catastrophes has made the world's governments pay attention – for example a severe drought in the US in 1988. They now talk a lot about global warming. But many are unwilling to take drastic action, because that means drastically cutting back on the use of fossil fuels – especially in the USA which is the main carbon dioxide polluter.

Fossil fuels keep an economy going – factories, transport systems, power stations, and so on. You could impose a huge tax increase on them to make people use less. But this could hurt the economy, and lose votes. Without alternative fuels in place, few governments will risk that.

At the same time many of us do not *really* feel climate change is our problem, or that we contribute to it in any way. Until we all accept that it is our problem, action to fight it will be too slow.

Most countries do now accept that our present use of fossil fuels is not **sustainable**: it cannot be continued at this level without harm. The pressure is on to develop clean renewable sources of energy, like wave power and solar power, as quickly as possible.

Carbon dioxide production per person

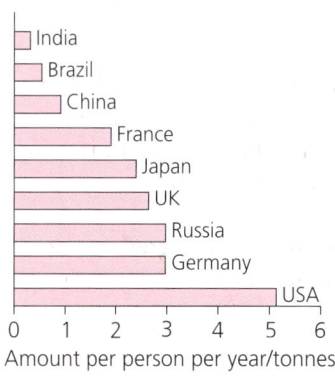

India
Brazil
China
France
Japan
UK
Russia
Germany
USA

0 1 2 3 4 5 6
Amount per person per year/tonnes

Questions

1 What does global warming mean?
2 Carbon dioxide acts as a greenhouse gas. Explain what this means.
3 Name two other greenhouse gases.
4 Explain the connection between climate change and our use of fossil fuels.
5 List four possible consequences of global warming.
6 Look at the graph above. The values are worked out by adding up all the carbon dioxide emissions from power stations and so on, and dividing by the population.
   a Suggest reasons why the USA tops the list.
   b Suggest reasons why the figure is so low for India.
7 In climate change, poorer countries are penalized for actions of the richer countries. Explain this.

6.11 The atmosphere past and present

What is the atmosphere?

The **atmosphere** is the layer of gas around the Earth. It may feel light, but when you lie flat out at the seaside to sunbathe, there are about 13 000 kilograms of gas pressing down on your body!

This gas layer has four parts. We live in the **troposphere** with trips into the **stratosphere** now and then. Only the astronauts have passed out of the atmosphere, on the way to the international space station which is 390 km above the Earth, or the moon, 356 000 km away.

The gas is at its most dense at sea level, but thins out rapidly as you rise through the troposphere. It is the mixture we call **air**.

What's in air?

This pie chart shows the gases that make up air:

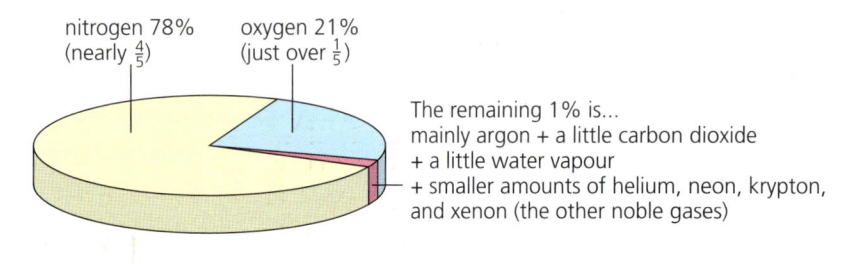

nitrogen 78% (nearly $\frac{4}{5}$)
oxygen 21% (just over $\frac{1}{5}$)

The remaining 1% is...
mainly argon + a little carbon dioxide + a little water vapour + smaller amounts of helium, neon, krypton, and xenon (the other noble gases)

The composition of air is not exactly the same everywhere. It changes a little from day to day and place to place. For example there is more water vapour in the air on a damp day. And over busy cities and industrial areas there is more carbon dioxide, as well as poisonous gases such as carbon monoxide and sulphur dioxide. (These spread everywhere in time.)

Where did the atmosphere come from?

The Earth was formed about 4600 million years ago, when gravity caused a hot, dense band of gas and dust around the Sun to collapse in on itself. Particles got pulled in and squashed together. In time the huge mass of material that formed broke up into the Earth and other planets. And then the Earth's atmosphere developed in stages.

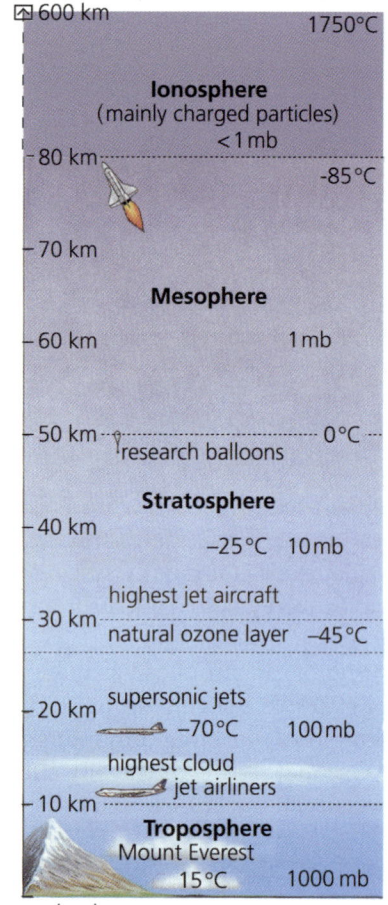

---

600 km — 1750 °C
Ionosphere (mainly charged particles) <1 mb
80 km — −85 °C
70 km
Mesophere
60 km — 1 mb
50 km — 0 °C
research balloons
Stratosphere
40 km — −25 °C 10 mb
highest jet aircraft
30 km — natural ozone layer −45 °C
supersonic jets
20 km — −70 °C 100 mb
highest cloud
jet airliners
10 km —
Troposphere
Mount Everest
15 °C 1000 mb
sea level

The four layers in the Earth's atmosphere. Note where the ozone layer is. The 'mb' figures show the pressure, which falls the higher you go.

The Earth from space – and beside it, a closer view.

The atmosphere is the thin band along the horizon.

First stage: the volcanic gases

Inside the young Earth there was intense activity. Hot gases seeped out through the crust and burst from volcanoes. Over billions of years our atmosphere developed from these gases.

- Scientists can't be certain which gases formed the early atmosphere. (It was a long time ago!) All agree it was mainly **carbon dioxide** and **water vapour**. Many think **ammonia** and **methane** were the other key gases present. (Others think hydrogen, nitrogen, and carbon monoxide.) But all agree that *there was little or no oxygen*.
- When the temperature dropped below 100 °C, the steam cooled and condensed. It rained for centuries – and the oceans formed.
- Carbon dioxide dissolved in the falling rain, forming an acidic solution. This attacked minerals in the Earth's crust, forming carbonates that were washed into the ocean. Other gases and minerals dissolved too. The oceans were hot and salty.

Next stage: life begins

- Then, around 3500 million years ago, primitive life appeared. It may have started in hot pools, with lightning providing the spark. First, reactions between the dissolved compounds produced amino acids. These polymerized to form proteins. And finally, proteins combined to form units that could reproduce themselves.
- The first living things were simple bacteria. **Nitrifying** bacteria fed on ammonia, producing nitrates that then formed proteins. **Denitrifying** bacteria fed on ammonia, releasing nitrogen to the atmosphere.

And then – oxygen!

- Then around 2200 million years ago, the first green plants developed and photosynthesis began. It used up carbon dioxide and *produced oxygen*.
- The first oxygen reacted fiercely with the elements on Earth's surface, forming oxides. It reacted with ammonia, giving nitrogen and water, and with methane, giving carbon dioxide and water. But eventually it began to build up in the atmosphere. It poisoned the organisms that fed on other gases. Their descendants, **anaerobic** bacteria, are now found only in places where there's no oxygen. For example they live in waterlogged soil or by volcanic vents on the ocean floor.
- Once oxygen began to collect in the atmosphere, a protective layer of ozone formed. (Check the diagram on the last page.)
- Then around 600 million years ago, when there was enough oxygen and protective ozone, the first simple animals developed.
- Our own species (*Homo sapiens*) first appeared only about 200 000 years ago – and we have taken over the Earth since then.

The composition of air has been steady for the last 200 million years.

Questions

1 Make a table to describe the layers of the atmosphere. You should use these headings:
   name   height   temperature   pressure
2 Describe the present composition of the atmosphere.
3 Explain why the atmosphere now contains:
   a less water vapour   b less carbon dioxide
   c more oxygen   than 4000 million years ago.
4 At first oxygen acted as a pollutant. Explain.

6.12 The atmosphere in balance

How the atmosphere stays balanced

All living things, including humans, depend on the nitrogen, oxygen, and carbon dioxide in air. We also depend on the water vapour that condenses and falls as rain. But these substances don't just get used up and disappear. They are **recycled**.

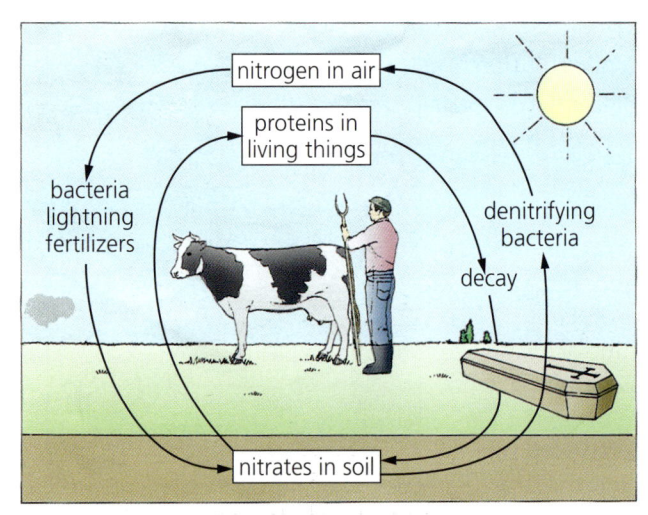

1 Nitrogen circulates between the air, the soil, and living things in the **nitrogen cycle**. You'll learn more about this in biology class.

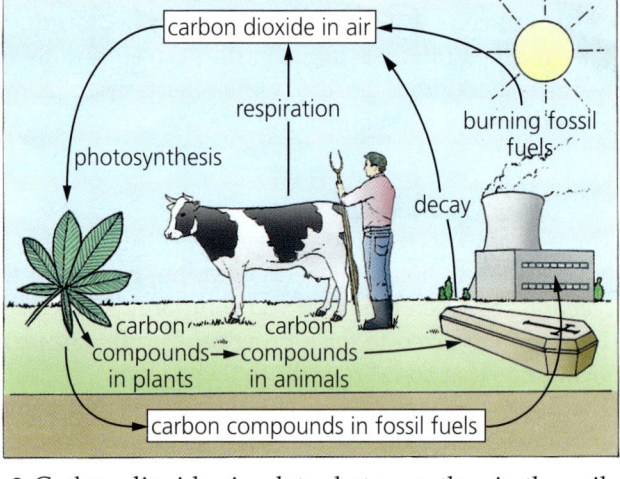

2 Carbon dioxide circulates between the air, the soil, and living things – and in and out of the ocean – in the **carbon cycle**.

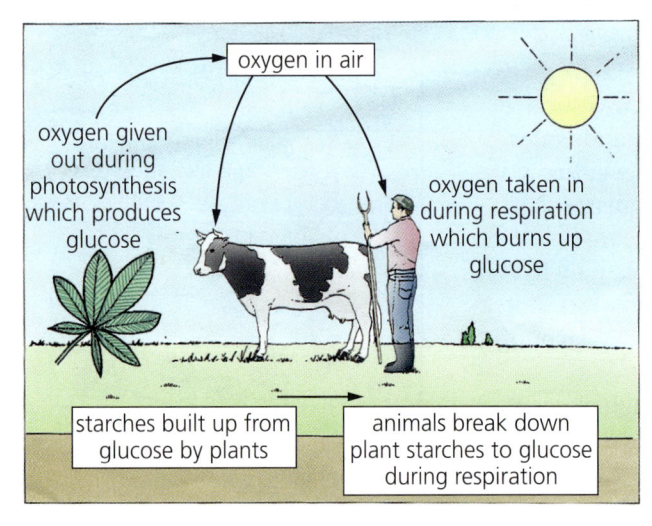

3 The process of photosynthesis, followed by respiration, also recycles **oxygen**.

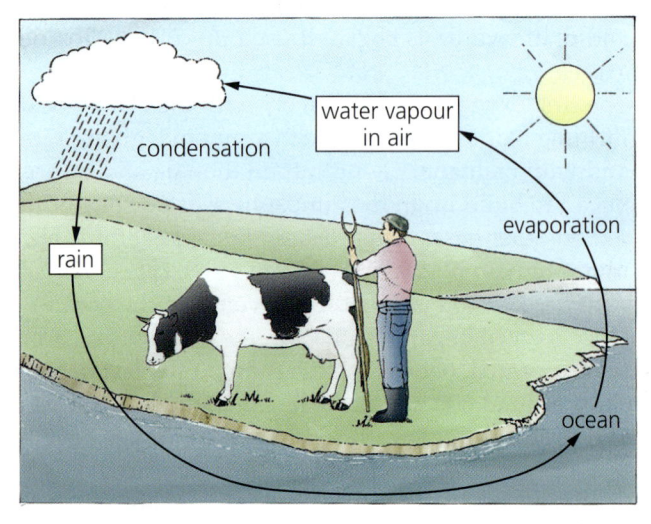

4 And finally, water circulates between the air, the oceans, and living things in the **water cycle**.

These recycling processes have kept the composition of the atmosphere stable over the last 200 million years.

But in the last 200 years (ever since the Industrial Revolution) we have gradually upset the balance of carbon dioxide by burning so much fossil fuel. Most scientists agree that this is the cause of **global warming**.

How oxygen was discovered

For many centuries people thought air was a single gas. Then in 1771 the Swedish chemist Karl Scheele found that hot manganese dioxide gave off a gas. In 1774 in England Joseph Priestley found hot mercuric oxide did the same – and that a candle burned better in this gas than in air.

In France, Antoine Lavoisier took their work further. He heated mercury in air in a closed vessel. It turned into a red powder, and about one-fifth of the air was used up. It was the same no matter how much mercury he started with, or how much he heated it.

Lavoisier tested the remaining gas and found that a candle would not burn in it. He put a mouse in it. The mouse died. So the gas that got used up was also the one that supported life! He named it **oxygen**.

Measuring the percentage of oxygen in air

You too can do an experiment along the same lines as Lavoisier's.

The apparatus A tube of hard glass is connected to two gas syringes A and B. The tube is packed with small pieces of copper wire. One syringe contains 100 cm³ of air. The other is empty:

Lavoisier (1743–1794) in his lab. He met his death on the guillotine during the French Revolution.

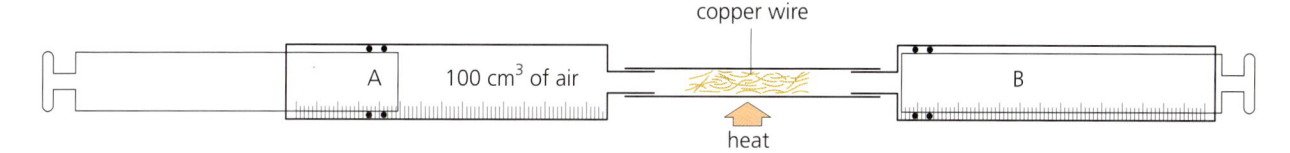

The method These are the steps:
1 The tube is heated by a Bunsen burner. When A's plunger is pushed in, the air is forced through the tube and into B. The oxygen in it reacts with the hot copper, turning it black. When A is empty, B's plunger is pushed in, forcing the air back to A. This cycle is repeated several times.
2 Heating is stopped after about 3 minutes, and the apparatus allowed to cool. Then all the gas is pushed into one syringe and its volume measured. (It is now less than 100 cm³.)
3 Steps **1** and **2** are repeated until the volume of the gas remains steady. This means all the oxygen has been used up. The final volume is noted.

The results Starting volume: 100 cm³. Final volume: 79 cm³.
So the volume of oxygen in 100 cm³ of air is 21 cm³.

The percentage of oxygen in air is therefore $\frac{21}{100} \times 100 =$ **21%**.

Questions

1 About what percentage of the air is oxygen?
2 6 billion humans on the Earth breathe in oxygen and covert it to carbon dioxide. Explain why the percentage of oxygen in the air does not fall.
3 a What is the carbon cycle?
  b Look at the carbon cycle on the opposite page. One step in this cycle is not 'natural'. Which one?
4 One gas in the air is less 'in balance' than it was 200 years ago. Which gas, and why?
5 Name *two* products you think Priestley obtained from heating mercuric oxide.
6 Name the red powder Lavoisier obtained.
7 In the experiment above, what happens to the oxygen in the gas syringe? What is the black powder that forms?

6.13 What have we done to the ozone layer?

Ozone, the Earth's sunscreen

High above you in the stratosphere, about 25 km up, is the **ozone layer**, the Earth's sunscreen. Ozone is a form of oxygen. It has the formula O_3. It protects us by absorbing harmful ultraviolet radiation from the Sun.

The ozone layer is not really a layer. It is a zone where ozone is more abundant than elsewhere. If the ozone gets depleted, harmful UV-B radiation will get through to the Earth and:

- kill off **plankton**, the tiny sea plants that are the basis of all food chains in the sea. So fish and other sea creatures will starve.
- stunt the growth of some crops, causing food shortages.
- damage eyesight and cause skin cancer in humans.

1928: The first of the new wonder products

80 years ago, fridges could kill. They used harmful gases like ammonia, sulphur dioxide, and methyl chloride as coolants. A leak could be fatal. So anyone who came up with a safe fridge gas would be onto a winner.

In 1928 an American chemist called Thomas Midgely did just that. He made a new chemical containing chlorine, fluorine, and carbon. It was a **chlorofluorocarbon** or **CFC**. Non-poisonous, unreactive, non-flammable, and cheap to make, it was a wonder product.

It was a great success. Soon a range of CFCs was developed. They found more and more uses: as coolants in fridges and air conditioning units, as propellants, for aerosols, as cleaning fluids. They were used to blow up plastics into rigid foam for packaging, and soft foam for pillows and furniture. By 1970 world production of CFCs was around a million tonnes a year.

1985: Not so wonderful after all

In 1985 Joe Harman and two colleagues from the British Antarctic Survey published a paper in the scientific journal *Nature*, warning that ozone was being depleted over the Antarctic, and suggesting that chlorine was the cause. Soon everyone was worrying about this 'hole' in the ozone layer. (It was not really a hole.)

Chlorine was indeed the cause – and CFCs the culprit. Every time an aerosol hair spray was used, or fridge dumped, or hamburger carton discarded, CFCs escaped into the air. Insoluble in rain and unreactive, they eventually rose into the stratosphere. There, at low temperatures, CFC molecules are broken down by UV radiation, giving chlorine atoms:

$$CFCl_3 \xrightarrow{\text{ultraviolet radiation}} CFCl_2 + Cl$$

The chlorine atoms then react with ozone molecules, converting them to oxygen. A single chlorine atom can destroy 100 000 ozone molecules.

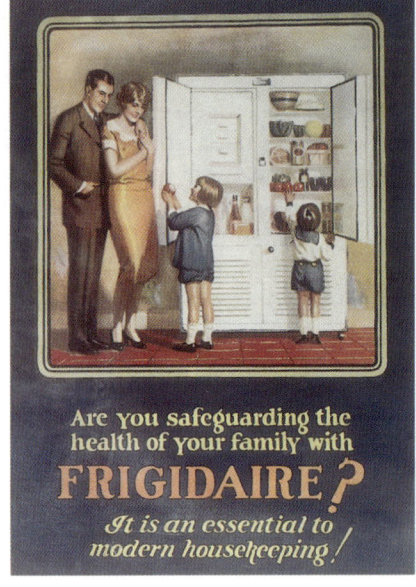

Trouble ahead! With the arrival of CFCs, sales of fridges soared.

A chain reaction

This is how a chlorine atom reacts with an ozone molecule:

$Cl + O_3 \longrightarrow O_2 + ClO$ then
$2ClO \longrightarrow Cl_2O_2$ then
$Cl_2O_2 \longrightarrow 2Cl + O_2$

So the chlorine atoms are freed again to attack more ozone. This is an example of a **chain reaction**.

Outdoor workers need to take special care, and use a high factor sun lotion.

CFCs leak away from discarded fridges and go all the way to the stratosphere unchanged.

1987: They have to go!

That report in *Nature* alarmed the world. In 1987, 27 countries signed a treaty called the **Montreal Protocol** to reduce their production of CFCs. In 1990 it was made tougher: production should be banned altogether by 2000. 150 countries have signed.

It's working. This graph shows how the concentration of one CFC, called CFC-11, increased over the South Pole until around 1993 and is now levelling off.

But CFC molecules take time to rise to the stratosphere. They are still being made in some developing countries. Many old cars still have air conditioning that runs on CFCs.

So it will take years for the graph to fall back to zero. In the meantime, depleted ozone levels have been reported over many countries, and skin cancer is on the increase.

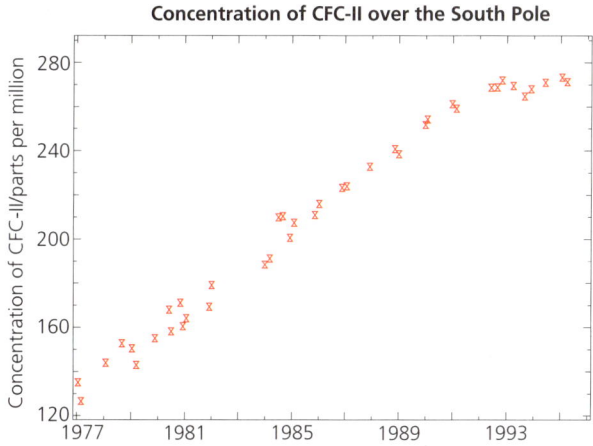

Concentration of CFC-II over the South Pole

A harsh lesson

The story of CFCs is a good example of how 'wonder' products created by scientists can have harmful results – usually detected by other scientists; and of how politicians respond. Now the hunt is on for CFC replacements. Whatever these turn out to be, one thing is certain: they must pass the ozone test.

Questions

1 a What is ozone?
   b Where is the ozone layer?
   c Why do we depend on it? Give two examples.
2 a What are CFCs?
   b Give four uses of CFCs.
   c Explain why they were so popular in industry.
3 Chlorine escapes from swimming pools – but it does not reach the stratosphere. Why do CFCs do so?
4 a How does a CFC damage the ozone layer?
   b The reaction is a *chain reaction*. What does that mean?
5 Has the Montreal Protocol been successful? Explain.

6.14 The oceans

The oceans: a vast solution

When the Earth's early atmosphere cooled below 100 °C, the water vapour condensed. It began to rain – and rained non-stop for centuries. Rain poured down the hot rocks, carrying eroded material into the hollows. It was the start of the oceans.

Today the oceans are like a vast solution, covering over 70% of the Earth. One expert estimates that they contain over 50 million billion tonnes of dissolved material. They probably contain every known element. But about 85% of the dissolved material is sodium and chloride ions. (The sodium ions make the water taste salty.) The rest is mainly magnesium, sulphate, calcium, and potassium ions.

The table on the right shows the amounts of different ions you would find, on average, in a kilogram of ocean water.

The oceans in balance

It is estimated that the world's rivers carry over 4 billion tonnes of dissolved material to the oceans each year. But the oceans remain in balance, because other processes remove dissolved material. Like this ...

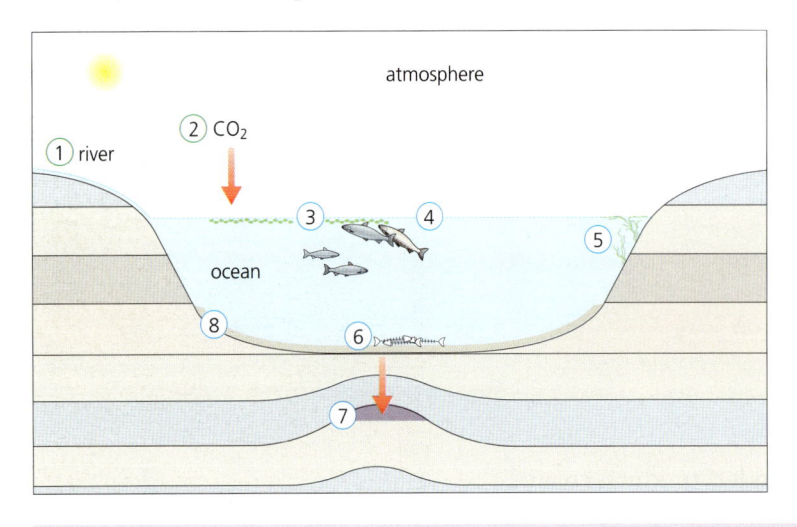

Sailing on a vast solution.

Main ions dissolved in ocean water	Amount/ g per kg of sea water
Chloride ions (Cl^-)	19.2
Sodium ions (Na^+)	10.7
Sulphate ions (SO_4^{2-})	2.7
Magnesium ions (Mg^{2+})	1.4
Calcium ions (Ca^{2+})	0.4
Potassium ions (K^+)	0.38
Hydrogencarbonate ions (HCO_3^-)	0.14
Bromide ions (Br^-)	0.07

Materials in
1. Rivers carry in dissolved substances from the weathering of rock (for example calcium ions).
2. Some carbon dioxide dissolves in the water.

Materials out
3. Phytoplankton (tiny plants) use the carbon dioxide in surface water for photosynthesis.
4. Other sea creatures obtain carbon by feeding on the phytoplankton. They extract calcium ions from the water. And they produce calcium carbonate for their shells and skeletons.
5. Other ions get extracted too – for example iodide ions by seaweed.
6. In time sea creatures die and fall to the bottom of the ocean. Over millions of years, skeletons and shells turn into limestone (calcium carbonate).
7. Over millions of years the soft remains of some dead things turn into fossil fuels: oil and gas.
8. Some ions react together to form insoluble compounds. For example calcium and carbonate ions form a deposit of calcium carbonate on the ocean floor.

These nodules of manganese have been deposited on the sea floor by complex biological and chemical reactions.

The oceans and carbon dioxide

The oceans play an important part in keeping the carbon dioxide in the atmosphere in balance.

- Rain removed much of the carbon dioxide from the early atmosphere, carrying it to the oceans as carbonic acid.
- The oceans continue to dissolve carbon dioxide. Some is used for photosynthesis and will eventually be locked up in limestone, oil, and gas. (We release it again by heating limestone to make quicklime, or burning oil and gas as fuel.)
- The rest reacts to produce insoluble carbonates (such as calcium carbonate) and soluble hydrogencarbonate ions.

Can the oceans solve the global warming problem?

The oceans already contain a huge amount of dissolved carbon dioxide: about 40 000 billion tonnes of it. But **equilibrium** is always reached between the amount already dissolved and the amount in the air. Because of this they can dissolve *some* of the extra we produce from burning fossil fuels, but they can *never* dissolve all of it.

Some scientists are seeking other ways to get carbon dioxide into the ocean, for example by pumping liquid carbon dioxide down to the ocean floor. (At the high pressure and low temperature found there, it forms an ice-like solid.) They have even tried adding iron to the ocean as a fertilizer for phytoplankton, to encourage more photosynthesis.

Other scientists disapprove, since there could be long-term harmful effects on the marine environment.

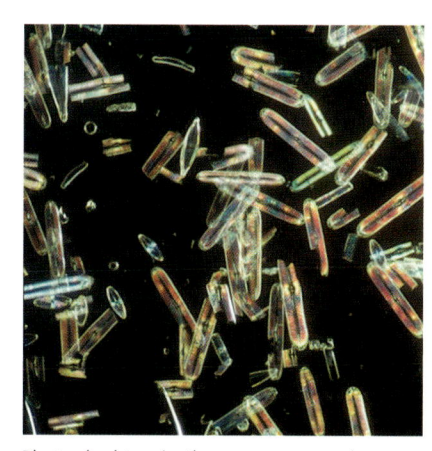

Phytoplankton in the ocean use carbon dioxide for photosynthesis. Perhaps they can help reduce global warming?

Questions

1 Name: **a** the top two positive ions
 b the top two negative ions present in ocean water.
2 How did these ions get into the ocean?
3 The ocean is a vast solution. But overall, its composition remains in balance. Explain why.
4 What are *phytoplankton*?
5 Describe two ways in which carbon dioxide from the air ends up locked in calcium carbonate on the ocean floor.
6 The ocean is vast, but it cannot dissolve all the extra carbon dioxide from burning fossil fuels. Why not?
7 Explain how some of the carbon dioxide locked up in the ocean is released again by humans.

6.15 The Earth's structure

The layers that make up the Earth

The Earth has three layers:
- The outer layer, where we live, is called the **crust**. It is made of solid rock. Compared with the other layers it is very thin.

- The middle layer is the **mantle**. It extends almost halfway to the Earth's centre. It is also made of rock, but much of this is soft enough to flow very slowly.

- The inner layer is the **core**. It has just over half the Earth's radius. It is made of metal (mainly iron and nickel). The outer core is liquid and the inner core is solid.

The upper part of the mantle is hard. This hard part and the crust together form the **lithosphere**, the Earth's hard outer part. Just below the lithosphere the rock gets softer.

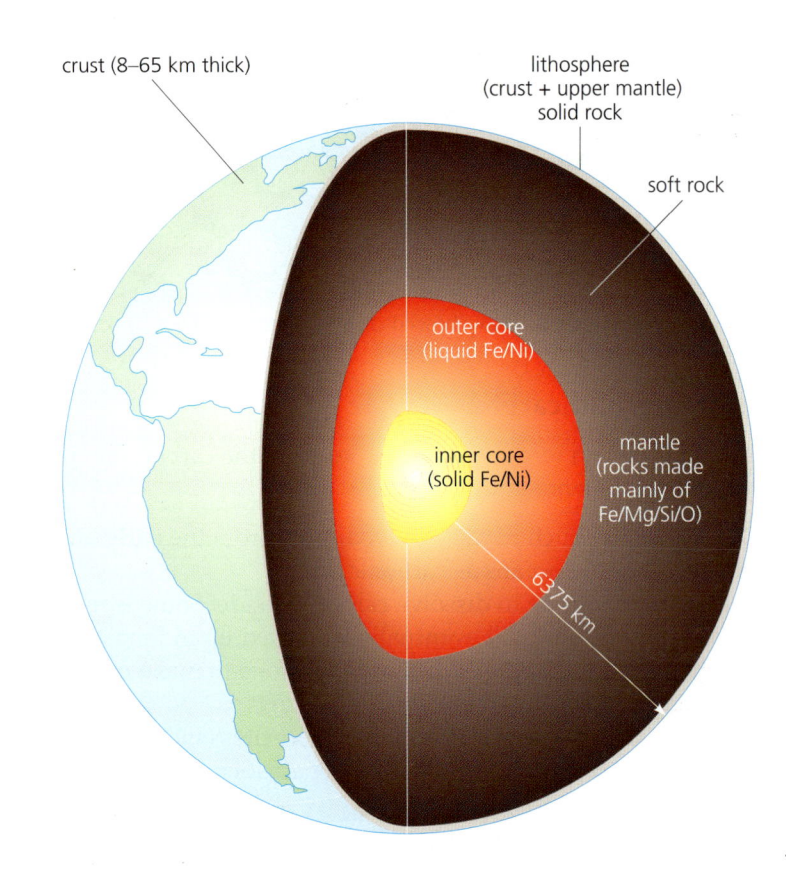

The deepest hole ever dug (8 km) did not even reach the mantle. So scientists have learned most of what they know about the Earth's structure by studying seismic waves from earthquakes.

The Earth's density

We think the young Earth was a molten ball that settled into its layers as it cooled. The lightest materials floated to the top to form the crust.

Density values support this idea. The average density of the Earth is calculated as 5.5 grams per cubic centimetre. But the average density of the rocks in the crust is around 2.7 g/cm³. So the materials inside the Earth must be different from and denser than those in crust.

Conditions inside the Earth

Temperature The deeper you go inside the Earth the hotter it gets. This is mainly due to the decay of radioactive isotopes inside the Earth – chiefly uranium. Radioactive decay gives out huge amounts of heat.

Pressure The closer you get to the centre of the Earth, the greater the pressure, because of the huge mass of rock pressing down. The pressure at the centre is about 2 million times that at the surface. This high pressure prevents the inner core from melting. (Melting and boiling points increase with pressure.)

More about the Earth's crust

The Earth's crust is made up of **oceanic crust** under the oceans, and **continental crust** which forms the continents:

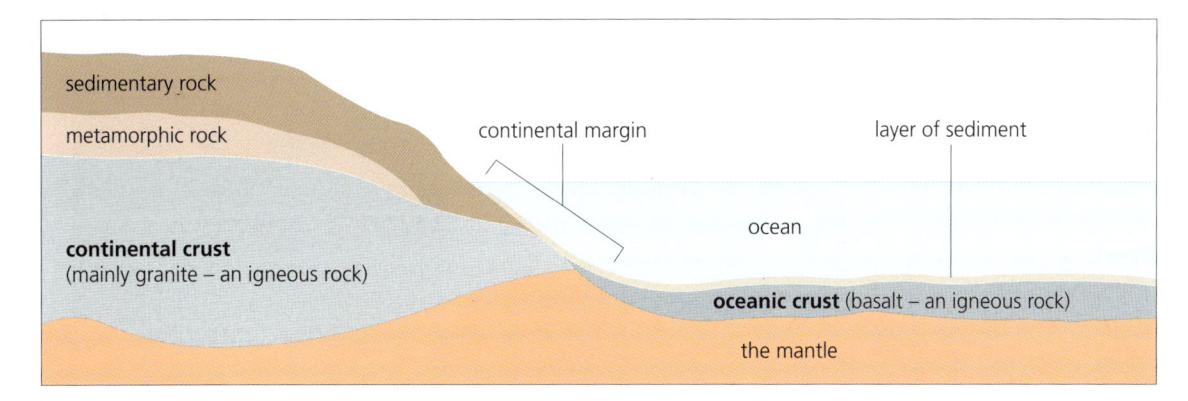

[Diagram showing cross-section of Earth's crust with labels: sedimentary rock, metamorphic rock, continental margin, layer of sediment, continental crust (mainly granite – an igneous rock), ocean, oceanic crust (basalt – an igneous rock), the mantle]

- Note that both types of crust are mainly **igneous rock**: granite in the continents and basalt under the ocean. (You can find out more about igneous rock below.)
- The continental crust is less dense than the oceanic crust. That is why the continents sit up above sea level!
- A layer of **sediment** covers the oceanic crust. It contains anything that drifts to the bottom of the ocean, such as shells or sand.
- The granite in the continents usually has some **metamorphic** and **sedimentary** rock above it – but not everywhere, nor spread evenly. They may be worn away in places leaving the granite exposed.

Igneous, sedimentary, and metamorphic rock

Different rocks are made of different minerals. But *all* rocks in the Earth's crust can divided into these three groups, depending on how they were formed:

Igneous rock is formed when rock melts and cools again, forming crystals.

Sedimentary rock is built up from particles compacted and cemented together.

Metamorphic rock is formed by the action of high temperatures and pressures on rock.

Questions

1 Draw a labelled diagram to show the structure of the Earth.
2 How did scientists learn about the Earth's layers?
3 The rock in the crust is less dense than the material inside the Earth. How did this happen?
4 The Earth's crust is made up of two parts.
   a Name them.
   b Both are mainly *igneous rock*. What is igneous rock?
   c Name the igneous rock in each part of the crust.
5 Name the two other rock groups, apart from igneous.

6.16 From weathering to deposition

Weathering

Weathering means the breaking down of rock at the Earth's surface – largely by the action of the weather! It happens to all rock, and is the first stage in its transformation into new rock. There are two types of weathering, physical and chemical, and they usually go on together.

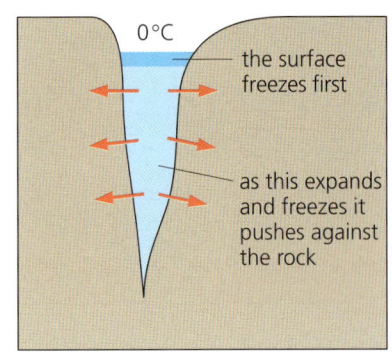

Physical weathering

This is where rock gets broken into fragments *but not chemically changed*. For example plant roots grow into cracks in rock and slowly prise them apart. (This is sometimes called **biological weathering**.) Or in winter, water in the cracks expands as it freezes, forcing the cracks wider. Eventually the rock breaks up.

In some places the temperature drops below freezing at night and rises during the day. So water in the cracks freezes and thaws cycle after cycle. This gives **freeze-thaw weathering**.

Freeze–thaw weathering of rock.

Chemical weathering

Here the rock is broken down by reactions with air and water. It is *chemically changed*. For example carbon dioxide dissolves in rainwater giving **carbonic acid**. This reacts with limestone rock (calcium carbonate) forming calcium hydrogencarbonate which is carried away in solution:

water + carbon dioxide + calcium carbonate ⟶ calcium hydrogencarbonate
H_2O (l) + CO_2 (g) + $CaCO_3$ (s) ⟶ $Ca(HCO_3)_2$ (aq)

Many rocks contain iron compounds. Water and oxygen react with iron ions to produce iron oxide, in a 'rusting' process. The rock turns red brown.

After weathering

Weathering breaks up the rock. Next the pieces get picked up and moved by a **transport agent**: a river, or the wind, or waves, or a glacier. It's a three-step process. This is how it works for a river:

1 Material is eroded.
The river picks up the weathered material. This is called **erosion**. The more energy the river has the faster it can erode.

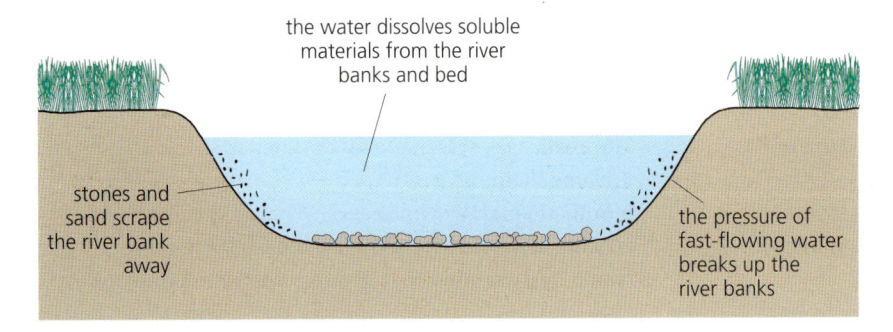

2 The material is transported.
The more energy the river has, and the more water, the more material it can transport. Trees, cars, and boulders can be carried along by floods.

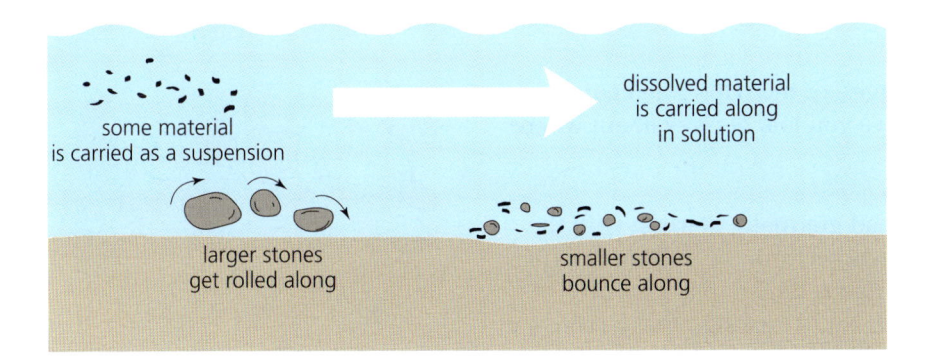

3 It is deposited again.
When the river slows down and runs out of energy it drops or **deposits** the undissolved material as **sediment**. Some is deposited in its flood plain. Some is carried out to sea and falls to the sea floor.

If the river loses energy gradually it deposits material gradually:

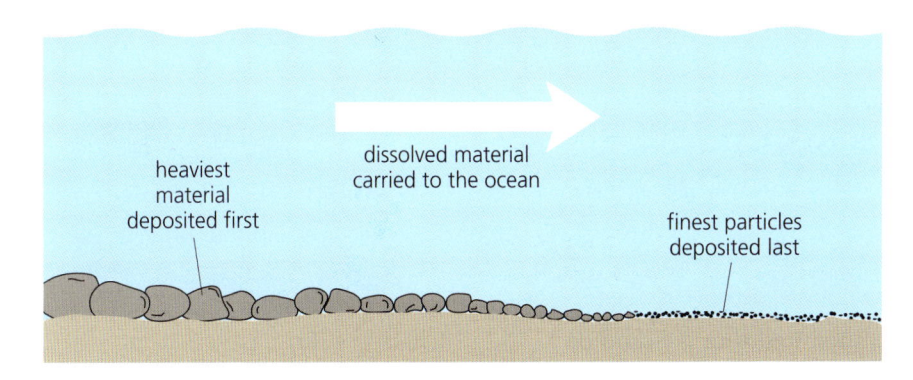

But if it suddenly loses energy, for example when a flood wave subsides, then it can drop everything at once.

What about the other transport agents?
Like rivers, the wind and waves erode and transport material and deposit it when they run out of energy, for example in sheltered bays or corners. Glaciers are different: the material they transport is trapped in ice, and deposited where the ice melts.

The sand formed by weathering of rock has been eroded and deposited as dunes by the wind.

Questions
1 What is the difference between physical and chemical weathering?
2 Give two examples of physical weathering.
3 Give one example of chemical weathering.
4 Weathered material is then generally *eroded*. What does this mean?
5 Name the four agents that can erode material and move it somewhere else.
6 Rivers can erode and transport larger pieces of material during floods. Why?
7 What makes a river deposit the material it is carrying?
8 The ocean contains many different minerals. Why?

6.17 From sediment to sedimentary rock

Clues from sediment

Sediment gets deposited when rivers, waves, or the wind lose energy, or a glacier melts. A study of sediment can give you lots of clues about where it came from, and how. Like these.

Particle size Usually a river drops its load gradually, starting with the heaviest material. This sorts the larger particles from the smaller ones, so the sediment is **well sorted**. In **poorly sorted** sediment particles of different sizes are all mixed up together. It could be a sign that a river was forced to slow down abruptly and drop its load.

Particle shape A fast-flowing river flings rocks and stones against each other. They get broken up and smoothed and rounded. The more vigorous the collisions, and the further a stone has been carried, the smaller and smoother it will be.

Poorly-sorted sediment.

The structure of the sediment You can usually pick out separate layers. This arrangement is called **bedding**. The top layer is the most recent one.

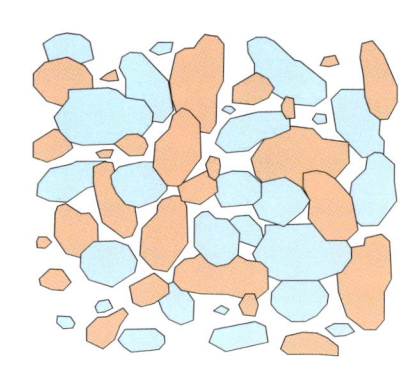

Each layer marks a change in the type of sediment or the energy with which it was carried. Here a heavy layer on top of a fine one could mean the river flooded recently.

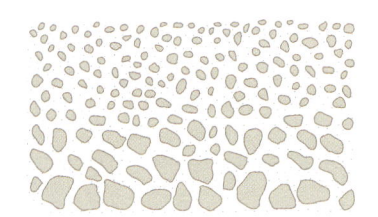

Graded bedding is found on ocean floors. River sediment builds up on the continental shelf. If it gets dislodged it tumbles forward into deep still ocean water and settles like this.

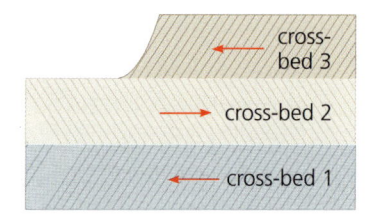

Cross-bedding. Layers lie at different angles, reflecting the direction of flow of the wind or water that deposited them. Wind-deposited sand dunes are often like this.

An individual layer of sediment may carry more clues:

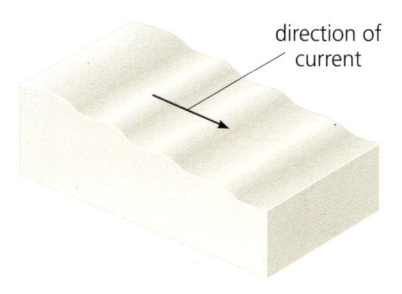

Ripple marks like these suggest that the sediment was shaped by a current of wind or water *after* it had been deposited.

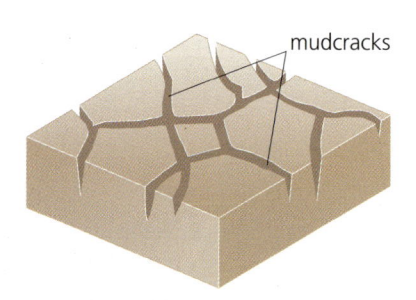

Mud cracks suggest that the watery environment in which this sediment was deposited dried up at some point.

How sedimentary rock is formed

Over millions of years, sediment can grow thousands of metres deep. Slowly the sediment turns into rock, like this:

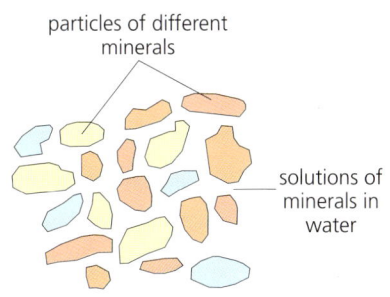

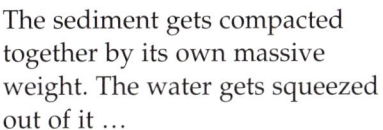

The sediment gets compacted together by its own massive weight. The water gets squeezed out of it …

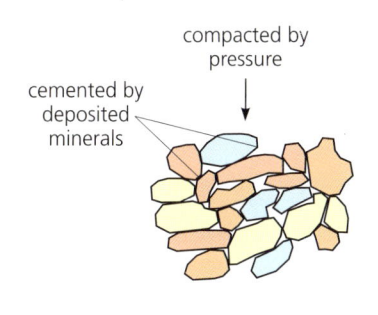

… depositing minerals such as calcium carbonate and silica as it goes. These act like cement, binding the particles together.

The result: sedimentary rock. An example is **sandstone** which is sand grains in a silica or carbonate cement.

And here is the exciting point: *this process does not upset the patterns in the sediment.* The bedding, ripple marks and so on will still be there – but now preserved in rock. Millions of years in the future a geologist will still be able to uncover all those clues. The remains of dead creatures trapped in the sediment will also end up embedded in the rock – or at least their shapes will. These **fossils** will give the geologist further clues.

Some examples of sedimentary rock

Here are some common types of sedimentary rock. The key difference between them is the size of the particles they are made from.

Ripple marks preserved in rock.

shale
made from clay
particle size:
<0.004 mm

siltstone
made from silt
particle size:
0.004–0.062 mm

conglomerate
made from pebbles
particle size: >2 mm

sandstone
made from sand grains
particle size: >0.063–2 mm

Questions

1. **a** Draw a sketch showing:
   i poorly sorted sediment **ii** well sorted sediment
   b Which type do you think would be deposited by a river in heavy flood? Explain why.
2. You would not find graded bedding in shallow water that has strong waves. Explain why.
3. Why is sedimentary rock called that?
4. What is the main factor that causes sediment to turn into rock?
5. Draw a sketch to show how sedimentary rock forms.
6. Which type of sedimentary rock will form from a layer of very fine sediment?

6.18 Evidence from fossils

What are fossils?

Fossils are the preserved remains or traces of ancient living things. The word is from the Latin for 'dig'. (Usually you have to dig them out of the rock.) Fossils are found in sedimentary rock …

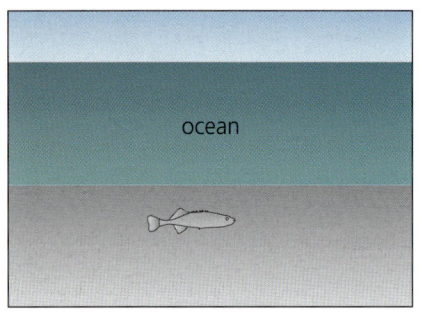

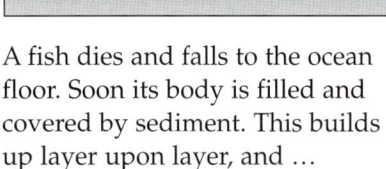

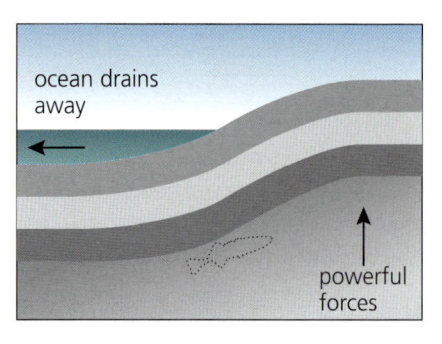

A fish dies and falls to the ocean floor. Soon its body is filled and covered by sediment. This builds up layer upon layer, and …

… in time, becomes sedimentary rock. Then, over millions of years, powerful forces inside the Earth raise the ocean floor to form land.

The rock gets weathered and eroded away… until one day, many millions of years after the fish died, its fossil is revealed.

By now there is nothing left of the original fish. But its shape has been preserved by the sediment that got cemented in and around it.

Why fossils are so useful

- They help you date rock. If that fish species lived between 60 and 70 million years ago, then the layer of rock dates from that period.
- They can tell you where it came from. If that fish species lived in warm tropical waters, then the rock was formed in warm tropical waters – even if it is now in Derbyshire!
- They help us trace how living things evolved. Hundreds of thousands of species that once lived are now extinct (like the dinosaurs). We only know about them through fossils.

Trapped in mid-flight: a fossil Archaeopteryx, one of the first bird-like animals.

A dinosaur footprint in sandstone, in Namibia. Size 6, would you say?

Example: fossils in the Peak District

This is the Peak District National Park in Derbyshire. The rock at its surface is sedimentary: mainly limestone, gritstone, and shale.

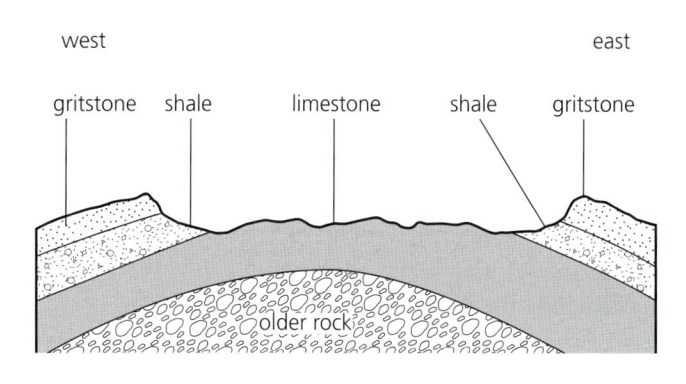

This is a cross-section through the Peak District from west to east. Note how different rocks are exposed. The older rock below is granite, an igneous rock.

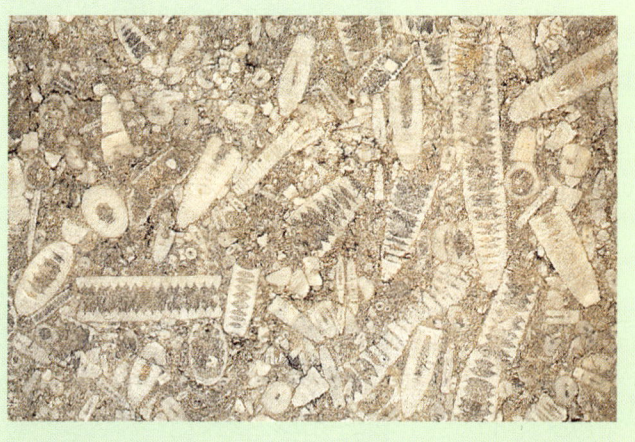

Some Peak District limestone. It is full of fossils of sea lilies or **crinoids** that once grew in warm tropical lagoons. The fossil stems look like nuts and bolts.

The story of the Peak District
- It once lay in the tropics, south of the equator (with the rest of the British Isles).
- It was covered by a warm shallow sea. Shells and skeletons of sea creatures piled up on the sea floor – and were turned into limestone rock.
- Then massive rock movements produced the Scottish Highlands, from which a huge river poured into the sea.
- The river deposited fine particles of clay on top of the limestone. This in time turned into shale.
- And then a coarser layer of sand and gravel, which in time formed gritstone.
- Meanwhile, over 300 million years, the whole area slowly drifted north to where it is today.
- At some point powerful forces caused the flat layers of rock to fold and bulge upwards.
- About 30 million years ago, the land was pushed right out of the sea. The rock began to weather and erode, exposing the layers we see today.

Fossils found in the limestone show that it is about 350 million years old – and that the Peak District once lay under a shallow tropical sea!

Using the evidence from the rocks and fossils, the Peak District story has been pieced together. Above are the key events.

Questions

1. **a** What is a fossil?
   b Fossils are found in sedimentary rock. Explain why.
2. Fossils give valuable information about rock. What kind of information?
3. How do we know that the Peak District once lay under a tropical sea?
4. Sediment is laid down in flat layers. Why are the rock layers in the Peak District not flat?

6.19 Igneous rock

How igneous rock is formed

Over 90% of the rock in Earth's crust is **igneous**. But most of it is hidden under a layer of sedimentary rock, or under the oceans. You can see it where sedimentary rock has been eroded away.

Igneous rock is formed when rock melts, cools, and recrystallizes again. The molten rock is called **magma**. It forms in the hot soft layer of rock below the lithosphere:

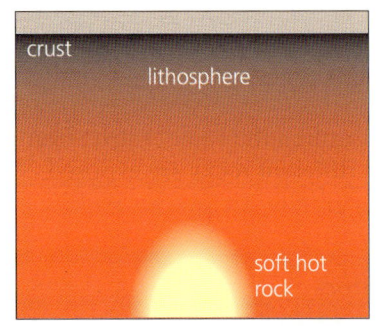

Just a small decrease in pressure (due to rock shifting), or a little water coming down (as steam), will make this rock melt.

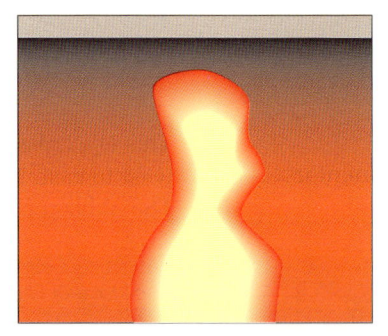

The magma pushes its way up. If it is very viscous, it moves very slowly. It cools as it travels. If it hardens *below* ground, it forms **intrusive** igneous rock.

Granite is the most common example. Slow cooling below ground gives large crystals. The crystals of different minerals interlock randomly.

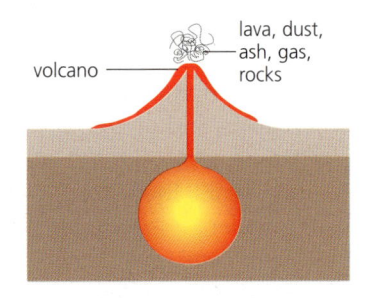

But some magma is very runny. It forces its way to the surface and spews out from a volcano as **lava**. It's often mixed with gases and solid rock.

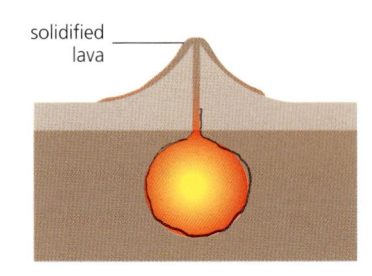

Out in the open the lava cools quickly. It solidifies around the volcano within days or even hours, forming **extrusive** igneous rock.

Basalt is the most common example. Its crystals are tiny because the magma cooled so fast. You'd need a microscope to see them properly.

What makes some magma more runny?

When rock melts, the minerals in it turn to liquid. How viscous (thick and sticky) this is depends on how hot it is and how much water it contains. But it also depends on the minerals.

The more silica the magma contains the more viscous and pale it will be. Granite is over 70% silica. The less silica the runnier it will be, and therefore the more likely to reach the surface. Basalt is 40–50% silica. It is darker than granite because of the iron and magnesium in it.

Silica

Silica is silicon dioxide, SiO_2. Quartz is mainly silica. So is sand.

The crystal size in igneous rock

When magma cools, the minerals in it start to crystallize one by one. The one with the highest melting point starts first. If cooling is slow, its crystals will grow large before the next crystals start forming. When these start, they grow until they meet the sides of the first crystals and then interlock with them.

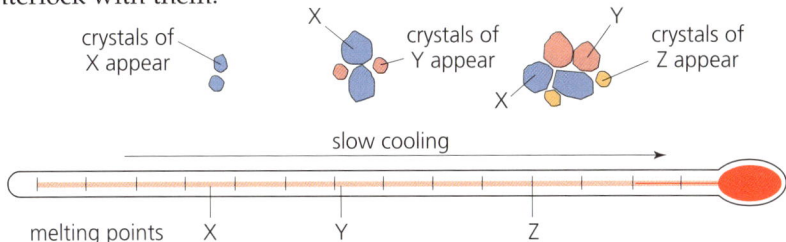

Clues that a rock is igneous

So how can you tell a rock is igneous? These are the clues:
- It is made of **crystals** which are arranged randomly. Their size gives further clues about how fast or slowly it cooled.
- It does not contain any fossils (since these are ancient remains or traces trapped in sediment).

But note that extrusive igneous rock *can* have different layers, formed at different periods of volcanic activity. By studying the rock on the slope of a volcano, geologists can tell a lot about its past eruptions.

Rock from volcanic ash

Volcanic ash can also form igneous rock. The ash is tiny fragments of lava that explode in a cloud from the volcano. Much of it blows away. But where it falls to the ground in a thick layer this compacts and cements to form a porous rock called **tuff**.

Igneous rock and metal ores

Ores of valuable metals such as gold, silver, copper, lead, tin, and zinc are often found in and around granite. Why?

When all the minerals have crystallized from magma, a solution of metal ions in water at several hundred °C is left behind. This is called a **hydrothermal solution**. The metal ions in it could not find a place in the crystal structures. The solution runs into cracks in surrounding rock and veins of metal ores are deposited

Quartz and copper are found together in this igneous rock.

Questions

1. What is *magma*? What causes it to form?
2. What's the difference between extrusive and intrusive igneous rock? Give an example of each.
3. Some magma is more viscous than others. Why?
4. Basalt is darker in colour than granite. Why?
5. A sample of igneous rock has large crystals. What can we deduce from this?
6. What is volcanic ash? What rock does it form?
7. Ores of valuable metals are often found in or near granite. Why?

6.20 Metamorphic rock

An in-between rock

Igneous rock forms at high temperatures and pressures deep inside the Earth. Sedimentary rock forms at the relatively low temperature and pressure on the Earth's surface. Metamorphic rock forms at conditions in between the two.

How metamorphic rock is formed

Rocks gets buried when new sediment forms above them, or during earthquakes and other earth movements. Around 10 km below Earth's surface, temperatures and pressures are high enough to change buried rock without melting it. The change is called **metamorphism**.

How conditions change
As you go down through the Earth:
• the temperature rises by about 30 °C per km
• the pressure rises by about 330 atmospheres per km

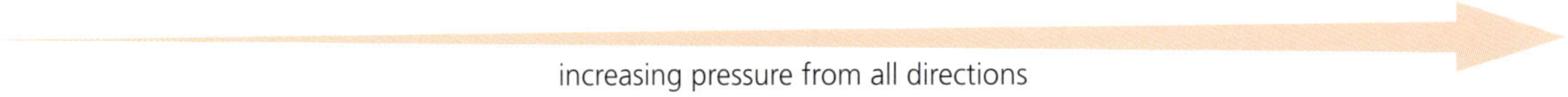

increasing pressure from all directions →

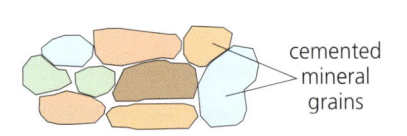

cemented mineral grains

Here mineral grains in the rock get pressed tightly together from all directions, to produce a more dense and compact structure.

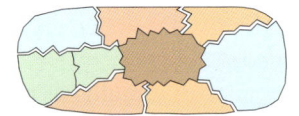

Circulating water speeds up the process. Substances dissolve and recrystallize. Ions migrate from one place to another.

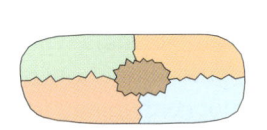

More extreme conditions may cause grains to fuse together to form bigger ones. Higher pressure means larger grains.

The result is metamorphic rock. **Marble** is an example. It is formed from limestone. But it is harder than limestone. Why?

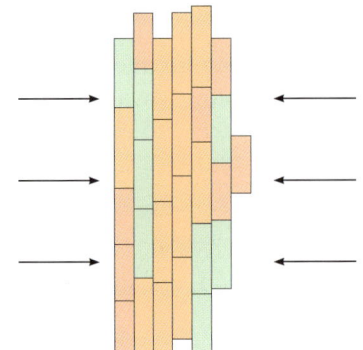

If pressure is **directed** like this, it forces grains to line up. If the grains are thin and flat, the rock takes on a layered look.

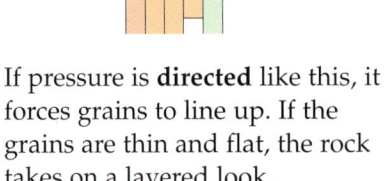

Slate is an example. It is formed from shale which is made up of flat particles of mica and other minerals.

The alignment of crystals in rocks due to directed pressure is called **foliation**. Slate is foliated. Marble is non-foliated.

The minerals in the parent rock determine the minerals in the metamorphic rock. Marble, like limestone, is calcium carbonate. But where a parent rock has several minerals, new minerals may form as ions move around.

Using metamorphic rocks
They provide us with:
• slate for floors
• marble for sculpting statues
• rubies and sapphires
• talc (used in talcum powder)

Which rocks can be metamorphosed – and where?

Any rock can be metamorphosed. Sedimentary rocks are the easiest to change. (Why?) Even metamorphic rock can go through further changes with increasing temperatures and pressures.

Slate is a **low-grade** metamorphic rock, which means it was formed at relatively low temperatures and pressures not far down in the Earth's crust.

With further pressure it is metamorphosed into **schist**. The grains grow larger (up to 1 cm across). If schist gets buried deeper in the crust, it turns into …

… **gneiss** (pronounced *nice*). This is a **high-grade** metamorphic rock. The higher pressures and temperatures have caused the minerals to migrate into layers.

If the temperature had risen any further when gneiss was formed, it would have melted to form igneous rock.

Metamorphic rock is usually found within mountain ranges. It is a sign of the high pressures and temperatures associated with mountain building.

In particular gneiss is found within ancient mountain ranges, or where mountains once stood. The gneiss that forms the Scottish islands of Lewis and Harris is the oldest rock in Europe.

Contact and regional metamorphism

Rocks *can* be metamorphosed by heat alone. This happens when molten magma pushes its way into the Earth's crust. The rocks in contact with it undergo **contact metamorphism**. The change can extend from a few centimetres to a few kilometres into the rock, depending on how big and hot the body of magma is.

But where rocks get buried under sediment, or trapped by large movements in the Earth's crust (as in mountain building), metamorphism may extend over thousands of square kilometres. This is called **regional metamorphism**.

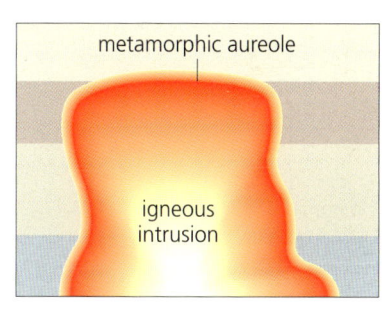

Contact metamorphism. The **aureole** is the metamorphosed zone around the intrusion. The result is low grade metamorphic rock.

Questions

1 How is metamorphic rock formed?
2 Name a metamorphic rock made from:
   a limestone **b** shale **c** slate
3 **a** What is *foliation* in metamorphic rock?
   b What causes it?
4 **a** What is the difference between:
   i slate and schist? **ii** schist and gneiss?
   b Do you think they contain different minerals?
5 Where would you expect to find examples of contact metamorphism?

6.21 The rock cycle

All change!

At this moment, on the ocean floor, some sediment is slowly turning into limestone. Come back a few million years from now, and you may find the limestone buried below ground, turning into marble. Check again a few million years later, and you may find its carbon in the carbon dioxide erupting from a volcano.

All rock in Earth's crust changes continually from one form to another over millions of years. It is recycled. This is summed up in the **rock cycle**.

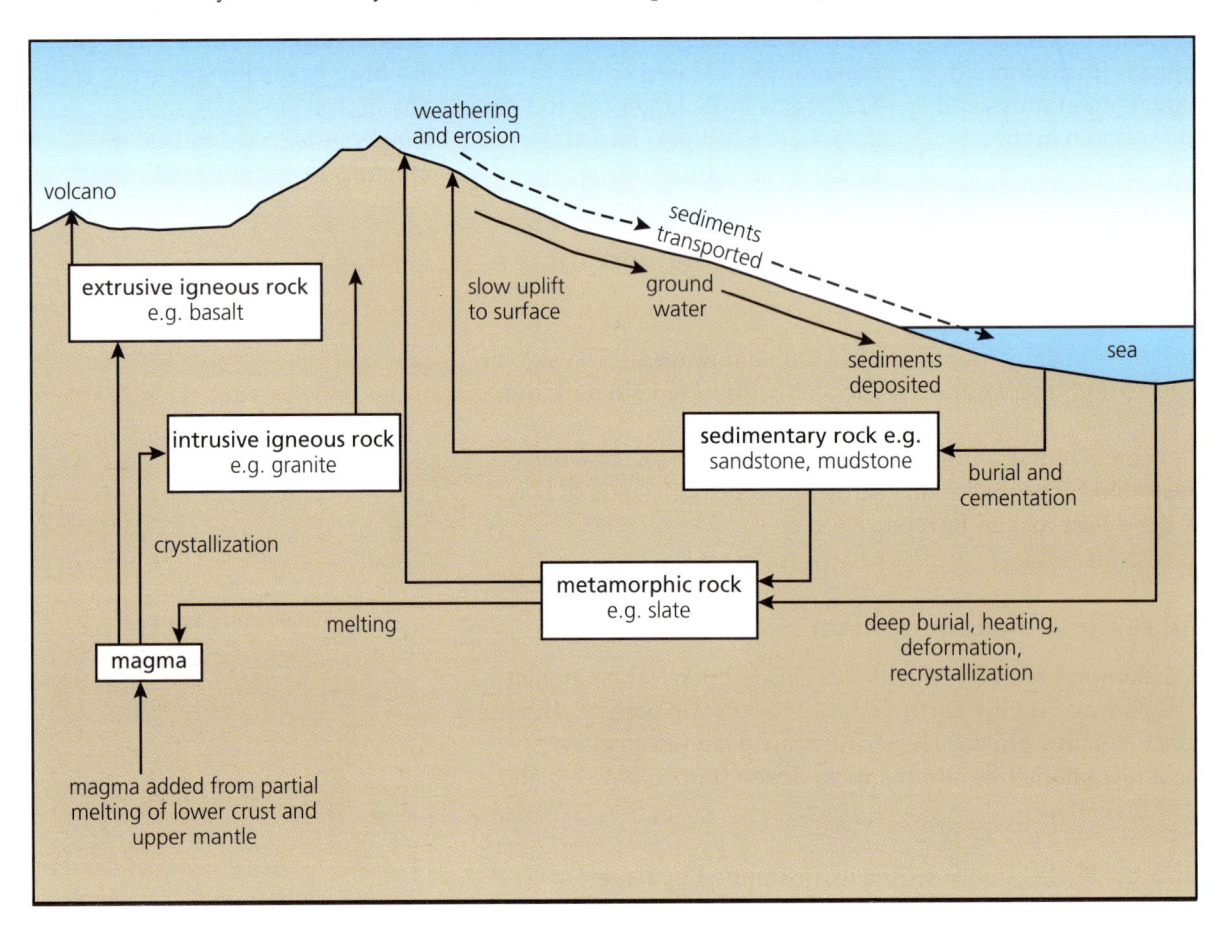

What drives the rock cycle?

This continual recycling of rock requires a great deal of energy.
- Energy from the sun drives the changes in the rock on the Earth's surface: its weathering and erosion. Without the sun's energy there would be no water cycle, and no wind or other weather.
- Sediment is deposited and transformed into rock through the action of gravity.
- Energy from within the Earth drives the rest of the cycle. It causes large slabs of the Earth called **tectonic plates** to move around. As a result, great masses of rock get buried and changed. You can find out more about tectonic plates in the next two spreads.

No movement – no change!

Note that if rock stays where it was formed, it does not change. For example igneous rock is stable as long as it remains deep underground. But at Earth's surface it is in a hostile environment: the temperature is low, and it is surrounded by oxygen, water, and weak acids. Under conditions like these, igneous rock breaks down.

Keeping track of changes

The continual recycling of rock means that geologists have to be detectives. They look for clues that will tell them where rocks have come from, and how they were formed. Below are examples of the clues they gather. You have met some of these already:

- Big crystals show that a rock cooled slowly from magma deep underground. Small crystals mean it cooled quickly on the surface.
- Cemented particles show it started as sediment.
- Sharp edges show that rock fragments did not suffer much chemical weathering, and were protected from collisions.
- If a rock contains shelly remains of sea animals it was created in a marine environment (even if you find it on a mountain top).
- Sediment is usually deposited in horizontal layers. If layers are slanted, it shows the rock has been folded or tilted at some stage by powerful forces.
- Usually, younger rock lies on top of older rock. If you find it *below* older rock, it show that powerful forces have turned the rock upside down.

When a volcano erupts, it gives off carbon dioxide and other gases. In this way, carbon dioxide that was removed from the air and locked up in limestone, via photosynthesis, is returned to the air millions of years later.

A sample of sedimentary rock called breccia.

Some powerful forces were at work here!

Questions

1 What is the *rock cycle*?
2 Explain how a grain of sand deposited on the ocean floor ends up in magma.
3 Rock changes only if it gets moved to a different environment. Explain this.
4 Look at the first photo on the left above. It shows a sedimentary rock called breccia. What conclusions can you draw from its appearance?
5 Study the second photo above, and then list your conclusions about the history of this rock.

6.22 The Earth's plates

Patterns on the Earth's surface

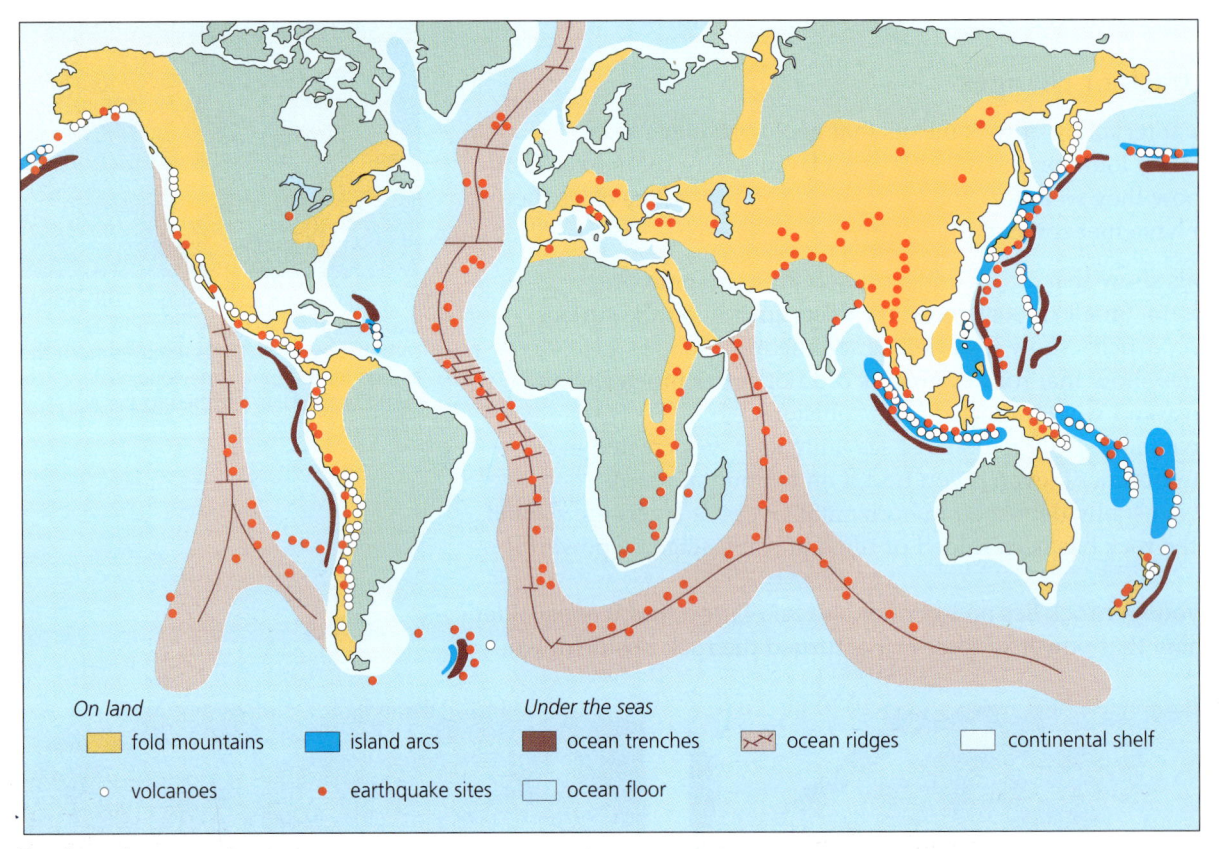

On land: fold mountains, island arcs, volcanoes
Under the seas: ocean trenches, ocean ridges, continental shelf, earthquake sites, ocean floor

Look at the map above.

- Note how earthquakes and volcanic eruptions seem to occur along lines – and often together.
- Note that there are more volcanoes in the ocean than on land.
- Lava also wells up continuously through the ocean floor, along the **ocean ridges**.
- On land the fold mountains often form long **mountain ranges**.
- Look at the **ocean trenches**. These are deep, V-shaped gashes in the ocean floor. They run parallel to lines of volcanoes, which suggests a link between the two.

Some of these patterns had puzzled scientists for decades. But all became clear about 35 years ago, with the theory of **plate tectonics.**

The theory of plate tectonics

This is what the theory says:

- The Earth's lithosphere (the crust + hard outer mantle) is cracked like an eggshell into big pieces called **tectonic plates**.
- The plates move around, dragged by currents in the soft rock below.
- They can pull apart, or push into each other, or slide past each other.
- This movement causes earthquakes, volcanic eruptions, and mountain building along or close to the **boundaries** (edges) of the plates. That explains the linear patterns of these on the map.

A volcano erupting on Big Island in Hawaii. Where there's a volcano, there's usually plate movement.

A map of the tectonic plates

Plates are around 80 km thick, on average. But they vary a lot in area. This map shows the main ones and the directions they move in:

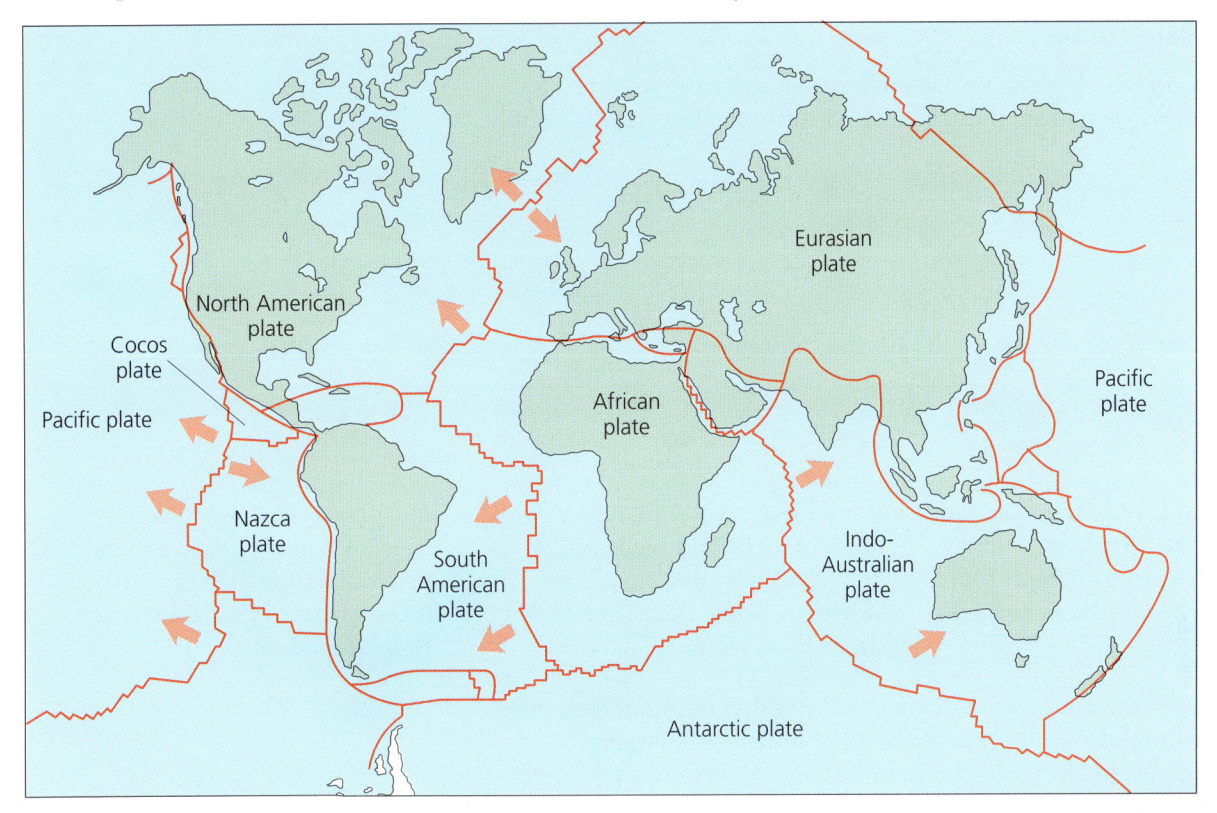

How fast do plates move?

We can track plate movements with the help of satellites. The fastest plates move at over 10 cm a year. Our plate and the North American plate are among the slowest. These two plates are moving apart – and the distance from New York to London increases by under 10 cm a year.

Some consequences of plate movements

Plate boundaries can be dangerous zones to live in. Earthquakes and volcanic eruptions have killed millions of people over the centuries, and cost billions in damage. For example Turkey suffers many serious earthquakes. (The small plate above Africa cuts right through it.)

On the positive side, plate movements help to bury sediment deep underground, so they help to form fossil fuels. And the lava that erupts from volcanoes forms very fertile soil when it weathers.

Part of the Turkish town of Adarpazari after the 1999 earthquake.

Questions

1 What is: **a** an ocean trench? **b** an ocean ridge?
2 What are the Earth's plates?
3 Which plate do you live on?
4 Name two plates that are pushing into each other.
5 Name three plates that are moving away from the African plate.
6 About how fast do plates move?
7 Plate boundaries can be dangerous places to live. Why?

97

6.23 A closer look at plate movements

Why do plates move?

Why exactly do the plates move? To answer that, we have to return to Earth's interior. There, deep in the mantle, heat from natural radioactivity makes rock soften.

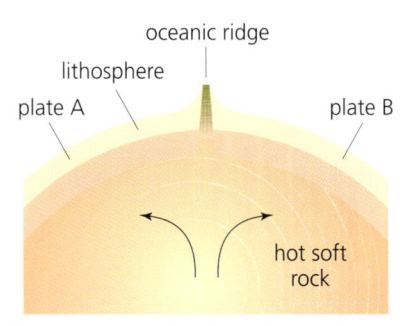

The hotter and softer the rock, the less dense it is. So it rises slowly, forming **convection currents** like you get when you heat very thick porridge.

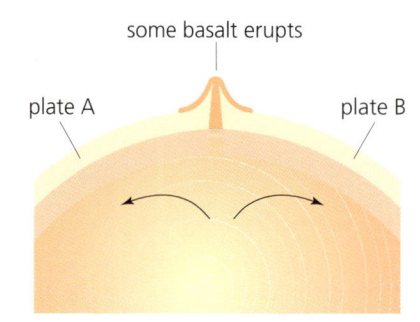

It reaches the solid plates. A small amount melts and wells up between them. But most is forced sideways below the plates and drags them along.

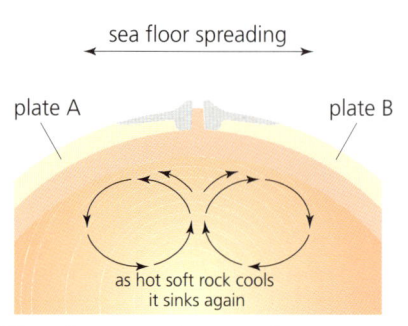

The plates continue to diverge over millions of years. The molten material that welled up between them hardens. The result is **sea floor spreading**.

In this example, the plates are moving apart. But that does not happen at every plate boundary – otherwise the Earth's surface would keep expanding! In fact plates can move in three ways relative to each other: they can move apart or **diverge**, they can push into each other or **converge**, and they can slide past each other. We'll now look at each of those movements.

1 Plates can diverge

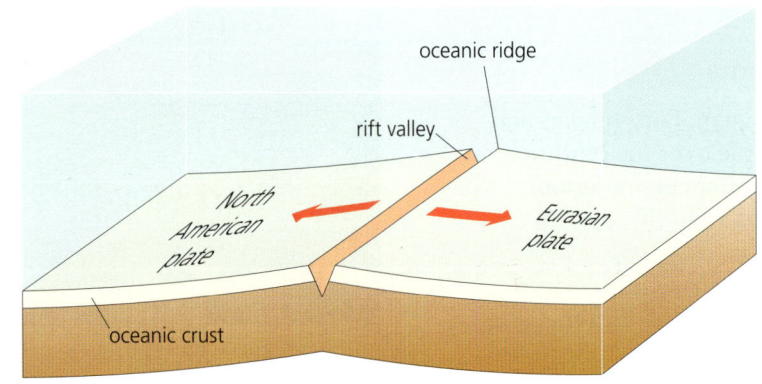

Two oceanic plates diverging.

The plates move apart, as you saw above, with new ocean crust forming at oceanic ridges. For example the Eurasian and North American plates are diverging along the Mid-Atlantic Ridge. There are numerous earthquakes along the ridge as a result.

Since new crust is being formed, the boundary between diverging plates is called a **constructive boundary**.

The jagged line in the ocean is the Mid-Atlantic Ridge, along which the Eurasian and North American plates diverge.

98

2 Plates can converge

They push up against each other. When an ocean plate and a continental plate converge, the thinner denser ocean plate is driven down or **subducted** below the thicker continental plate.

This diagram shows how the Nazca plate is being subducted under the South American plate, off the west coast of South America. Note the volcanoes that form in the Andes as a result.

Since crust is destroyed by subduction, this boundary is called a **destructive boundary**.

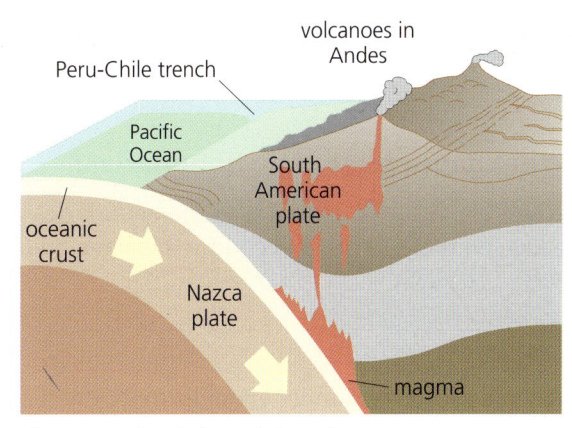

The Nazca plate being subducted.

Subduction takes place along ocean trenches. The massive rock movement causes earthquakes. The subducted rock melts and the magma forces itself to the surface somewhere close by, producing volcanoes. That explains why you find chains of volcanoes parallel to ocean trenches, as in the Andes.

But what if two continental plates converge? Neither is dense enough to be subducted. So rock gets folded upwards, and the result is **mountain ranges**. That's how the Alps and Himalayas were formed.

3 Plates can slide past each other

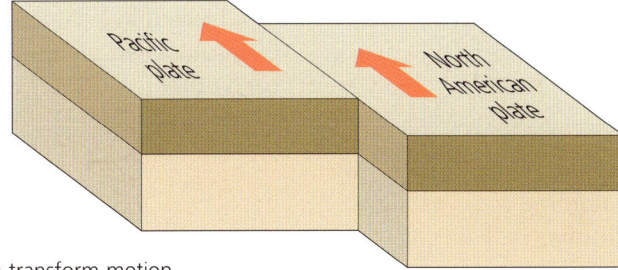

Two plates in transform motion.

The Pacific plate and North American plate are sliding past each other along the San Andreas Fault in California. (Both are moving in the same direction but the Pacific plate moves faster.)

The movement is called **transform motion**. No crust is created or destroyed, so the boundary between the plates is called a **conservative boundary**. Since there is no subduction, there are no volcanoes. But the friction between the plates can cause devastating earthquakes.

Questions

1 Explain in your own words why plates move.
2 What is happening at the Mid-Atlantic Ridge?
3 Draw a diagram to show what happens when:
   a an ocean plate and a continental plate converge
   b two continental plates converge
4 For each part of question 3, give two examples of where this is happening.
5 What is *transform motion*?
6 Why are there earthquakes but no volcanoes along the San Andreas fault in California?

6.24 Shaping the Earth's surface

What happens at plate boundaries?

When plates move, the rocks at the boundaries get pushed and pulled by powerful forces. What effect does this have on them? You will find out here.

Faults

When a rock is put under enough stress it will crack and shift. A **fault** is a crack in a rock along which movement has taken place. The direction of movement depends on the direction of the force, as these diagrams show:

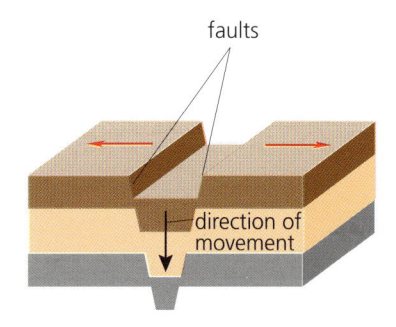

Here a block of rock has dropped down between two faults. This happens when rock is being pulled in opposite directions, for example where two plates are diverging.

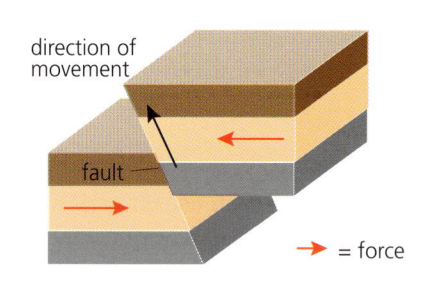

Here one block is pushed up over the other at the fault. This happens when rock is being squashed with great force, for example where two plates are converging.

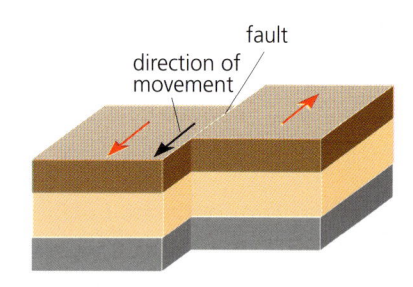

Here one block gets pushed past the other. This happens when rock is subjected to a shearing stress, for example when plates slide past each other.

When a rock is under stress, it stores up **strain energy**. When it gives way, this energy is released. It passes through the surrounding rocks as high-speed waves and causes the vibrations called **earthquakes**. Faulting always gives rise to earthquakes.

When the rock comes to rest, new forces may build up around it and cause it to move again. This means more earthquakes.

Folds

Folds are formed when compressed rocks deform *without* cracking, as shown on the right. It's like when you press your hands on a table cloth and slide it towards the centre of the table.

Folding mainly takes place at the boundary between two converging plates. The diagram shows how sedimentary layers fold. Extreme pressure will cause folds to **overturn**, as shown. Each part of a fold, and type of fold, has its own name. For example, the ridge of a fold is called an **anticline**, and the trough a **syncline**.

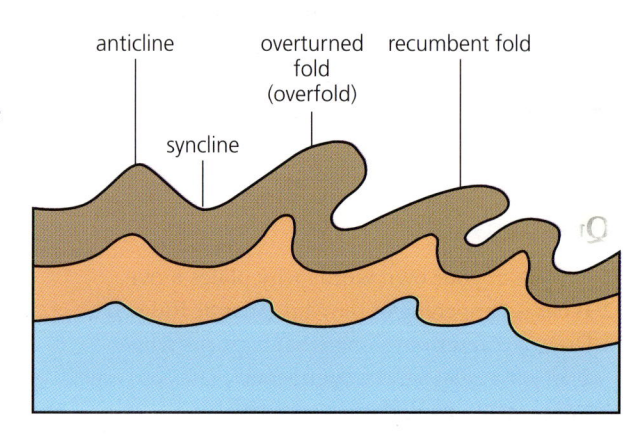

Fold mountains

Many of the world's highest mountain ranges are the result of plate movement. They started as sedimentary rock on the sea floor. As this was squashed by converging continental plates it folded upwards, as shown in this diagram.

For example, when the Indo-Australian plate converged on the Eurasian plate, an ancient sea called the **Tethys** got squashed upwards. The result is the Himalayas. The plates are still converging so the Himalayas are still growing, and causing major earthquakes in China.

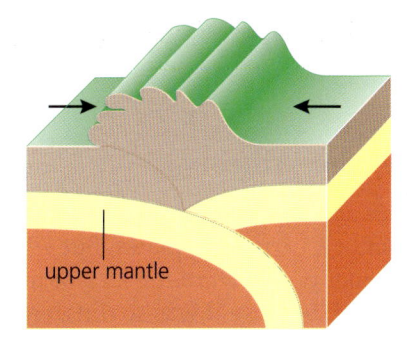

The making of a fold mountain.

The fold mountains of the Himalayas, including Everest, are growing slowly as the plates converge.

Volcanoes

Volcanoes are linked with plate movements.
- Underwater volcanoes erupt non-stop at diverging boundaries, making new ocean floor.
- Subduction at ocean trenches also leads to volcanoes. Subducting rock heats, softens, and melts as it slides below the other plate. Magma forces its way to the surface, forming a line of volcanoes parallel to the trench. They may appear in the ocean as **island arcs** like the islands of the Caribbean. They may be on a continent, like the volcanoes of the Andes.

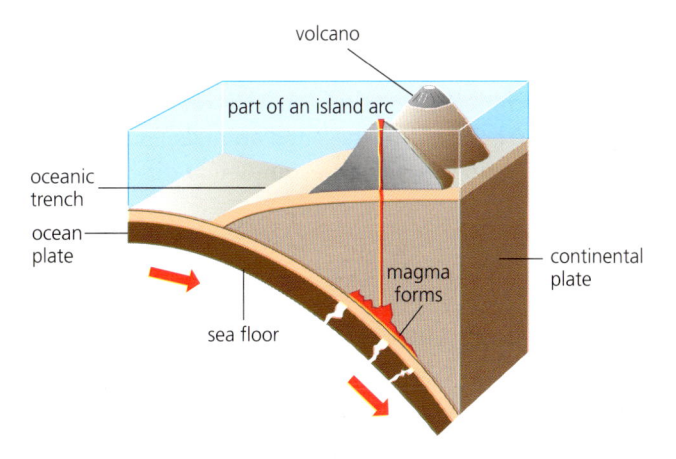

An island arc forming. The islands lie along a curve (arc).

Note that volcanic eruptions return to the air some of the carbon dioxide once locked up in limestone. So they can add to global warming. But volcanic dust and ash can have the opposite effect – by blocking out sunlight they can cause temperatures around the Earth to fall.

Questions

1. What is a fault?
2. **a** Give two examples of movements that can take place along a fault.
   b Explain why they cause earthquakes.
3. What causes folds in rock?
4. Draw a diagram to show how a fold mountain forms and give two examples of mountain ranges that have been formed in this way.
5. Draw a diagram to show how subduction can lead to the formation of an island arc.

6.25 Wegener: a scientist ahead of his time

Wegener's theory

Since around 1970, we have accepted the idea that the continents are on the move, carried on plates. But when the German geophysicist Alfred Wegener put forward the idea of moving continents in 1915 people just laughed, because it was so different from what they believed then.

Wegener suggested that around 200 million years ago the continents were all joined as one supercontinent, which he named **Pangaea**, meaning *all lands*. Over millions of years Pangaea broke up and the continents moved apart. He called their movement **continental drift**.

The evidence

This is the kind of evidence Wegener used to support his ideas.

Continental fit   The continents look like pieces of a jigsaw puzzle. For example, if you move South America it will fit snugly into the curve of Africa. Wegener played around with the shapes of the continents until he got Pangaea.

Linearity   Chains of mountains and volcanoes lie roughly along straight lines. Earthquakes also tend to occur along lines. This suggests an underlying pattern such as you might find on a jigsaw.

Fossils   Animals and plants survive only in climates and habitats that suit them. Yet fossils of the same animals and plants have been found on different continents, often in completely different climates.

For example, the small reptile *Mesosaurus* lived 240 million years ago. From its fossil, scientists can tell that it swam only in shallow waters. Yet it has been found in Brazil and South Africa, separated by 5000 km of ocean. So South America and Africa may once have been joined. And look where the plant species *Glossopteris* is found, on the map below.

Rocks   Similar rocks, 390 million years old, have been found in mountains in eastern North America, western Europe, and West Africa. This suggests that North America, Africa, and Europe were once joined, with a long line of mountains running through them.

Wegener's supercontinent, Pangaea

Why they laughed

In 1915, people believed that:

- After Earth was formed, its crust wrinkled as it cooled.
- The wrinkling gave us mountains and other features.
- Once it had cooled and hardened, the crust changed very little.
- There was no way continents could move through the hard dense rock of the ocean floors.

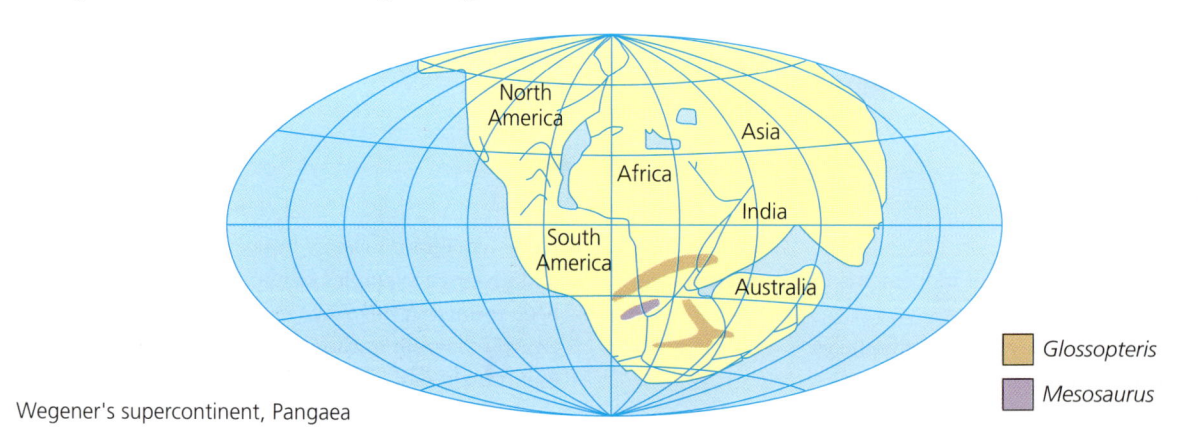

Alfred Wegener (1880–1930), who was proved right in the end.

There has been a huge increase in marine exploration in the last thirty years. This is the British Antarctic research vessel *James Clark Ross* at work in Antarctica.

The proof

Wegener's ideas were laughed at. But in 1968 a deep sea drilling project was launched. It led to several amazing discoveries:

- The rock in the ocean floor is much younger than the rock in the continents. The oldest continental rock found so far is 3.96 billion years old. The oldest ocean rock is only 0.18 billion years old.
- The further it is from the oceanic ridges, the older the ocean rock. The rock at the top of the ridges is almost new.
- As the basalt in the ocean floor hardens, the iron crystals in it line up along the Earth's magnetic field, just like iron filings around a magnet. Once the rock has hardened, the crystals stay lined up, giving a permanent record of the Earth's magnetic field.
- But as you move out from an oceanic ridge, the basalt shows stripes of different magnetization, matching reversals in the Earth's magnetic field. This proves that new basalt is formed at ocean ridges, pushing older basalt aside. In other words, the sea floor is spreading.

These discoveries shattered all the earlier beliefs that the Earth's crust was ancient and unchanging. If the sea floor is spreading, then *the continents are moving too!* The theory of tectonic plates took over.

In fact, scientists have used the magnetic strips to retrace the plates' paths from Pangaea. It appears to have broken up over 200 million years ago, in the age of the dinosaurs. So Alfred Wegener was right after all.

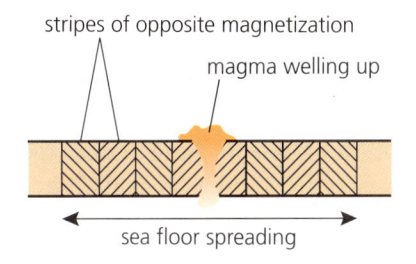

Questions

1 In 1915, how did people think mountains had been formed?
2 List four pieces of evidence Wegener used to support his theory of continental drift.
3 What did Wegener call the supercontinent?
4 Basalt records the direction of the Earth's magnetic field. Explain how.
5 Magnetic stripes in the basalt prove that the ocean floor is spreading. Explain why, in your own words.
6 Oceanic rock is younger than continental rock. Why?

6.26 Can we predict earthquakes and eruptions?

Prediction would save lives

Large numbers of people live in danger zones, along or near plate boundaries. Cities at risk from earthquakes include Tokyo, San Francisco, Los Angeles, and Mexico. The people of Naples live in the shadow of Mount Vesuvius, which could erupt at any time.

If we could predict earthquakes and eruptions to the nearest day or two, people could be moved to safety and lives saved. But we are not so good at prediction yet. Scientists are still working on this challenge.

Predicting earthquakes

Scientist can tell where earthquakes are *likely* to occur. For example the risk is high along the San Andreas fault in California, where the Pacific plate slides past the North American plate. It is particularly high where sections of these plates are locked together. The rock here is storing up enormous strain energy.

But earthquakes occur very suddenly. Predicting them in the short term – to within a day or two – is very difficult. These are some of the instruments scientists use and the clues they look for:

- **A global positioning system (GPS)** can be used to measure how fast plates are moving, if at all. It bounces signals off satellites.
- A **strainmeter** is used to measure strain energy building up in rock.
- When a quake is getting closer, tiny cracks form in the rock. These cause the rock to bulge. A **tiltmeter** will check for bulges.
- The cracks fill with groundwater, so the **water level** in any nearby wells will fall. Radon gas escaping from the cracks may bubble from the well.
- There will be small **foreshocks** before the main earthquake. A **seismometer** will record these.
- And finally, a living clue. Before an earthquake **animals** act strangely. Snakes and rats crawl out of their holes. Dogs howl.

To pick up these clues you have to be there on the spot, with the right equipment. But earthquakes can occur anywhere on a plate boundary or in the many fractures (faults) off to each side. You can't be everywhere!

Any success?

The most successful earthquake prediction ever was in Haicheng city in China in 1975. The public had been taught to watch out for signs of earthquakes. Early in the year they reported gas bubbling from wells and snakes crawling from holes, even though the weather was icy.

On 4 February scientists warned of an earthquake within hours. They told everyone to leave home and go to the parks. At 7.30 in the evening an earthquake destroyed most of the homes in Haicheng – but over 100 000 lives were saved.

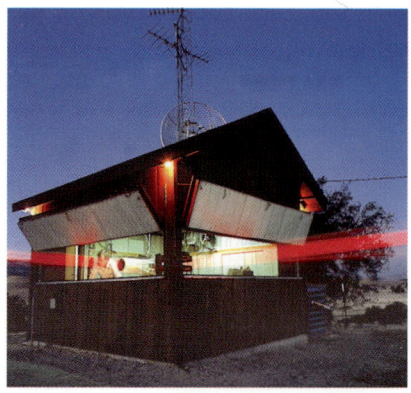

Monitoring the risk of earthquakes along the San Andreas Fault.

Predicting volcanic eruptions

It is much easier to predict volcanic eruptions than earthquakes. For a start, earthquakes can occur anywhere but volcanoes sit still. And second, volcanoes don't suddenly waken, erupt, and go back to sleep. They may belch dust and ash months before an eruption, as warning signals. Let's look at other clues scientists can pick up:

- When magma is on the move inside a volcano, it causes showers of tiny earthquakes. **Seismometers** will detect these.
- The moving magma warms up the ground. **Infrared cameras** on **satellites** can detect these heat changes.
- The magma causes the volcano to swell and bulge. A **global positioning system** or a **tiltmeter** will detect this.
- The moving magma gurgles and belches. Scientists are now studying these **sounds** to see if they can be used as warning signals.
- The more **sulphur dioxide** a volcano gives out, the closer the eruption. So scientists regularly collect gas samples on the volcano's slopes.

You can also collect sulphur to monitor the activity of a volcano. Sulphur is deposited around the volcano as sulphur-containing gases condense out onto the rocks.

Any success?

On 2 April 1991, Mount Pinatubo in the Philippines, after 500 years of dormancy, started belching dust and steam. Scientists moved in to monitor it. In early June they warned of a large eruption within weeks. Over 75 000 people were moved to safety. On 15 June a violent eruption of gas and lava, followed by mudflows, destroyed many of their homes.

In the future

Prediction methods will improve. But that does not mean no lives will be lost. Monitoring equipment is expensive and a poorer country might not be able to afford it. And many places still do not have good emergency plans for warning people, or moving them quickly to safe zones.

Fleeing to safety at Mount Pinatubo. Notice how everything is covered in volcanic dust.

Questions

1 Explain why it is difficult to predict earthquakes.
2 a Water wells can give clues that an earthquake is on the way. Explain how.
  b List three other clues scientists look for.
3 Eruptions are easier to predict than earthquakes. Why?
4 Give four clues that an eruption is on the way, and say how scientists will detect these clues.
5 What part do satellites play in helping us predict earthquakes and eruptions?
6 Can earthquakes and eruptions be *prevented*?

Questions for Module 6

1 a Copy this diagram and complete it with:
  i the common names
  ii the chemical formulae

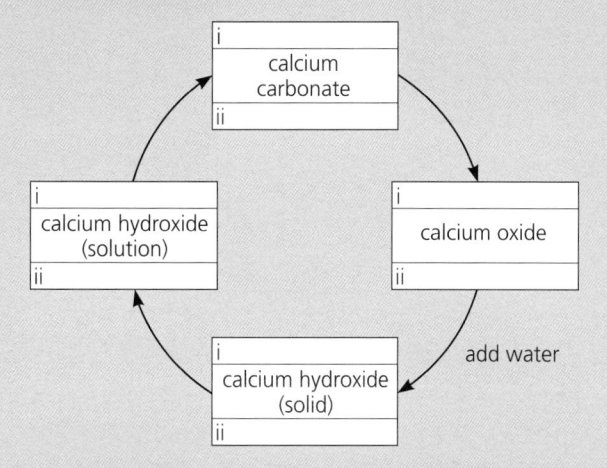

[Diagram: cycle with four boxes — calcium carbonate → calcium oxide → calcium hydroxide (solid) → calcium hydroxide (solution) → calcium carbonate. Arrow from calcium oxide to calcium hydroxide (solid) labelled "add water". Each box has spaces labelled i and ii.]

b Beside each arrow say how the change would be carried out. One example is shown.
c Give three reasons why limestone is an important raw material.

2 Limestone is calcium carbonate, $CaCO_3$. It is quarried on a huge scale.
  a Which elements are present in it?
  b Much of the limestone that's quarried is used to make quicklime (CaO) for the steel industry.
    i What is the chemical name for quicklime?
    ii How is quicklime made from limestone?
  c Powdered limestone is used to improve the water quality in acidified lakes.
    i Suggest how the water might have become acidified in the first place.
    ii How does limestone improve its quality?
    iii Why is the limestone powdered?
  d You live in a limestone area. A new limestone quarry is being proposed for the area. List the advantages and disadvantages to local people.

3 When calcium carbonate is heated, a white solid is formed and carbon dioxide is given off.
  a Write a word equation for this reaction.
  b Describe a test for carbon dioxide.
  c A few drops of water were added to the white solid. Some of the water turned to steam.
    i Classify the reaction between water and the white solid as endothermic or exothermic.
    ii Name the substance formed when water was added to the white solid.
    iii Give an industrial use for the white solid.

4 Fossil fuels are the world's main source of energy.
  a What is a *fossil fuel*? Name three.
  b Explain how oil was formed. Draw a diagram to show where it is found in rocks.
  c Is wood a fossil fuel? Explain.
  d 'Fossil fuels are stores of solar energy.' Is this true? Explain.
  e Fossil fuels are considered a non-renewable resource. Why is this?
  f 'We must become less dependent on fossil fuels in the 21st century.' Give two reasons why, and suggest a list of steps to achieve this aim.

5 A crude oil sample was separated into four fractions. These were collected in the temperature ranges shown here:

Fraction	Temperature range / °C
A	25 – 70
B	70 – 115
C	115 – 200
D	200 – 380

Which fraction:
  a has the lowest range of boiling points?
  b burns most readily?
  c has molecules with the longest chains?
  d would be the most viscous?

6 The diagram below represents the process used to separate crude oil into different fractions.

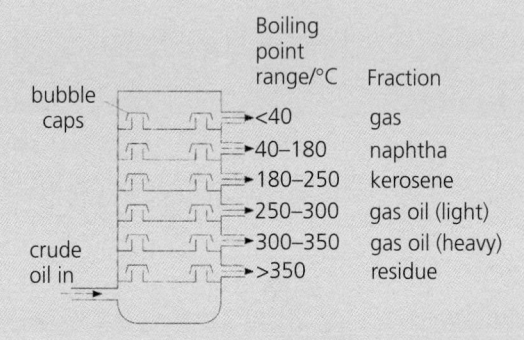

[Diagram: fractionating column with bubble caps; crude oil in at base. Fractions: <40 gas; 40–180 naphtha; 180–250 kerosene; 250–300 gas oil (light); 300–350 gas oil (heavy); >350 residue.]

  a Name the process.
  b This process is based on the fact that different compounds have different ____ ____. What are the missing words?
  c i Using the terms *evaporation* and *condensation*, explain how naphtha is produced.
    ii Is naphtha likely to be a single compound or a group of compounds? Explain your answer.
  d How do the bubble caps help in the process?

106

e Write down one use for each fraction obtained.
f i A hydrocarbon has a boiling point of 200 °C. In what fraction will it be found?
  ii Are the carbon chains in its molecules shorter or longer than those found in naphtha?
  iii Is it more, or less, viscous than naphtha?

7 This shows how hydrocarbons are cracked:

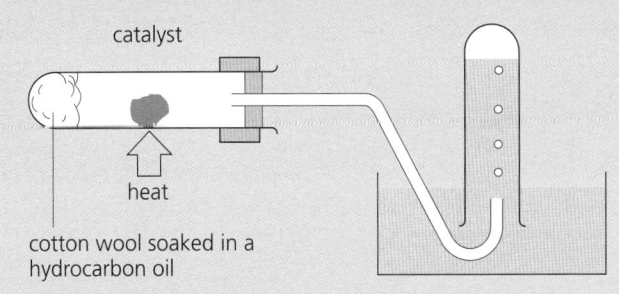

a What is *cracking*?
b What two things are needed to crack hydrocarbons?
c Why should the first tube of gas that is collected be discarded?
d Is the gaseous product soluble or insoluble in water? Explain your answer.
e Ethane, C_2H_6, can be cracked to give ethene, C_2H_4, and hydrogen. Write an equation for this reaction.

8 Propene is one of the many important hydrocarbons made from oil. Like *propane*, it is made up of molecules which contain three carbon atoms. Like *ethene*, it has a double bond.
a Draw a molecule of propene.
b How does it differ from a molecule of propane?
c To which group of hydrocarbons does:
  i propane   **ii** propene   belong?
d Write formulae for propane and propene.
e Which of the two is a *saturated* hydrocarbon?
f i Explain why propene reacts immediately with bromine water, while propane does not.
  ii What would you see in the reaction?
g Name a hydrocarbon with **four carbons** in the chain that will react with bromine in water in the same way that propene does.
h Propene is obtained by breaking down longer-chain hydrocarbons. What is this process called?
i Propene is the monomer for making an important plastic. Suggest a name for the polymer it forms.

9 a Which of these molecules could be used as monomers for making plastics? Explain your choice.
ethene, $CH_2=CH_2$ ethanol, C_2H_5OH
propane, C_3H_8 styrene, $C_6H_5CH=CH_2$
chloropropene, $CH_3\,CH=CHCl$
b Suggest a name for each polymer obtained.

10 'Teflon' is the commercial name given to the polymer obtained from the monomer tetrafluoroethene.
The monomer has this structure:

$$\begin{array}{c}F\\F\end{array}\!\!C=C\!\!\begin{array}{c}F\\F\end{array}$$

a What feature of the monomer makes polymerization possible?
b Name the type of polymerization that occurs.
c Show the structure of the repeating unit in the polymer.
d What is the chemical name for this polymer?
e Write an equation for the polymerization reaction.

11 Two plastics have these structures:
  A long-chain molecules
  B long-chain molecules with cross-linking between the chains
a Draw diagrams to show these structures.
b Which diagram could represent:
  i a thermosetting plastic?
  ii a thermosoftening plastic?
c For each choice in **b**, explain why the structure gives rise to that particular property.

12 Look at the table below. It shows information about fossil fuels that can be used for heating and cooking at home.

Fuel	Heat produced kJ/g	kJ/cm^3	Relative cost per kJ
Natural gas	56	0.001	1.4
LPG (liquefied petroleum gas)	50	30	2.0
Gas oil	48	35	1.8
Coal	34	76	1.6

a What is the compound that makes up natural gas?

107

b Name the products when this compound burns in:
  i plenty of oxygen
  ii a limited supply of oxygen
c When coal is burned, other gases are formed which cause acid rain. Name an impurity in coal that forms one of these gases when burned.
d Which fuel in the table is the most expensive?
e **i** Which fuel gives out the most energy per gram when burned?
  ii Which gives out the most energy per cm^3?
  iii Compare the ways these two fuels are supplied to homes.

13 Hydrocarbons with large molecules, obtained from crude oil, can be cracked. The result is a mixture of smaller hydrocarbons including alkanes and alkenes.
  a Why is cracking economically important?
  b What conditions are needed for cracking?
  c Give the structural formulae and names for the four hydrocarbons shown below. (In a structural formula, you show a line for each bond.)

  $$CH_4 \quad C_2H_6 \quad C_2H_4 \quad C_3H_8$$

  d Classify each hydrocarbon above as an alkane or an alkene.
  e Which hydrocarbon above undergoes an addition reaction to form a polymer?
  f Draw part of the polymer molecule and name the polymer.

14 The mixture of gases we call air developed over millions of years from the gas that burst from volcanoes.
  a Most of the volcanic gas consisted of water vapour. But now there is only a tiny percentage of this in air. Why?
  b The volcanic gas contained a large percentage of carbon dioxide. Where did most of this go?
  c Some scientists believe that the volcanic gas also contained hydrogen chloride. Why is there none of this gas in air?
  d Hydrogen is also absent from air, even though it may have been a constituent of volcanic gas. Where has it gone to?
  e Which important gas in air was not present in the volcanic gas? Name the process that produced this gas over millions of years.
  f Which gas makes up approximately 78% of the air?
  g Only one gas in the mixture will allow things to burn in it. Which gas is this?
  h Which gas containing sulphur is a major cause of air pollution?
  i Name two other gases which contribute to air pollution.

15 Copy and complete:
The carbon cycle and the water cycle are both driven by _____ from the _____. In the _____ cycle, energy in the form of _____ causes ocean water to _____. When it cools, the vapour forms _____ which collect as _____. Eventually this falls as _____. This soaks into the ground and is taken in by the _____ of plants which need it for a process called _____. This process is the start of the _____ cycle. It requires energy in the form of _____.

16 Imagine slicing Earth in half and taking a journey from the crust to the centre.
  a Draw a diagram showing the cross-section through the centre, and label the layers.
  b What happens to:
    i temperature **ii** density **iii** pressure
    as you move from the surface to the centre?
  c Why is the inner core solid and not liquid?
  d Why is the overall density of Earth much greater than the average density of the rocks that form the crust?

17 Erosion and transport are part of the rock cycle.
  a What is erosion?
  b What name is given to the movement of rock particles that causes erosion of other rocks?
  c List the four transport agents.
  d Explain what would cause each transport agent to drop the load of particles it carries.
  e A sediment deposited by wind or water may be *well sorted*. What does that mean?
  f Explain how a stream sorts particles.

18 Sedimentary rocks often contain rock fragments that have been transported by rivers and deposited on river and sea beds. The rock fragments can be classified as sand, mud, or pebbles according to their size. All three are affected to different extents by moving water.
  a Which of the tree types of particle requires most energy to transport it? Why?

b Place the following observations in order of increasing water speed.
  A the small pebbles move further downstream
  B sand is lifted from the sea bed
  C mud settles on the river bed
  D large pebbles are moved from one place to another
  E only pebbles are found
  F sand is dropped on the sea bed
c Explain why the opposite banks of a river do not always look the same.

19 Water containing dissolved minerals seeps into tiny pores in rocks. Crystals form in these pores when the water evaporates. Over time, the crystals grow until eventually they exert enough pressure to force a surface layer of rock to break and flake off.
  a Decide whether this is an example of physical weathering or chemical weathering.
  b Limestone (a sedimentary rock) is greatly affected by the process described above. Marble (a metamorphic rock) is affected less than the limestone. Granite (an igneous rock) is almost unaffected.
    i Suggest a feature of the structure of granite which enables it to resist the process described.
    ii Describe the conditions under which limestone changes to marble. Suggest why marble is affected less than limestone by the process described.

20 A laboratory demonstration was carried out to show how igneous rock forms. The following results were obtained by recrystallizing equal amounts of the chemical salol at 0 °C and at 50 °C.

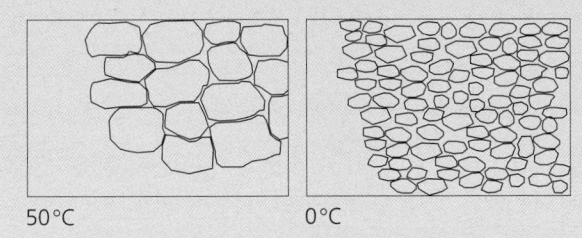

50 °C   0 °C

  a What is meant by *recrystallizing* and how is it carried out in the laboratory?
  b What effect does the temperature have on:
    i the speed of crystallization?
    ii the size of the crystals?
  c Explain how these results relate to the two main types of igneous rock, extrusive and intrusive.

21 Gabbro is an *intrusive* igneous rock. Andesite is an *extrusive* igneous rock.
  a Explain the terms in italics.
  b In what way(s) would you expect these rocks to look: **i** similar? **ii** different?
  c Which of them appears at Earth's surface only after erosion of other rocks?

22

This is a sample of the igneous rock **obsidian**. It is a volcanic glass, which means it is not crystalline.
  a What is the source of the rock obsidian?
  b What type of structure (giant or molecular) does it have?
  c How would you describe the arrangement of the atoms in obsidian: *ordered* or *disordered*?
  d What kind of cooling produced obsidian: rapid or slow? Explain your answer.
  e How do you think obsidian would respond to being struck with a hammer? Explain.
  f Suggest how obsidian may have been used by early civilizations.

23 The following diagram represents a geological feature found in many areas.

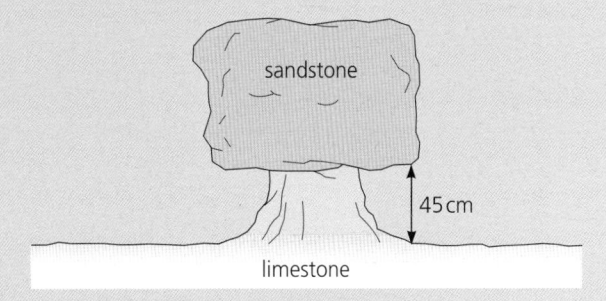

sandstone

45 cm

limestone

  a Which type of rock is:
    i sandstone? **ii** limestone?
  b Explain the sequence of events that led to sandstone being deposited on the limestone.
  c Why is the limestone under the sandstone a different height from the rest?

109

d Assume that the rate of weathering of limestone is about 0.025 mm per year.
   i How long has it been since the limestone was level?
   ii What assumption have you made in this calculation?

24 Dolerite is an igneous rock found in Cumbria. In one area it is found in contact with the sedimentary rock sandstone.

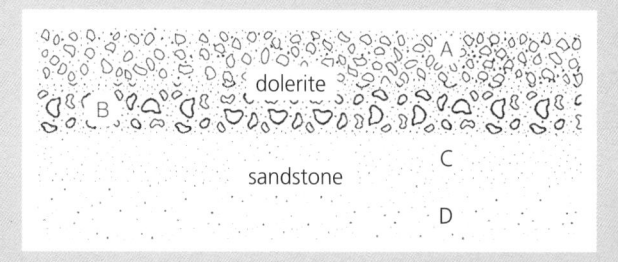

a The crystals of dolerite at A are smaller than those at B. Suggest a reason.
b The sandstone at C has become 'baked' and is much tougher than the sandstone at D. Why?
c From this evidence, which rock is older? Explain.
d Sometimes dolerite forms a **sill**, where lava pushes its way between the planes of sedimentary rock. Make a sketch of a sill.
e Is a sill younger or older than the sedimentary rock above it?
f What characteristic(s) might you expect the sedimentary rock on top of the sill to have?

25 The diagram below shows a section through the Earth, showing layers labelled **P**, **Q**, **R**, and **S**.

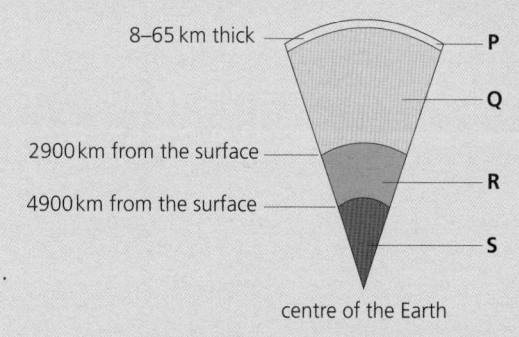

a Give the name of:
   i layer **P**   **ii** layer **Q**
b What do layers **R** and **S** make up?
c Give one difference between layers **R** and **S**.

The outer part of the Earth is made up of plates which move relative to one another. Two such plates are shown on the diagram below.

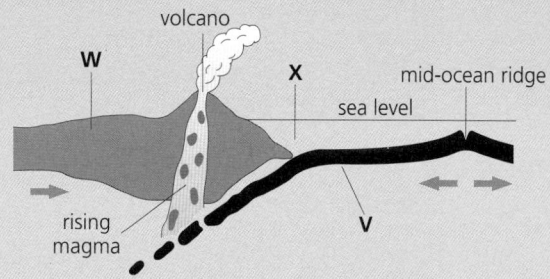

d What is the name given to these plates that make up the outer part of the Earth?
e Name the two types of crust making up plates **V** and **W**.
f A volcano is shown in the diagram. Describe and explain the difference between igneous rocks that form inside the Earth and those that form at the surface of a volcano.
g Explain what is happening at point **X** where the two plates meet.

26 Plate tectonics is the driving force behind the recycling of rocks. The key to plate tectonics is the heat generated in the Earth's interior.
   a What is meant by the *recycling* of rocks?
   b i Name an important source of heat energy in Earth's interior.
      ii Explain how this heat energy leads to the movement of plates.
   c Sedimentary rock can be recycled as metamorphic rock.
      i How does sedimentary rock get pushed down through Earth's crust?
      ii Explain how it becomes metamorphic rock.
   d Sedimentary rock can also be recycled as igneous rock. What further conditions would be needed to turn it into igneous rather than metamorphic rock?

27 When continents collide, rock sometimes becomes folded and sometimes broken.
   a Which of these two processes always results in:
      i land being uplifted?   **ii** a fault forming?
   b Draw diagrams to illustrate the difference between the two processes.
   c What is: **i** an anticline?   **ii** a syncline?
   d Identify an anticline and a syncline on your diagrams for **b**.

28

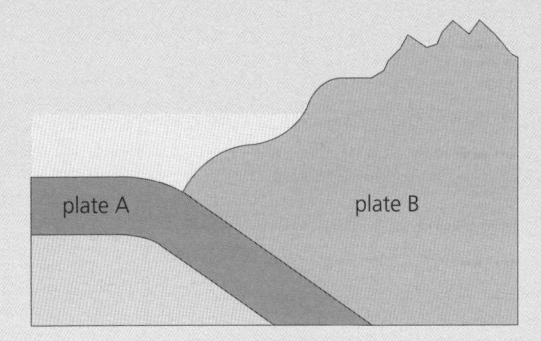

a This diagram shows a cross-section through two plates, A and B. Copy it and add these labels:
ocean   oceanic crust   continental crust mountains   oceanic trench   sediments continental margin
b i Draw an arrow to show the movement of A.
  ii Guess how far A would move in 12 months.
  iii What other name is given to the crust in A? Explain.
c i Why does plate A sink beneath plate B?
  ii What scientific name is given to this process?
d Name a mineral you'd find in the sediment.
e Mark on your diagram an area where metamorphic rock could form.
f Mark a point on your diagram where you think a volcano might erupt. Explain your reasoning.

29 As the basalt in the ocean floor hardens, the iron crystals in it line up along the Earth's magnetic field. The Earth's magnetic field reverses from time to time over millions of years. 'Stripes' of magnetic reversal appear in the oceanic crust.

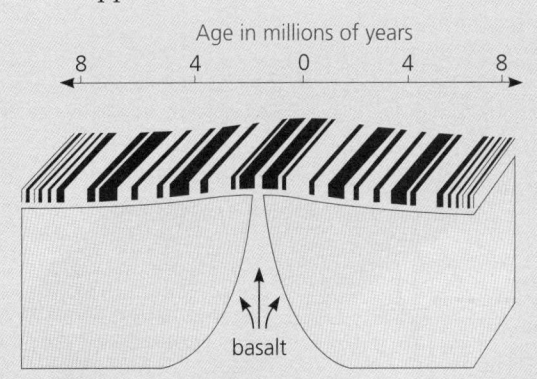

a i Where on the diagram is the oceanic ridge?
  ii What process happens here?
  iii Where is the oldest rock?
  iv What do the black and white stripes represent?
  v Why is the pattern symmetrical?

b Would you expect this pattern of magnetic reversals to continue in the future? Explain.

30

Above is a map of the North Atlantic ocean. Greenland and Europe were once joined. Now the North Atlantic ocean is widening by about 2 cm per year.
a What is the cause of the widening?
b Is the plate boundary constructive or destructive?
c What do you find along the mid-Atlantic ridge?
d How was Iceland formed? From what type of rock?
e i Estimate how long ago the two continents were joined.
  ii Why are distances from the edges of the continental shelves used in the calculation rather than distances from the coasts?
  iii What other assumption have you made in doing the calculation?
f What other tectonic process must be taking place to balance out the fact that the crust is getting larger between Europe and Greenland?

31 Which of the following statements provide support for the idea that South America and Africa were once joined together?
A They have similar rock along their adjacent coastlines.
B They are of a similar shape.
C Today, many of the same crops are grown on both continents (for example coffee).
D The adjacent coastlines have shapes that seem to fit snugly together.
E Similar fossils have been found in each of the continents.
F Satellite images show that South America is moving slowly away from Africa.
G They experience similar weather conditions.

Module 7
Patterns and chemical change

In chemistry class you make small amounts of a substance. But in industry, huge quantities of plastics, paint, dyes, fertilizers, and other products are made on a daily basis. A factory might make 1000 tonnes of fertilizer a day, for example.

A factory has to make a profit. This means a lot of careful planning: about how much of each reactant to use, for example, and the rate of the reaction. The faster it goes, the faster you get the product. The more product you can make in a day, the more profitable your factory.

Luckily, we can control the rate of most reactions: speed them up or slow them down to give a rate that suits. But usually it means a compromise between speed and cost. Speeding up a reaction can mean a larger fuel bill, for example.

This section covers many aspects of reactions: the quantities that react; ways to change the rate of a reaction, and why these work; reactions that pose a special problem because they can go backwards; and catalysts that speed up reactions by just being there. You'll also look in more detail at one important group of products: fertilizers. ∎

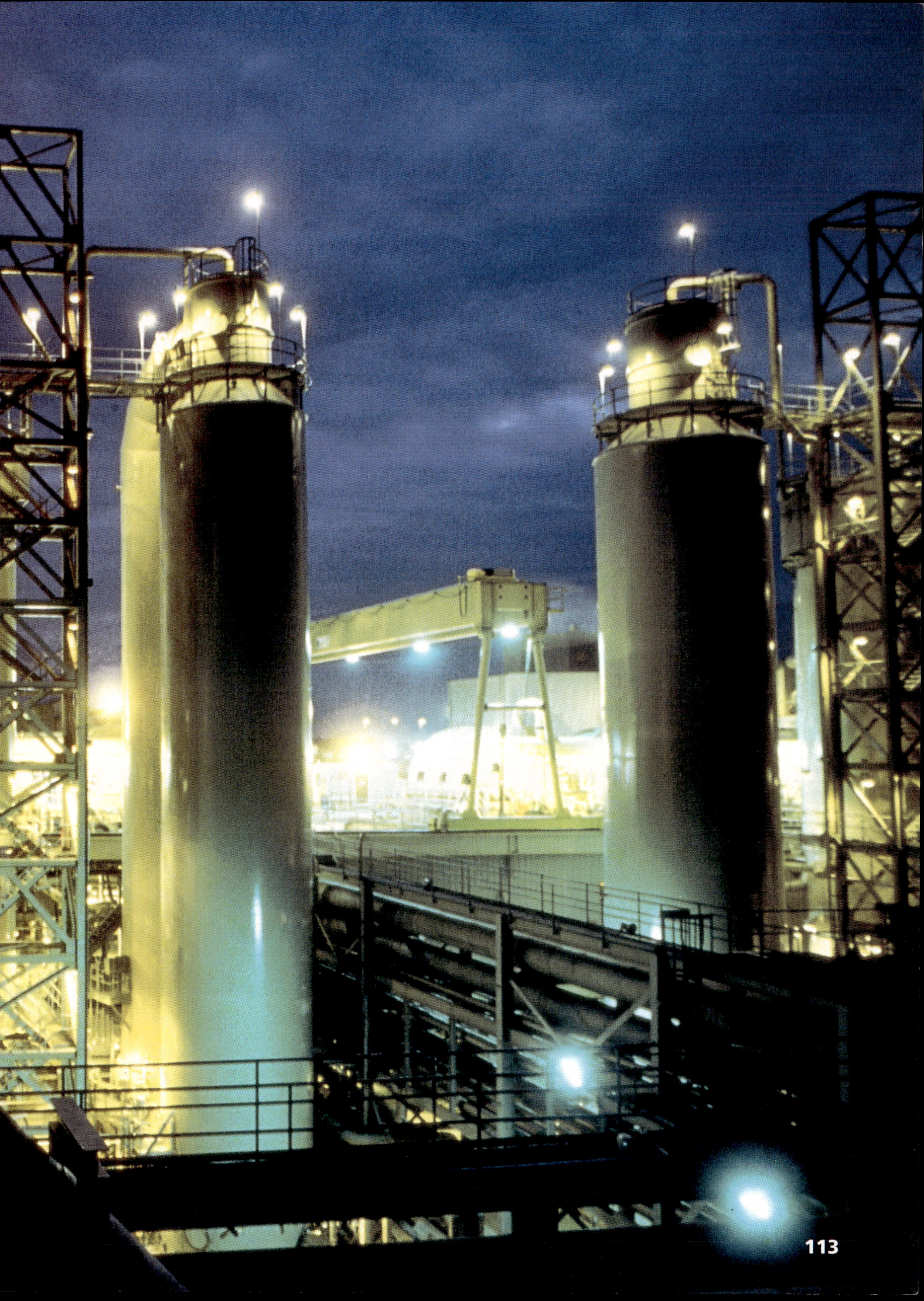

7.01 Rates of reaction

Fast and slow

Some reactions are **fast** and some are **slow**. Look at these examples:

Silver chloride precipitating, when solutions of silver nitrate and sodium chloride are mixed. This is a very fast reaction.

Concrete setting. This reaction is quite slow. It will take a couple of days for the concrete to harden.

Rust forming on a heap of scrap iron in a scrapyard. This is usually a very slow reaction.

It is not always enough to know just that a reaction is fast or slow. For example, in a factory that makes products from chemicals, the chemical engineers need to know *exactly* how fast each reaction is going, and how long it takes to complete. In other words, they need to know the **rate** of each reaction.

What is rate?

Rate is a measure of how fast or slow something is. Here are some everyday examples.

This plane has just flown 2000 kilometres in 1 hour. It flew at a **rate** of 2000 kilometres per hour.

This petrol pump can pump petrol at a **rate** of 50 litres per minute.

This machine can print newspapers at a **rate** of 10 copies per second.

From these examples you can see that:

Rate is a measure of the change that happens in a single unit of time.

Any suitable unit of time can be used – a second, a minute, an hour, even a day.

Rate of a chemical reaction

When zinc is added to dilute sulphuric acid, they react together. The zinc disappears slowly, and a gas bubbles off.

After a time, the bubbles of gas form less quickly. The reaction is slowing down.

Finally, no more bubbles appear. The reaction is over, because all the acid has been used up. Some zinc remains behind.

In this example, the gas that forms is hydrogen. The equation for the reaction is:

zinc + sulphuric acid $\longrightarrow$ zinc sulphate + hydrogen
$Zn\ (s)$ + $H_2SO_4\ (aq)$ $\longrightarrow$ $ZnSO_4\ (aq)$ + $H_2\ (g)$

Both zinc and sulphuric acid get used up in the reaction. At the same time, zinc sulphate and hydrogen form.
You could measure the rate of the reaction, by measuring either:

- the amount of zinc used up per minute *or*
- the amount of sulphuric acid used up per minute *or*
- the amount of zinc sulphate produced per minute *or*
- the amount of hydrogen produced per minute.

For this reaction, it is easiest to measure the amount of hydrogen produced per minute – the hydrogen can be collected as it bubbles off, and its volume can then be measured.

In general, to find the rate of a reaction, you should measure:

**the amount of a reactant used up per unit of time *or*
the amount of a product produced per unit of time.**

Questions

1 Here are some reactions that take place in the home. Put them in order of decreasing rate (the fastest one first).
   a gloss paint drying
   b fruit going rotten
   c cooking gas burning
   d a cake baking
   e a metal bath rusting
2 Which of these rates of travel is slowest?
   5 kilometres per second
   20 kilometres per minute
   60 kilometres per hour
3 Suppose you had to measure the rate at which zinc is used up in the reaction above. Which of these units would be suitable?
   a litres per minute
   b grams per minute
   c centimetres per minute
   Explain your choice.
4 Iron reacts with sulphuric acid like this:
   $Fe\ (s) + H_2SO_4\ (aq) \longrightarrow FeSO_4\ (aq) + H_2\ (g)$
   a Write a word equation for this reaction.
   b Write down four different ways in which the rate of the reaction could be measured.

7.02 Measuring the rate of a reaction

A reaction that produces a gas

The rate of a reaction is found by measuring the amount of a **reactant** used up per unit of time, or the amount of a **product** produced per unit of time. Take, for example, this reaction:

magnesium + hydrochloric acid ⟶ magnesium chloride + hydrogen
Mg (s) + 2HCl (aq) ⟶ MgCl$_2$ (aq) + H$_2$ (g)

Here hydrogen is the easiest substance to measure. This is because it is the only gas in the reaction. It bubbles off and can be collected in a **gas syringe**, where its volume is measured.

Some reactions are so fast that their rates would be very difficult to measure, like this detonation of an old mine.

The method

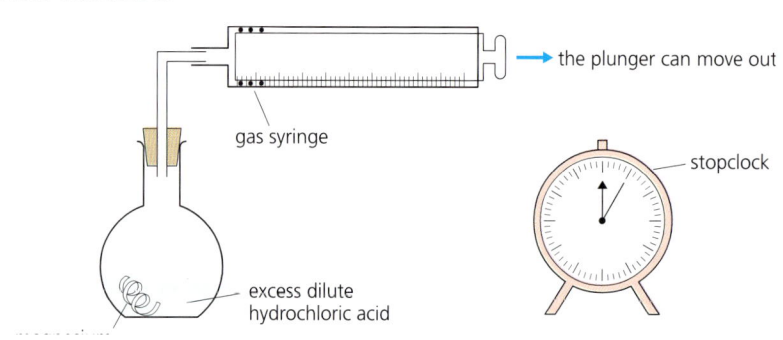

The magnesium is cleaned with sandpaper to remove impurities. Dilute hydrochloric acid is put into the flask. The magnesium is dropped in and the stopper plus syringe inserted immediately. The clock is started at the same time. Hydrogen begins to bubble off. It rises up the flask and into the gas syringe, pushing the plunger out:

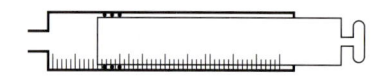

At the start the plunger is fully in. No gas has yet been collected.

Now the plunger has moved out to the 20 cm^3 mark. 20 cm^3 of gas have been collected.

The volume of gas in the syringe is noted at intervals, for example at the end of each half-minute. How will you know when the reaction is complete?

Typical results

Time/minutes	0	$\frac{1}{2}$	1	$1\frac{1}{2}$	2	$2\frac{1}{2}$	3	$3\frac{1}{2}$	4	$4\frac{1}{2}$	5	$5\frac{1}{2}$	6	$6\frac{1}{2}$
Volume of hydrogen/cm^3	0	8	14	20	25	29	33	36	38	39	40	40	40	40

These results can be plotted on a graph, as shown on the next page.

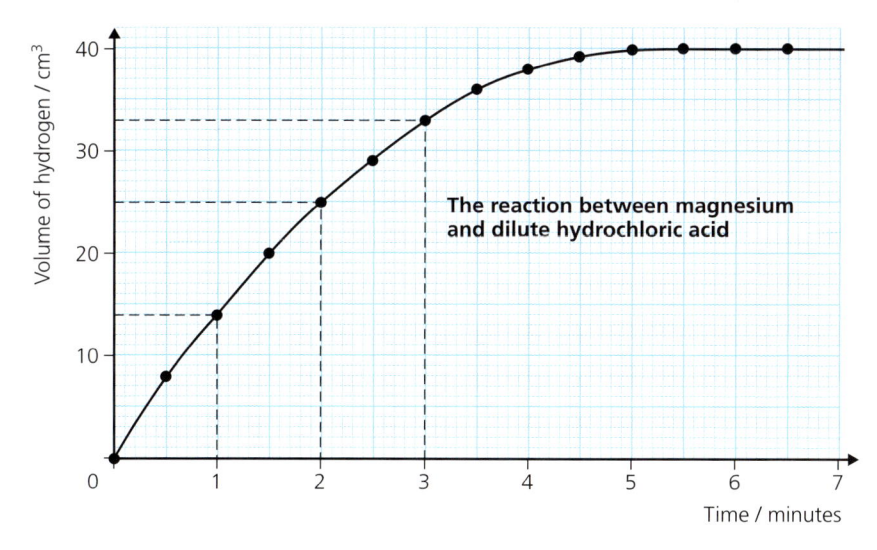

The reaction between magnesium and dilute hydrochloric acid

Notice these things about the results:

1 In the first minute, 14 cm³ of hydrogen are produced.
   So the rate for the first minute is 14 cm³ of hydrogen per minute.
   In the second minute, only 11 cm³ are produced. (25 − 14 = 11)
   So the rate for the second minute is 11 cm³ of hydrogen per minute.
   The rate for the third minute is 8 cm³ of hydrogen per minute.
   So you can see that the rate decreases as time goes on.
   The rate changes all through the reaction. It is greatest at the start, but gets less as the reaction proceeds.

2 The reaction is fastest in the first minute, and the curve is steepest then. It gets less steep as the reaction gets slower.
   The faster the reaction, the steeper the curve.

3 After 5 minutes, no more hydrogen is produced, so the volume no longer changes. The reaction is over, and the curve goes flat.
   When the reaction is over, the curve goes flat.

4 Altogether, 40 cm³ of hydrogen are produced in 5 minutes.

The *average* rate for the reaction $= \dfrac{\text{total volume of hydrogen}}{\text{total time for the reaction}}$

$= \dfrac{40 \text{ cm}^3}{5 \text{ minutes}}$

$= $ **8 cm³ of hydrogen per minute**.

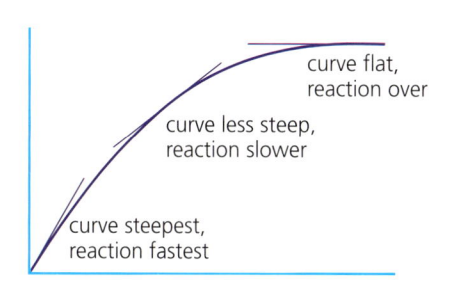

Note that this method can be used to measure the rate of *any* reaction in which one product is a gas.

Questions

1 For this experiment, can you explain why:
  a the stopper is replaced immediately?
  b the magnesium ribbon is first cleaned?
  c the clock is started the moment the reactants are mixed?

2 From the graph above, how can you tell when the reaction is over?

3 Look at the graph at the top of the page.
  a How much hydrogen is produced in:
     i 2.5 minutes?   ii 4.5 minutes?
  b How many minutes does it take to produce:
     i 10 cm³   ii 20 cm³ of hydrogen?
  c What is the rate of the reaction during:
     i the fourth minute?   ii the fifth minute?

117

7.03 Changing the rate of a reaction (I)

The effect of concentration

A reaction can be made to go faster or slower by changing the **concentration** of a reactant.

Suppose the experiment with magnesium and excess hydrochloric acid is repeated twice (A and B below). Everything is kept the same each time, *except* the concentration of the acid:

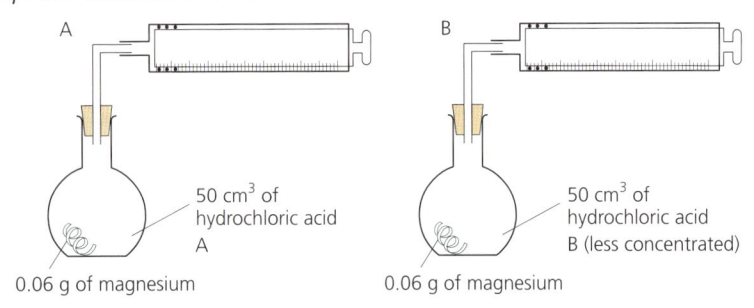

A — 50 cm^3 of hydrochloric acid A — 0.06 g of magnesium

B — 50 cm^3 of hydrochloric acid B (less concentrated) — 0.06 g of magnesium

The acid in A is *twice as concentrated* as the acid in B.
Here are both sets of results shown on the same graph.

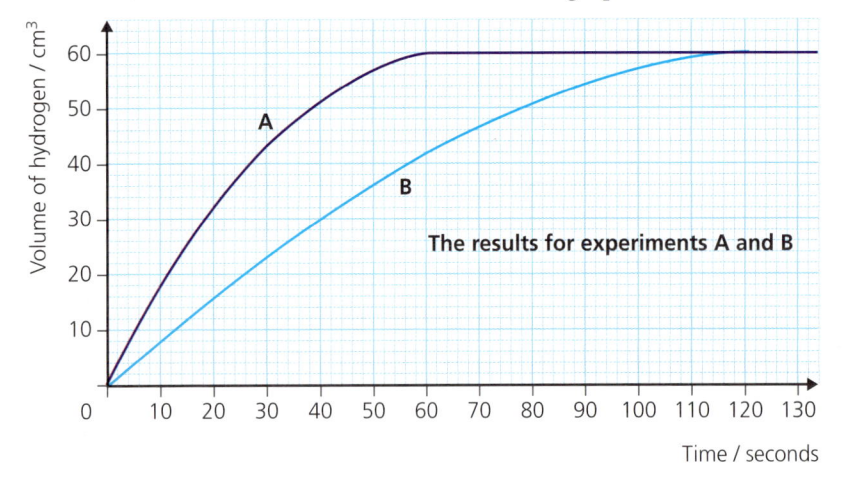

The results for experiments A and B

Notice these things about the results:

1. Curve A is steeper than curve B. From this you can tell straight away that the reaction was faster in A than in B.
2. In A, the reaction lasts 60 seconds. In B it lasts for 120 seconds.
3. Both reactions produced 60 cm^3 of hydrogen. Do you agree?
   In A it was produced in 60 seconds, so the average rate was 1 cm^3 of hydrogen per second. In B it was produced in 120 seconds, so the average rate was 0.5 cm^3 of hydrogen per second.
   The average rate in A was twice the average rate in B.

These results show that:
A reaction goes faster when the concentration of a reactant is increased.

For the reaction above, the rate doubles when the concentration of acid is doubled.

This stain could be removed with a solution of bleach. The more concentrated the solution, the faster the stain will disappear.

118

The effect of temperature

A reaction can also be made to go faster or slower by changing the **temperature** of the reactants.

This time, a different reaction is used: when dilute hydrochloric acid is mixed with sodium thiosulphate solution, a fine yellow precipitate of sulphur forms. The rate can be followed like this:

1. A cross is marked on a piece of paper.
2. A beaker containing some sodium thiosulphate solution is put on top of the paper. The cross should be easy to see through the solution, from above.
3. Hydrochloric acid is added quickly, and a clock started at the same time. The cross grows fainter as the precipitate forms.
4. The clock is stopped the moment the cross can no longer be seen from above.

The low temperature in the fridge slows down decomposition reactions.

View from above the beaker:

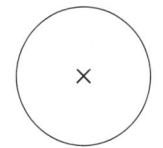

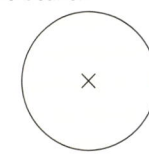

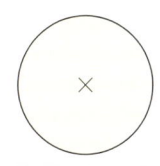

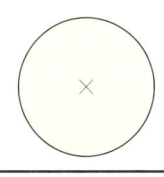

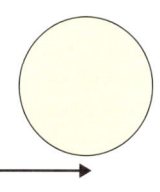

The cross grows fainter with time

The experiment is repeated several times. The quantity of each reactant is kept exactly the same each time. Only the temperature of the reactants is changed. This table shows the results:

Temperature/°C	20	30	40	50	60
Time for cross to disappear/seconds	200	125	50	33	24

The higher the temperature, the faster the cross disappears

The cross disappears when enough sulphur forms to blot it out. Notice that this takes 200 seconds at 20 °C, but only 50 seconds at 40 °C. So the reaction is *four times faster* at 40 °C than at 20 °C.

A reaction goes faster when the temperature is raised. When the temperature increases by 10 °C, the rate approximately doubles.

This fact is used a great deal in everyday life. For example, food is kept in the fridge to slow down decomposition reactions and keep it fresh for longer. Can you think of any other examples?

Comparing rates for this reaction

$$\frac{\text{average}}{\text{rate}} = \frac{\text{mass of sulphur (g)}}{\text{time to produce it (s)}}$$

The mass of sulphur needed to hide the cross is the same each time. Suppose it is 1 gram. Then:

At 20 °C av. rate = $\frac{1}{200}$ g/s
= 0.005 g/s

At 40 °C av. rate = $\frac{1}{50}$ g/s
= 0.02 g/s

So the reaction is 4 times faster at 40 °C than at 20 °C.
(4 × 0.005 = 0.02)

Questions

1. Look at the graph on the opposite page.
   a. After two minutes, how much hydrogen was produced in:
      i experiment A?   ii experiment B?
   b. From the shape of the curves, how can you tell which reaction was faster?
2. Explain why experiments A and B both produce the same amount of hydrogen.
3. Copy and complete: A reaction goes ...... when the concentration of a ...... is increased. It also goes ...... when the ...... is raised.
4. Why does the cross disappear, in the experiment with sodium thiosulphate and hydrochloric acid?
5. What will happen to the rate of a reaction when the temperature is *lowered*? Use this to explain why milk is stored in a fridge.

7.04 Changing the rate of a reaction (II)

The effect of surface area

In many reactions, one of the reactants is a solid. The reaction between hydrochloric acid and calcium carbonate (marble chips) is one example. Carbon dioxide gas is produced:

$$CaCO_3\,(s) + 2HCl\,(aq) \longrightarrow CaCl_2\,(aq) + H_2O\,(l) + CO_2\,(g)$$

The rate can be measured using the apparatus on the right.

The method The marble chips are placed in the flask and the acid is added. The mass of the flask is measured immediately and the clock started. The flask is plugged with cotton wool to stop any liquid splashing out. The mass is noted regularly until the reaction is complete.

Since carbon dioxide can escape through the cotton wool, the flask gets lighter as the reaction proceeds. (It is a heavy gas so the loss of mass is appreciable.) So by finding the mass of the flask at regular intervals you can follow the rate of reaction.
The experiment is repeated twice. Everything is kept exactly the same each time, except the *surface area* of the marble chips.

For experiment 1, large chips are used. Their surface area is the total area of exposed surface.

For experiment 2, the same *mass* of marble is used – but in small chips, so the surface area is greater.

The results The results of the two experiments are plotted below:

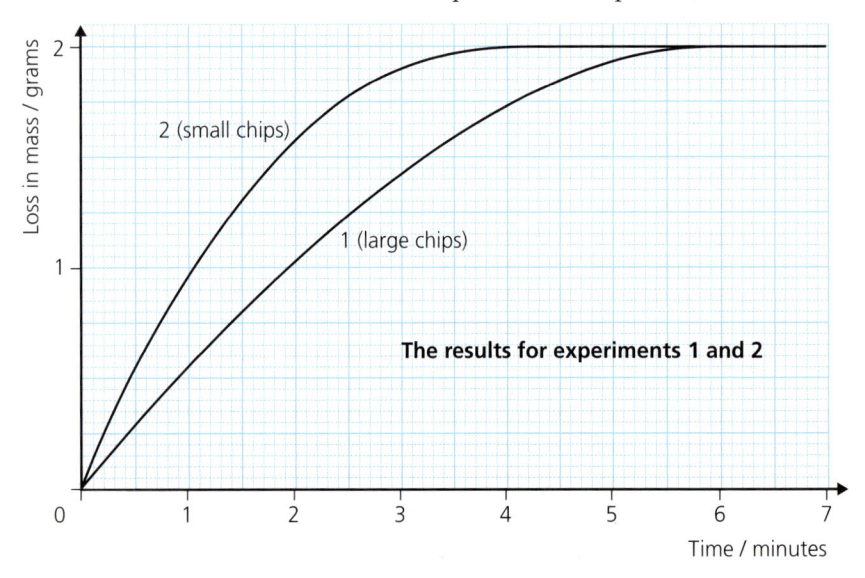

The results for experiments 1 and 2

How to draw the graph

First you have to find the *loss in mass* at different times:

loss in mass at a given time = mass at start − mass at that time

Then you plot the values for loss in mass against the times.

You should notice these things about the results:

1. Curve 2 is steeper than curve 1. This shows immediately that the reaction is faster for the small chips.
2. In both experiments, the final loss in mass is 2.0 grams. In other words, 2.0 grams of carbon dioxide are produced each time.
3. For the small chips, the reaction is complete in 4 minutes. For the large chips, it lasts for 6 minutes

These results show that:
The rate of a reaction increases when the surface area of a solid reactant is increased.

The effect of a catalyst

Hydrogen peroxide is a clear, colourless liquid with the formula H_2O_2. It can decompose to water and oxygen:

hydrogen peroxide ⟶ water + oxygen
$2H_2O_2\ (aq)$ ⟶ $2H_2O\ (l)$ + $O_2\ (g)$

The rate of the reaction can be followed by collecting the oxygen:

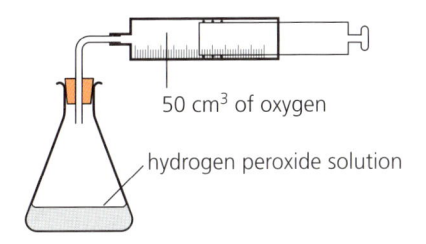

The reaction is in fact very slow. It could take 500 days to collect 50 cm³ of oxygen.

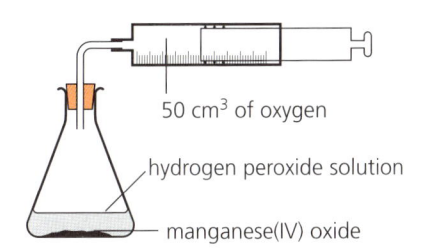

However, if 1 gram of black manganese(IV) oxide is added, the reaction goes much faster. 50 cm³ of oxygen are produced in a few minutes.

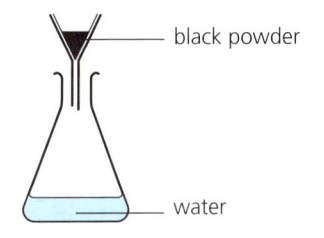

After the reaction, the black powder is removed by filtering. It is then dried, weighed, and tested. It is still manganese(IV) oxide, and still weighs 1 gram.

The manganese(IV) oxide speeds up the reaction without being used up itself. It is called a **catalyst** for the reaction.

A catalyst is a substance that changes the rate of a chemical reaction but remains chemically unchanged itself.

Catalysts have been discovered for many reactions. They are usually **transition metals** or **compounds of transition metals**. For example, iron speeds up the reaction between nitrogen and hydrogen, to make ammonia. There are also many biological catalysts, called **enzymes**. For example, the pancreatic juice made by your pancreas contains enzymes that speed up digestion.

Questions

1. This question is about the graph on the opposite page. For each experiment find:
   a. the loss in mass in the first minute
   b. the mass of carbon dioxide produced during the first minute
   c. the average rate of production of the gas, for the reaction.

2. What is a catalyst? Give two examples of reactions and their catalysts.

3. Look again at the decomposition of hydrogen peroxide, above. How would you show that:
   a. the reaction goes even faster *if more than* 1 gram of catalyst is used?
   b. the catalyst is not used up?

7.05 Explaining rates

A closer look at a reaction

Magnesium and dilute hydrochloric acid react together:

magnesium + hydrochloric acid ⟶ magnesium chloride + hydrogen
$Mg\ (s)\ +\ 2HCl\ (aq)\ \longrightarrow\ MgCl_2\ (aq)\ +\ H_2\ (g)$

In order for the magnesium and acid particles to react together:

- **they must collide with each other.**
- **the collision must have enough energy.**

This is shown by the drawings below.

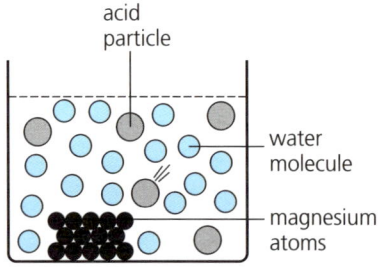

The particles in the liquid move around continually. Here an acid particle is about to collide with a magnesium atom.

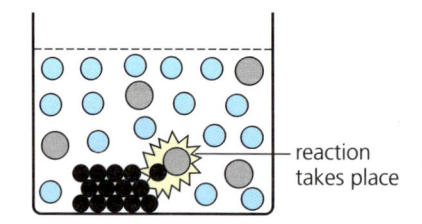

If the collision has enough energy, reaction takes place. Magnesium chloride and hydrogen are formed.

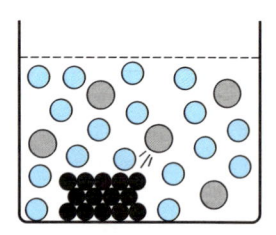

If the collision does not have enough energy, no reaction occurs. The acid particle bounces away again.

If there are lots of successful collisions in a given minute, then a lot of hydrogen is produced in that minute. In other words, the reaction goes quickly – its rate is high. If there are not many, its rate is low.
The rate of a reaction depends on how many successful collisions there are in a given unit of time.

> **In a successful collision:**
> - old bonds are broken (this needs energy)
> - new bonds are formed (this gives out energy).

Changing the rate of a reaction

Why rate increases with concentration If the concentration of the acid is increased, the reaction goes faster. It is easy to see why:

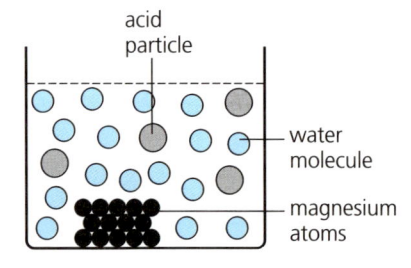

In dilute acid, there are not so many acid particles. This means there is not much chance of an acid particle hitting a magnesium atom.

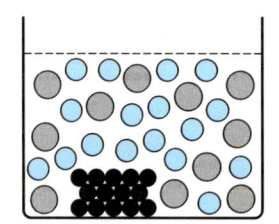

Here the acid is more concentrated – there are more acid particles in it. There is now more chance of a successful collision occurring.

> **Reactions between gases**
> - Increasing the pressure on two reacting gases has the same effect as increasing the concentration.
> - It means you squeeze more gas molecules into a given space.
> - The more molecules in a given space, the greater the chance of successful collisions.
> - So if pressure ↑ then rate ↑ for a gaseous reaction.

The more successful collisions there are, the faster the reaction.

122

This idea also explains why the reaction between magnesium and hydrochloric acid slows down as time goes on:

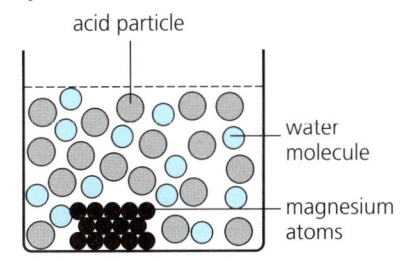

At the start, there are plenty of magnesium atoms and acid particles. But they get used up during successful collisions.

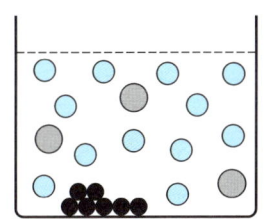

After a time, there are fewer magnesium atoms, and the acid is less concentrated. So the reaction slows down.

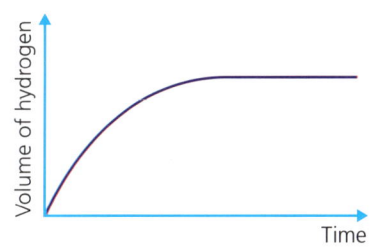

This means that the slope of the reaction curve decreases with time, as shown above.

Why rate increases with temperature At low temperatures, particles of reacting substances do not have much energy. However, when the substances are heated, the particles take in energy. This causes them to move faster and collide more often. The collisions have more energy, so more of them are successful. Therefore the rate of the reaction increases.

Why rate increases with surface area The reaction between the magnesium and acid is much faster when the metal is powdered:

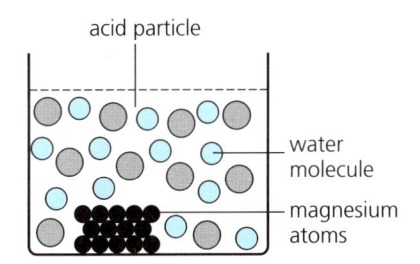

The acid particles can collide only with those magnesium atoms in the outer layer of the metal ribbon.

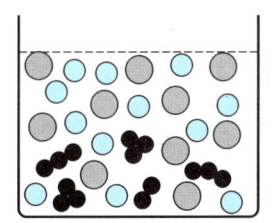

When the metal is powdered, many more atoms are exposed. So there is a greater chance of successful collisions.

Why a catalyst increases the rate Some reactions can be speeded up by adding a catalyst. *In the presence of a catalyst, a collision needs less energy in order to be successful.* The result is that more collisions become successful, so the reaction goes faster. Catalysts are very important in industry, because they speed up reactions even at low temperatures. This means that less fuel is needed, so money is saved.

This airborne detonation test of an industrial explosive shows clearly the rapid rate of reaction.

Questions

1 Copy and complete: Two particles can only react together if they …… and if the …… has enough …… .
2 What is:
  a a successful collision?
  b an unsuccessful collision?
3 In your own words, explain why the reaction between magnesium and acid goes faster when:
  a the temperature is raised
  b the magnesium is powdered
4 Explain why a catalyst can speed up a reaction, even at low temperatures.

7.06 More about catalysts

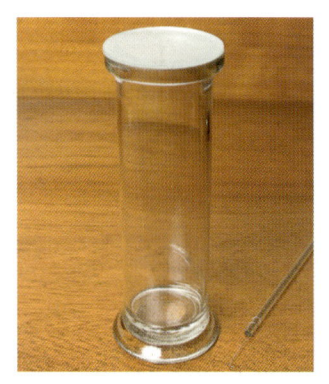

This gas jar contains a mixture of hydrogen and oxygen. Even if you leave it for hours, the two gases won't react together.

But hold a platinum wire in the mouth of the jar, and the gases explode immediately with a pop, producing water.

So platinum is a **catalyst** for the reaction between hydrogen and oxygen. It speeds up the reaction, while remaining chemically unchanged itself. The diagram on the right shows how it works.

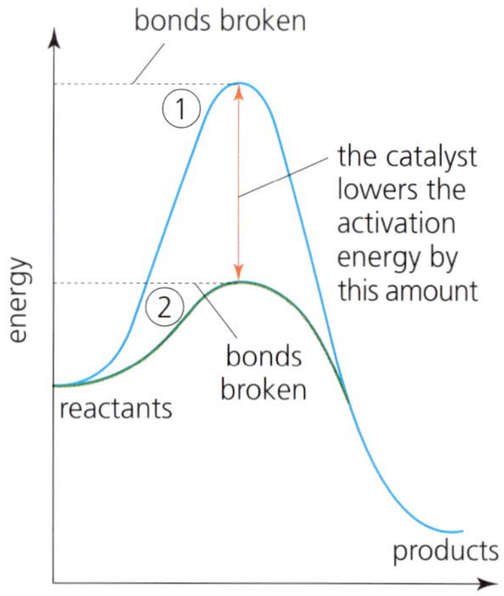

① This curve shows the situation in jar 1. The reaction can't start until the reacting molecules collide with enough energy to break bonds. The energy they need for successful collisions is called the **activation energy**. In jar 1 they don't have enough energy – so there's no reaction.
② When platinum is present, the activation energy is lowered. So some collisions have enough energy to be successful, and reaction begins. It is exothermic – it gives out energy. This leads to *more* successful collisions. Soon the reaction goes so fast it's explosive.

A catalyst lowers the energy needed for a reaction to take place. In other words, it lowers the activation energy.

A closer look at the catalyst

In the reaction above, platinum acts as a **surface catalyst**:

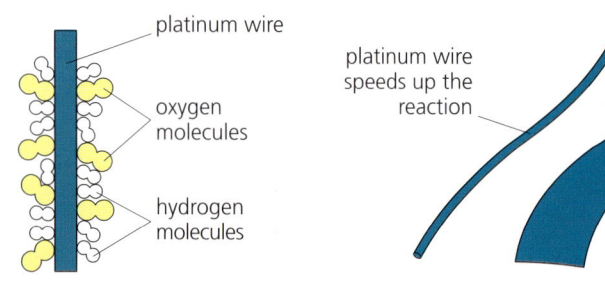

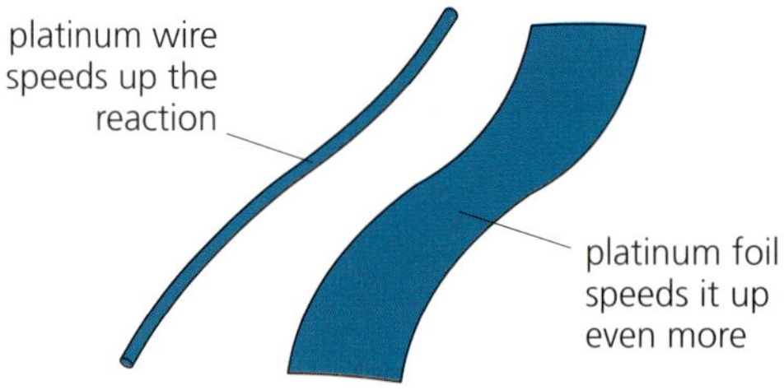

Instead of meeting on collision, gas molecules are adsorbed on to the metal surface. They get very close together, making it much easier for reaction to take place.

The larger its surface area, the better the platinum works, because more gas can be adsorbed. Many other catalysts work in the same way. For example . . .

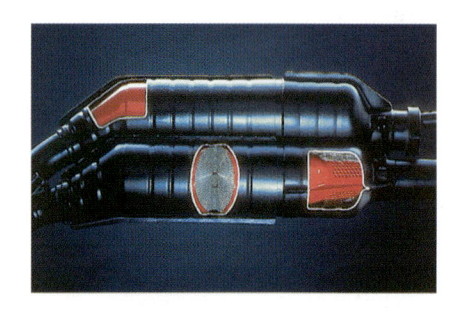

. . . in catalytic converters in car exhausts, harmful gases are adsorbed on to a rhodium catalyst, where they react with oxygen to make less harmful products.

Catalysts in industry

In industry, many reactions are endothermic. Energy must be put in to make them happen. The energy may come from oil, gas, coal, or electricity. But in any case it's expensive. It's one of the biggest costs industry faces.

With a catalyst, a reaction goes faster *at a given temperature.* That means you get your product faster, saving time and therefore money. Even better, it may go fast enough at a lower temperature. And that means a lower fuel bill and higher profits.

So catalysts can turn an uneconomic process into a profitable one. This makes them very important in industry – so important that over 10 million tonnes of catalysts a year are produced worldwide. There are different ones for different reactions. For example:
- **vanadium pentoxide**, the catalyst for converting sulphur dioxide into sulphur trioxide during the manufacture of sulphuric acid
- **iron**, a catalyst for the reaction between nitrogen and hydrogen to make ammonia
- **ZSM-5**, a catalyst for converting methanol into hydrocarbons to use in petrol. It is a compound of silicon, aluminium, and oxygen.

Although a catalyst is not itself chemically changed in a reaction, it does adsorb impurities. This eventually stops it being active. So the beds of catalyst in a chemical plant are replaced from time to time.

Catalysts are made in a variety of shapes and sizes.

Fitting a platinum–rhodium catalyst gauze to a reactor where ammonia will be oxidized to make nitric acid.

A close-up of the platinum–rhodium gauze. Can you think of three advantages of using the catalyst in this form?

Questions

1 Which does a catalyst *not* change?
   a the speed of a reaction   b the products formed
   c the total amount of each product formed
2 Explain how a surface catalyst works.
3 Give two reasons why a catalyst can turn an uneconomic process into a profitable one.
4 Try to think of a reason why the same catalyst will not work for all reactions.

7.07 Enzymes

What are enzymes?

Enzymes are **proteins** that act as **catalysts**. They are different from other catalysts in one big way: *they are made by living cells.*

Every living cell in every plant and every animal – including you – is like a busy chemical factory, humming with non-stop reactions. For example reactions help you move, see, and digest your food. But body temperature is quite low. To make the reactions go fast enough to keep you alive, the cells use enzymes as catalysts. Enzymes can make the reactions go over a million times faster.

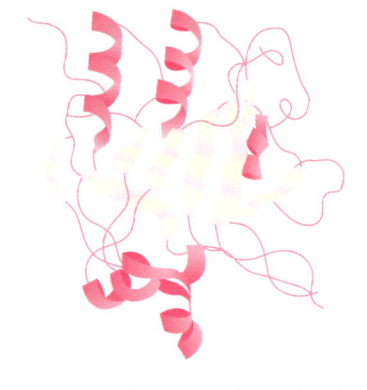

Enzymes are divided into groups depending on the job they do.
- **Amylases** break down starch. Cells in your mouth produce amylase and release it into saliva, where it catalyses the breakdown of starch in food when you chew.
- **Proteases** break down protein (for example when you digest food).
- **Lipases** break down fats and oils (lipids).

Studying an enzyme in the lab

Hydrogen peroxide is a colourless liquid that decomposes like this:

hydrogen peroxide $\longrightarrow$ water + oxygen
$2H_2O_2\ (l) \longrightarrow 2H_2O\ (l)\ +\ O_2\ (g)$

You can use a glowing splint to confirm that oxygen is produced.

Some hydrogen peroxide is poured into four large measuring cylinders. One is left as a control. Manganese(IV) oxide, raw liver, and cooked liver are added to the others. Compare what happens:

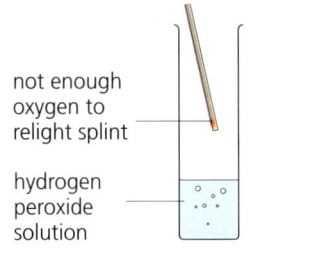

Enzymes have large complex molecules. This is a model of an enzyme molecule.

The test for oxygen

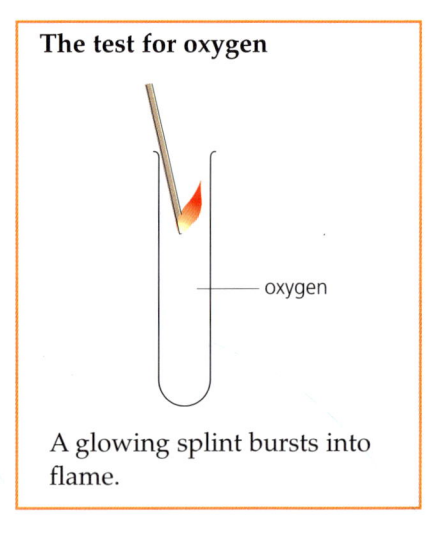

A glowing splint bursts into flame.

Hydrogen peroxide decomposes very slowly on its own, into water and oxygen.

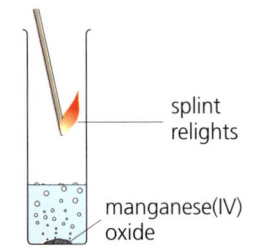

Manganese(IV) oxide makes the reaction go thousands of times faster.

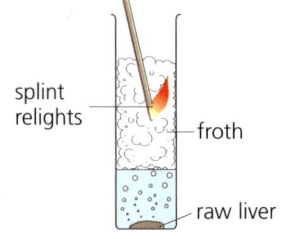

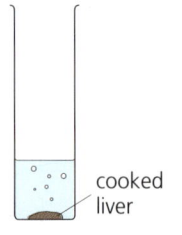

Raw liver also speeds it up. The liquid froths as the oxygen bubbles off.

But the cooked liver has no effect on the reaction rate.

So something in the raw liver acts as a powerful catalyst for the reaction. It is the enzyme **catalase**, made by liver cells.

Catalase is vitally important to *you*. It stops you being poisoned! Many reactions in your cells produce hydrogen peroxide, which is toxic. But many body cells also produce catalase. One molecule can catalyse the breakdown of five million hydrogen peroxide molecules in one minute flat.

How enzymes work

This simplified model shows how an enzyme catalyses the breakdown of a molecule (such as a hydrogen peroxide molecule):

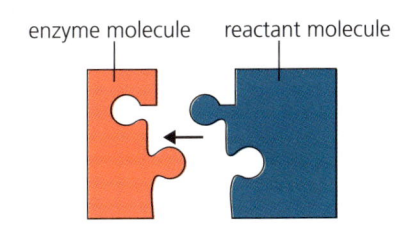

First, the two must fit together like jigsaw pieces. For this, the reactant molecule has to be the right shape.

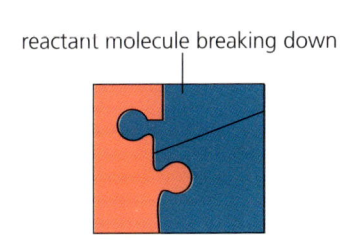

The 'complex' that forms makes it easier for the reactant molecule to break down. When decomposition is complete ...

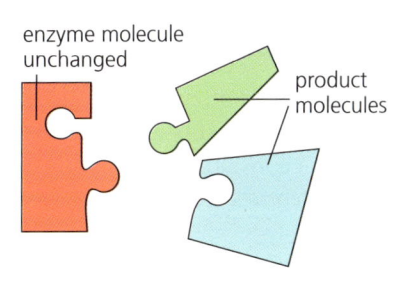

... the product molecules break away. Another molecule of reactant will take their place ... as long as its shape is right!

Note these important points:
- As you'd expect, an enzyme works best in the conditions the living cell created it for!
- For most enzymes that means in a temperature range of 25–45 °C, since most living things operate in that range. (Body temperature for humans is 36 °C for example.)
- At higher temperatures an enzyme loses its shape so stops working. It becomes **denatured**. At lower temperatures it becomes inactive.
- This graph shows how the rate of one enzyme-catalysed reaction changed with temperature. This particular enzyme is at its most active at 40 °C. That's its **optimum** temperature.
- An enzyme also works best in a particular pH range, again depending on the cell that made it. You can denature it by adding acid or alkali.

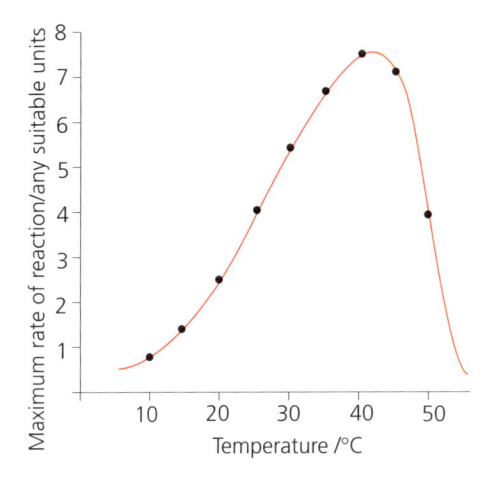

Enzymes in our food

As you saw, enzymes are produced in the cells of all plants and animals, and each has a specific job. For example some help fruit and vegetables to ripen. The enzyme activity continues after they are picked, and in meat after animals are slaughtered. It helps make food go 'off' (along with bacteria and other microorganisms).

- Keeping food in the fridge helps slow down decay. Why?
- Enzymes are completely inactive (but not destroyed) around –18 °C. At 0 °C they are very slightly active. So raw vegetables are often blanched (immersed quickly in boiling water) before freezing. Why?

The enzymes that ripen fruit don't know when to stop!

Questions

1 What are enzymes?
2 What function does amylase have in the body?
3 a What gas is produced when hydrogen peroxide decomposes?
   b How would you test for this gas?
4 What is the name of the enzyme found in liver?
5 Explain how an enzyme works.
6 a What is meant by *denatured*?
   b How do enzymes become denatured?
7 Why does the cooked liver not catalyse the decomposition of hydrogen peroxide?
8 What happens to enzymes *below* –18 °C?

7.08 Some traditional uses of enzymes

They go back a long way!
We have used enzymes for thousands of years without knowing it, in making wine and beer and bread. All three processes depend on **yeast**, a fungus consisting of millions of tiny living cells. In fact the word enzyme comes from the Greek for *'in yeast'*.

Then in 1926 an American biochemist called James Sumner extracted crystals of an enzyme called urease from beans. And that led to the big question: could enzymes be extracted and used as catalysts in factories?

The answer was yes. Today enzymes are very widely used in industry. Sometimes the living cells are present as part of the process (for example yeast in making bread). But often the cells are elsewhere, and the enzymes are brought to the factory in powdered form.

In this spread we look at some traditional uses of enzymes.

Making wine in the fifteenth century.

Yeast and fermentation
Yeast is a fungus consisting of millions of tiny living cells. They feed on sugar, using an enzyme to break it down into ethanol and carbon dioxide. The process is called **fermentation**. You can show it in the lab:

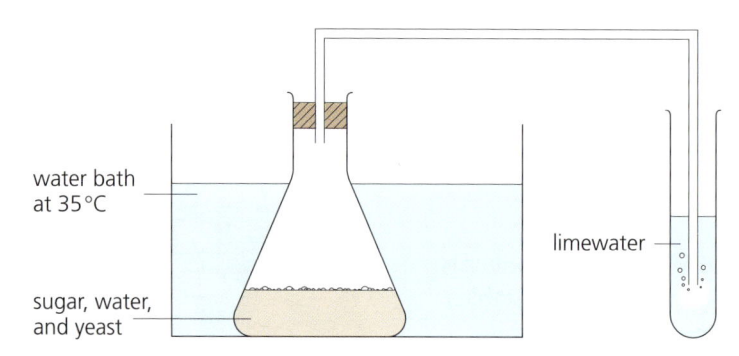

The limewater turns milky which confirms that carbon dioxide is produced. Since ethanol is a liquid it remains in the flask.
The equation for the fermentation is:

glucose $\xrightarrow{\text{enzymes in yeast}}$ ethanol + carbon dioxide

$C_6H_{12}O_6$ (aq) $\longrightarrow$ $2C_2H_5OH$ (l) + $2CO_2$ (g)

As you'll see next, each of those products is the basis of an industry.

Bread – using the carbon dioxide from fermentation
Bread dough contains yeast and sugar. When the dough is left to stand in a warm place for an hour or so, the yeast feeds on the sugar and carbon dioxide is produced. This makes the dough rise. Then it is baked in a hot oven. The heat makes the carbon dioxide expand, so the dough rises still further. The heat also kills off the yeast.

Beer and wine – using the ethanol from fermentation

Ethanol is the alcohol in alcoholic drinks. It is the ingredient that makes people drunk.

Beer It is made from barley and yeast, by fermentation. First the barley grains are soaked in hot water and then mashed to give a liquid called **wort**. It contains sugars made by the breakdown of starch in the grains. Yeast is added to the wort and fermentation begins, giving an alcoholic liquid. When the alcohol content has reached the desired level the yeast is removed.

Wine Grapes are crushed to extract the juice. Yeast is added and fermentation starts. It carries on until either the yeast is removed or the alcohol has 'poisoned' it.

A starting point for ethanol!

Making yoghurt

In making yoghurt, as in making bread, the living cells are present. First milk is **pasteurized** – heated up in order to kill any bacteria in it. Then a **starter culture** of special bacteria is added. These produce enzymes which convert the sugar (lactose) in the milk into lactic acid and other substances, and make the proteins thicken.

Making cheese

Cheese is made from milk too, using enzymes. But this time the cells that produce the enzymes are *not* present.

In the past cheese was made using **rennet**, a substance obtained from a calf's fourth stomach. It contained the enzymes chymosin and pepsin. The farmer would kill the calf, remove and wash the fourth stomach, and hang it out to dry. When it was needed for cheese, strips of the stomach were added to milk, or the stomach was soaked in salty water and the solution added.

Today it is not easy to get a good supply of calves' stomachs. So now the enzymes are usually made, far more cheaply, by bacteria living in tanks! The bacteria have been genetically modified for this task.

The enzymes (mainly chymosin) are added to pasteurized milk and soon a rubbery solid called **curd** is formed, and a liquid called **whey**. The whey is drained off. Further treatment turns the curd into cheese.

Yoghurt with **bio-** in the name contains the bacterium *Lactobacillus acidophilus*, which aids digestion. Other yoghurt is heat-treated to kill off all bacteria.

Questions

1 What is yeast?
2 Name the process in which yeast breaks down sugar, and write a word equation for it.
3 What test is used to identify carbon dioxide?
4 Explain how enzymes are used:
   a in making bread   b in making wine
5 a Name the main enzyme used in the making of cheese.
   b Where was this enzyme obtained from in the past?
   c Where is it obtained today?
6 Name one process in the food industry where:
   a living cells make the enzymes on the spot
   b enzymes taken from absent cells are used

7.09 Some modern uses of enzymes

Suddenly they're everywhere

Today enzymes have hundreds of new uses. For example:
- in **biological detergents**. Stains like sweat, grease, grass stains, and blood contain proteins, starches, and fats. So **proteases**, **amylases**, and **lipases** are added to detergents – and the stains vanish in the wash.
- in making baby food. **Proteases** 'pre-digest' protein in the food to make it easier for babies to digest.
- in making soft-centred chocolates. Have you ever wondered how they got the runny fillings into chocolates? Easy! The filling starts off as an almost solid paste. It contains some cane sugar (sucrose) and a little of the enzyme **invertase**. It is coated with chocolate and left for a couple of weeks. Inside, the enzyme gets to work on the sucrose and breaks it down into simpler sugars, making the paste go runny.
- to convert starch syrup (from potatoes for example, and not sweet) into sugar syrup (which is sweet) for use in baking, soft drinks, and making sweets. A **carbohydrase** is used for this.
- to convert glucose (a sugar) into fructose (another sugar) which is much sweeter. So you need less, which is why it is used in slimming foods. An enzyme called **isomerase** is used for this.
- to stonewash blue jeans for that washed-out look. They used to put the jeans in a washing machine with pumice stones, to knock some of the indigo dye from the denim. A **cellulase** enzyme does the job now.

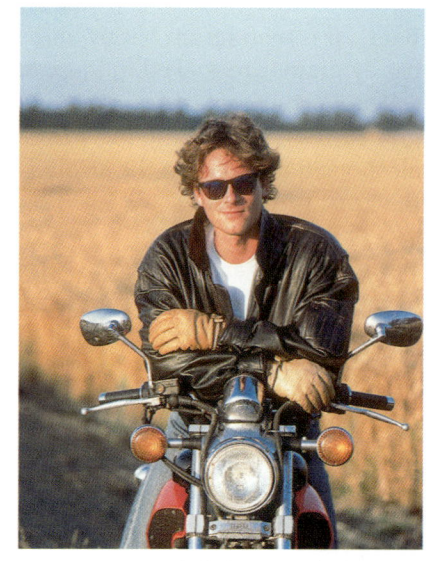

Enzymes remove hair from cow hide to make leather, and enzymes also give jeans the stonewashed look.

Why so popular?

Enzymes are becoming more widely used as industrial catalysts because:
- they usually work best at quite low temperatures. That saves money on heating.
- they operate in mild conditions (in water and usually close to pH 7) that do not harm fabrics or food. No harsh chemicals are needed!
- you can control them easily, and stop them working altogether, by changing the temperature or pH.
- they are biodegradable. Disposing of waste enzymes is no problem.

But first, get your enzymes

Today, most enzymes used in industry are made by bacteria living in tanks, in a broth that contains all the nutrients they need. When the required amount of enzyme is made, the broth is centrifuged to remove solids. The enzyme is separated from the liquid, and sold to factories.

A bacterium may produce over 1000 different enzymes. So the bacteria are usually genetically modified to give just the enzyme that's needed.

Today, scientists scour the world looking for **extremophiles** – bacteria that live in very hot or very cold places. Their enzymes can obviously function in harsh conditions. So genes from these bacteria could be used to modify 'ordinary' bacteria in order to produce 'tougher' enzymes. (For example enzymes might work in boiling water, or at a low pH.) It could mean an enzyme for every industrial process one day.

Some bacteria live at up to 350 °C in 'black smokers' – deep vents in the ocean floor. Their enzymes could be very useful to us one day.

Using enzymes: batch or continuous flow?

Compare these two processes that use enzymes:

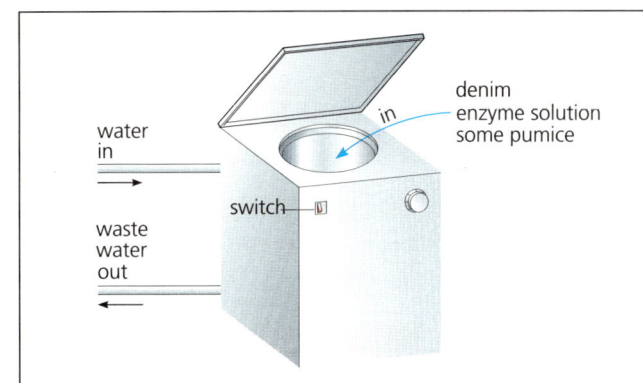

To 'stonewash' denim, you put it in a large washing machine with a cellulase enzyme, and perhaps a small amount of pumice stone. Press the 'on' switch.

When the wash and rinse is complete you drain the water, open the machine, and take the denim out to dry.

This is an example of a **batch process**. You put a batch in the machine and start the process. It has a definite start and end.

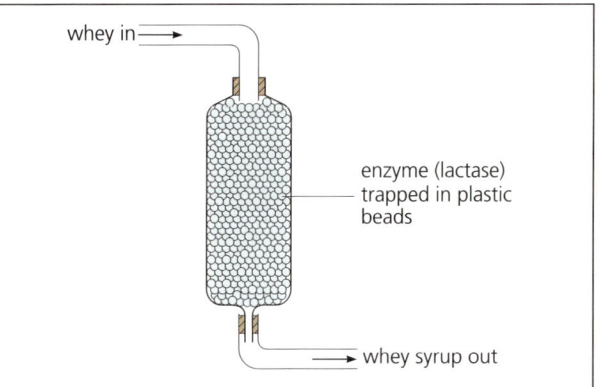

Here sour whey (from making cheese) is being turned into sweet whey syrup (for making sweets and cakes) by the enzyme lactase.

This time the enzyme is trapped in plastic beads. The sour whey is poured in at the top of the column. As it runs over the beads the enzyme goes to work. The syrup flows out at the bottom.

This is called a **continuous flow process**. Can you see why?

The continuous flow process has many advantages:
- Because the enzyme is trapped or **immobilized** it can be reused for as long as it is active. This saves money. (Enzymes can cost a lot.) But with batch processes the enzyme is usually not reused, since it would be too hard to recover from the liquid.
- With continuous flow you can control the reaction quickly and easily by adjusting the flow rate. You can't control a batch process so easily.
- You don't have to stop and start, like you do with a batch process. The system can keep running. And it can be computer controlled, with automatic testing of the product as it flows out, to make sure it is working well. (If there is a problem, the computer will turn off the flow straight away.)

But the enzyme needs to be very stable, for the process to work well. If it is destroyed by a small change in pH for example, then you have to get new beads and repack the column every time this happens.

Questions

1 Enzymes are added to some washing powders.
  a Explain why, and name the enzymes.
  b These powders are not recommended for washing silks and woollens. Why not?
  c You should not use them in very hot water. Why?
2 Explain how the runny centres get into some chocolates.
3 Some denim jeans were called *stonewashed*. Why? Why might this name no longer be appropriate?
4 Describe how most enzymes for industry are made.

5 Give the main difference between a batch process and a continuous flow process, and one example of each.
6 Is it batch or continuous flow?
  a using biological detergents in the washing machine
  b making bread   c making yoghurt
7 In a continuous flow process, enzymes are *immobilized*. What does that mean?
8 Give two advantages of a continuous flow process.

7.10 Exothermic and endothermic reactions

Energy changes in reactions

In most chemical reactions there is an energy change. Energy is taken in or given out, often in the form of heat. So reactions are divided into two groups, **exothermic** and **endothermic**. (*therm* is from the Greek for heat.)

Exothermic reactions

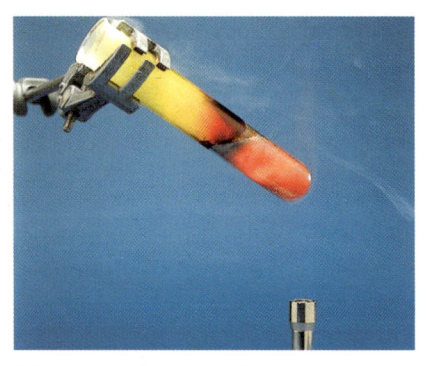

When iron and sulphur react together it's easy to see that energy is given out. The mixture glows – with the Bunsen off!

Mixing silver nitrate and sodium chloride solutions gives a white precipitate – and a temperature rise!

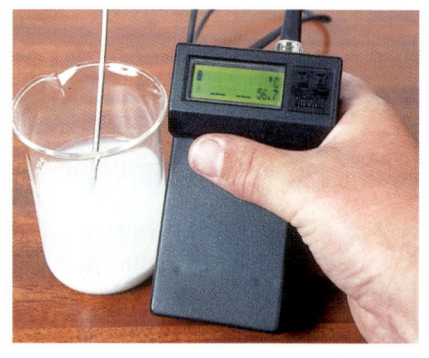

When you add water to quicklime (calcium oxide) heat is given out and the temperature rises.

All three reactions are **exothermic.** They give out heat energy. (*Exo* means *out*.) This can be summarized as:

  reactants ⟶ products + *heat energy*

This heat energy is transferred to the surroundings.

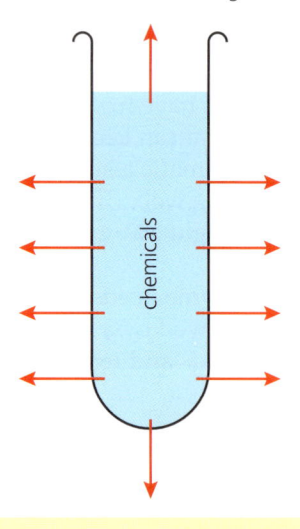

The energy level diagram

If overall a reaction gives out energy, it means the products have *less energy* than the reactants.

You can show that on an **energy level diagram** like this one.

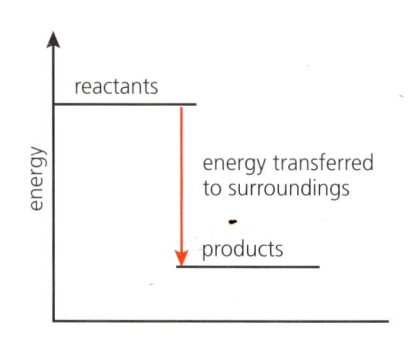

exothermic: energy transferred out to surroundings

Other examples of exothermic reactions

- The neutralization of an acid by an alkali.
- The combustion of fuels. If they didn't give out heat energy they'd be no good as fuels!
- Respiration in your body cells. It provides the energy to keep your heart and lungs working, and for warmth and movement.

The joule
The unit for energy is the joule (J) (just like in physics).

1 kilojoule (kJ) = 1000 joules

132

Endothermic reactions

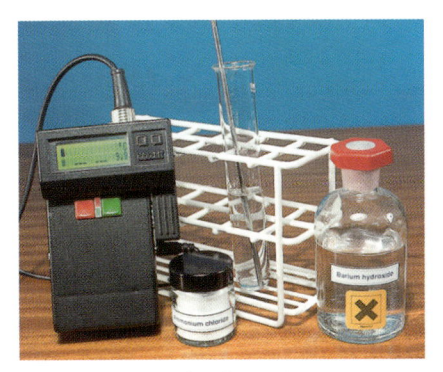

When barium hydroxide reacts with ammonium chloride the temperature falls.

Neutralizing ethanoic acid with sodium carbonate also produces a fall in temperature.

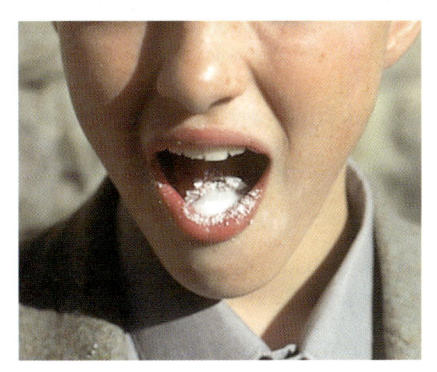

Sherbet is citric acid plus sodium hydrogencarbonate. The neutralization cools your tongue.

This time the reactions *take in* heat energy. They are **endothermic**. (*Endo* means *in*.) The reactions can be summarized as:

reactants + *heat energy* → products

The heat energy is transferred from the surroundings.

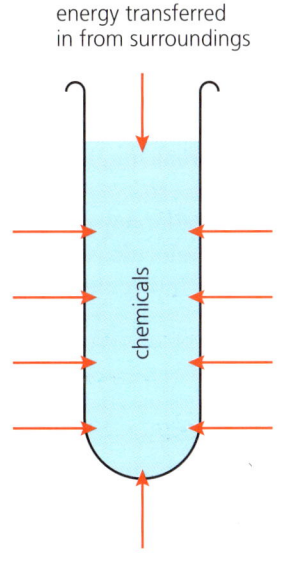

The energy level diagram

If overall a reaction takes in energy, it means the products have *more energy* than the reactants. You can show it on an energy level diagram like this one.

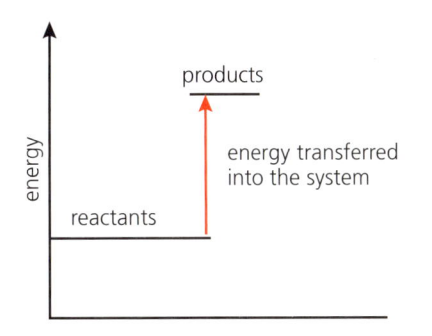

The reactants must climb the energy gap for the reaction to go ahead – so they steal energy from the surroundings.

endothermic: energy transferred in from surroundings

Other examples of endothermic reactions

The reactions above all occur *spontaneously* when you mix the chemicals. There are many other endothermic reactions where energy must be supplied to start, and keep, the reaction going. For example:

- the reactions that take place within food during cooking.
- the polymerization of ethene to polythene.
- photosynthesis. It takes in energy from sunlight.
- electrolysis. Energy is provided in the form of electricity.

Questions

1 What is: **a** an endothermic reaction?
   b an exothermic reaction?
2 Is it exothermic or endothermic?
   a the burning of a candle
   b the reaction between sodium and water
   c the change from raw egg to scrambled egg
3 What unit is used to measure energy?
4 In which type of reaction is heat energy transferred from the chemicals to the surroundings?
5 Draw an energy level diagram for:
   a an endothermic reaction
   b an exothermic reaction

133

7.11 Explaining energy changes

Breaking and forming bonds

A mixture of hydrogen and chlorine will explode in sunshine:

hydrogen + chlorine ⟶ hydrogen chloride

The explosion is a sign that a lot of energy is given out. But where does it come from? Let's look more closely at what happens:

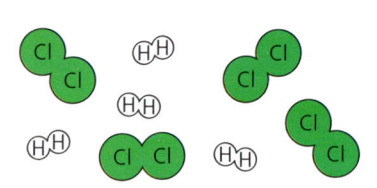

1 Molecules of hydrogen and chlorine are mixed together. The atoms in each molecule are held together by covalent bonds.

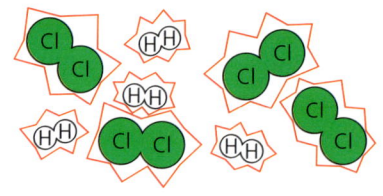

2 These bonds must be broken before reaction can take place. This step takes in energy. It is **endothermic**.

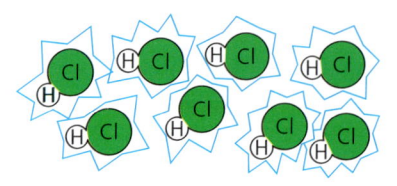

3 Now new bonds form between the hydrogen and chlorine atoms. This step releases energy. It is **exothermic**.

The energy given out in step **3** is greater than the energy taken in in step **2**. And that is why the reaction is exothermic overall.
If the energy released in bond making *is greater than* the energy needed for bond breaking, then the reaction is exothermic.

Now look what happens if you heat ammonia strongly. This time we show the molecules in a simpler way, with a line for each bond. (A nitrogen molecule has three lines since it has a triple bond.)

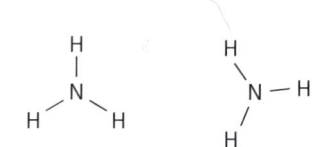

1 Take some ammonia gas. Its molecules are made of hydrogen and nitrogen atoms held together by covalent bonds.

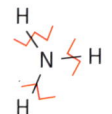

2 For reaction to take place these bonds must break. This step is **endothermic**. You supply the energy for it by heating.

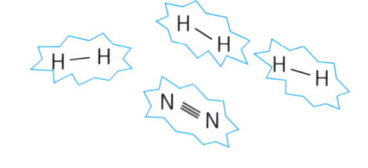

3 Now the hydrogen atoms bond together. So do the nitrogen atoms. This step releases energy. It is **exothermic**.

But this time the energy released in step **3** is less than the energy supplied for step **2**. So overall, the reaction is endothermic.
If the energy released in bond making *is less than* the energy needed for bond breaking, then the reaction is endothermic.

Bond energies

Chemists have worked out how much energy it takes to break the bonds in molecules. This is called the **bond energy**. It is given in kilojoules (kJ).

Look at the bond energies on the right. 242 kJ must be supplied to break the bonds in chlorine molecules to give chlorine atoms. But if the atoms join again to form molecules, 242 kJ of energy are given out again.
The bond energy is the energy needed to break bonds, or released when bonds form.

Bond energy/kJ

H—H	436
Cl—Cl	242
H—Cl	431
C—C	346
C=C	612
C—O	358
O=O	498
O—H	464
N≡N	946
N—H	391

It is worked out for the formula mass of the substance in grams.

Calculating the energy changes in reactions

For the reaction between hydrogen and chlorine
H—H + Cl—Cl → 2 H—Cl

Energy in to break bonds:
1 × H—H	436 kJ
1 × Cl—Cl	242 kJ
Total energy in	**678 kJ**

Energy out from bonds forming:
2 × H—Cl   2 × 431 = **862 kJ**

Energy out – energy in   862 kJ
                                – 678 kJ
                                 184 kJ

Overall the reaction gives out **184 kJ** of energy.

You can show this on an energy level diagram:

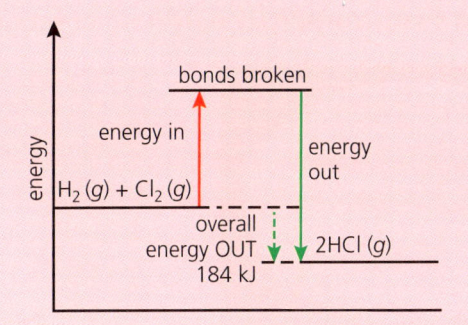

[Energy level diagram showing $H_2(g) + Cl_2(g)$ at higher level with bonds broken above, energy in (red arrow up), energy out (green arrow down) to $2HCl(g)$, overall energy OUT 184 kJ]

For the decomposition of ammonia

2 N(—H)(—H)(—H) → N≡N + 3 H—H

Energy in to break bonds:
6 × N—H   6 × 391 = **2346 kJ**

Energy out from bonds forming:
1 × N≡N		946 kJ
3 × H—H	3 × 436 =	1308 kJ
Total energy out		**2254 kJ**

Energy in – energy out   2346 kJ
                                – 2254 kJ
                                   92 kJ

Overall the reaction takes in **92 kJ** of energy.

You can show this on an energy level diagram:

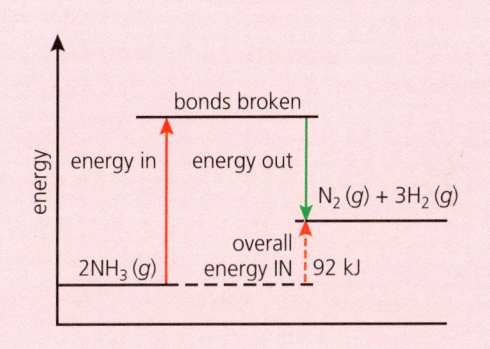

[Energy level diagram showing $2NH_3(g)$ at lower level, bonds broken above, energy in (red arrow up), energy out (green arrow down) to $N_2(g) + 3H_2(g)$, overall energy IN 92 kJ]

Note that for *every* reaction, even an explosive one, some energy must be supplied to break some bonds to start it off. This is called the **activation energy**. It could come from a lit match or Bunsen, or even from sunlight.

The burning of fuels

Fuels give out energy when they burn – combustion is exothermic.

Fuel	Used for	Equation for burning in oxygen	Energy (kJ) given out by burning the formula mass in grams
Methane	heating, cooking	$CH_4(g) + 2O_2(g) \rightarrow CO_2(g) + 2H_2O(g)$	890
Hydrogen	rocket fuel	$2H_2(g) + O_2(g) \rightarrow 2H_2O(g)$	486
Glucose	your body fuel	$C_6H_{12}O_6(aq) + 6O_2(g) \rightarrow 6CO_2(g) + 6H_2O(g)$	2803

Questions

1. When chlorine and hydrogen react:
   a. which part of the reaction is endothermic?
   b. which is exothermic?
   c. why does the reaction *give out* energy overall?
2. Write the reaction between hydrogen and oxygen with lines for bonds, as above. Any double bonds?
3. Now calculate the energy change for the reaction in **2**. Do you agree with the value given in the table above?
4. Which is the cleanest fuel in the table above?
5. Draw an energy diagram for the combustion of North Sea gas (methane) in a gas cooker. How would you provide the activation energy for this reaction?

7.12 Reversible reactions

When you heat copper(II) sulphate crystals …

When you heat blue crystals of copper(II) sulphate, they break down into *anhydrous* copper(II) sulphate, a white powder.

$CuSO_4 . 5H_2O\ (s) \longrightarrow$
$\quad CuSO_4\ (s) + 5H_2O\ (g)$

The reaction is easy to reverse: just add water! The white powder turns blue again. In fact this is used as a test for water.

$CuSO_4\ (s) + 5H_2O\ (l) \longrightarrow$
$\quad CuSO_4 . 5H_2O\ (s)$

Water of crystallization
- The water in blue copper(II) sulphate crystals is called **water of crystallization**.
- The blue compound is **hydrated**.
- The white compound is **anhydrous**.

The above reaction is reversible: it can go in either direction. Reaction 1 is the **forward** reaction. Reaction 2 is the **back** reaction. We use the symbol $\rightleftharpoons$ to show that the reaction is reversible.

$CuSO_4 . 5H_2O\ (s) \rightleftharpoons CuSO_4\ (s) + 5H_2O\ (g)$

Many chemical reactions are reversible.

Two tests for water
- It turns white anhydrous copper(II) sulphate blue.
- It turns blue cobalt chloride paper pink.

Both changes are reversible.

What about the energy change?

In reaction 1 above you must heat the blue crystals to obtain the white powder – so the reaction is **endothermic**. In reaction 2 the white powder gets hot and spits when you drip water on – this reaction is **exothermic**. In fact it gives out *the same amount of heat* as reaction 1 took in!

If a reversible reaction is endothermic in one direction it is exothermic in the other. The same amount of heat is transferred each time.
This is easy to see from the energy level diagram on the right.

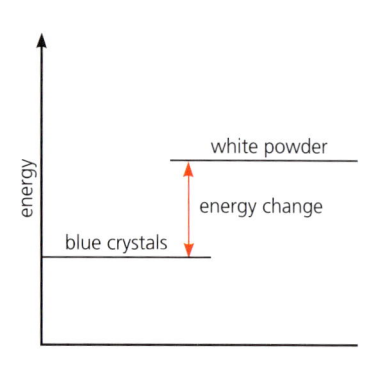

The thermal decomposition of ammonium chloride

Thermal decomposition means breaking a substance down into simpler substances by heating it.

The thermal decomposition of ammonium chloride is another example of a reversible reaction:

$NH_4Cl\ (s) \underset{cool}{\overset{heat}{\rightleftharpoons}} NH_3\ (g) + HCl\ (g)$

This diagram shows what happens.

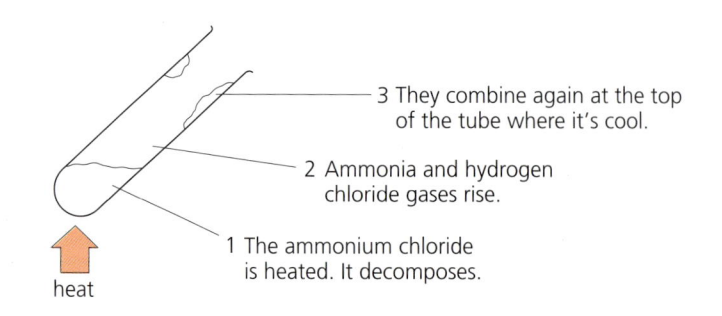

3 They combine again at the top of the tube where it's cool.
2 Ammonia and hydrogen chloride gases rise.
1 The ammonium chloride is heated. It decomposes.

136

Reversible reactions and dynamic equilibrium

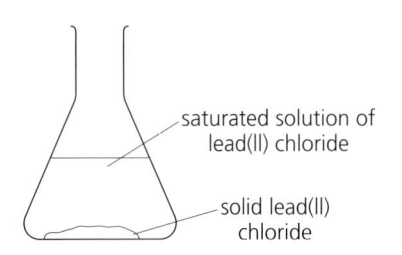

Lead(II) chloride is not very soluble. Only 1.1g will dissolve in 100 g of water:
$PbCl_2 (s) \longrightarrow Pb^{2+}(aq) + 2Cl^-(aq)$
If you add more lead chloride it will sink to the bottom.

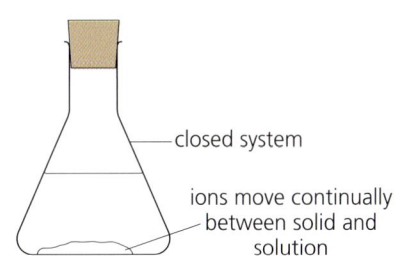

With a stopper on, the flask becomes a closed system. Inside, it's all action. Ions move continually from the solid to the solution and from the solution to the solid.

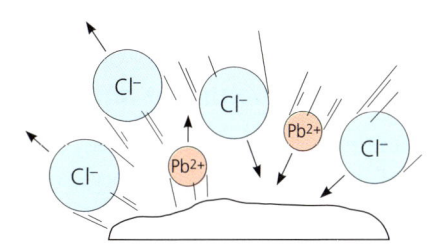

This is happening *at the same rate in both directions*, so no extra lead(II) chloride is dissolving. We can describe the situation like this:
$PbCl_2 (s) \rightleftharpoons Pb^{2+}(aq) + 2Cl^-(aq)$

Inside the flask, the solid and its saturated solution have reached a state of **dynamic equilibrium** (often just called **equilibrium**).

Equilibrium means there is no overall change. But *dynamic* means that change is in fact taking place continuously. You can prove this by adding lead(II) chloride containing radioactive lead ions to the flask above. The solid falls to the bottom as you'd expect. After 20 minutes its mass is unchanged, but there are radioactive lead ions in the solution.

Dissolving is a physical change. But exactly the same thing happens with a reversible chemical reaction.

In a closed system, a reversible reaction eventually reaches a state of dynamic equilibrium. The forward and back reactions take place at the same rate, so no overall change occurs.

Dynamic equilibrium and industry

Many very important reactions in industry are reversible. An example is the reaction between nitrogen and hydrogen to make ammonia:

$N_2 (g) + 3H_2 (g) \rightleftharpoons 2NH_3 (g)$

This reaction eventually reaches a state of dynamic equilibrium: while some molecules of ammonia are forming, others are breaking down. So the reaction *never goes to completion*.

This is a problem for companies that make ammonia. They want the yield to be as large as possible. So what steps can they take to make more ammonia? You will find out on the next spread.

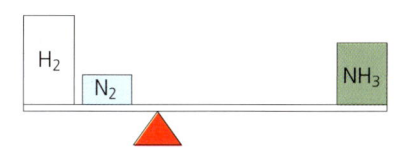

Nitrogen, hydrogen, and ammonia in equilibrium. The mixture is in balance.

Questions

1 What is a *reversible* reaction?
2 Write an equation for the reversible reaction between copper(II) sulphate and water.
3 What does *dynamic equilibrium* mean?
4 Explain how you could establish a state of dynamic equilibrium between table salt and water.

7.13 Shifting the equilibrium

The manufacture of ammonia

Imagine you run a factory that makes ammonia from nitrogen and hydrogen. The reaction is **reversible**:

$$N_2\,(g) + 3H_2\,(g) \rightleftharpoons 2NH_3\,(g)$$

Let's look more closely at this reaction:

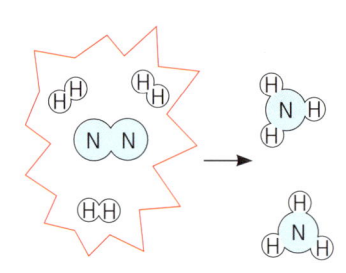

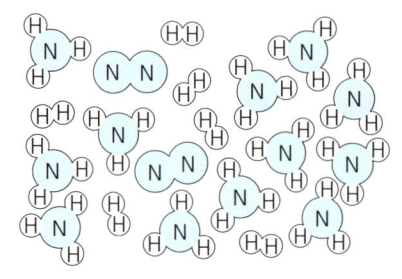

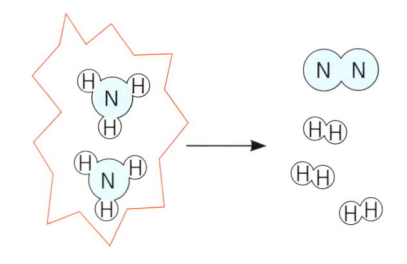

Three molecules of hydrogen react with one of nitrogen to form two of ammonia. So if you mix the right amounts of nitrogen and hydrogen …

… will it all turn into ammonia? No! Once a certain amount of ammonia is formed the system reaches a state of dynamic equilibrium. From then on …

… every time two ammonia molecules form, another two break down into nitrogen and hydrogen. So the level of ammonia remains unchanged.

But the more ammonia that forms, the better your profits. So how can you increase the yield?

This idea, called Le Chatelier's principle, will help you:
When a reversible reaction is in equilibrium and you make a change, it will do what it can to oppose that change.

You can't make a reversible reaction go to completion. It *always* ends up in a state of equilibrium. But by changing the conditions you can *shift equilibrium to the right* and obtain more product. So let's see what changes you can make to obtain more ammonia.

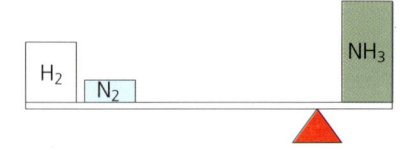

Shifting the equilibrium to the right will give you more ammonia.

1 Increasing the temperature

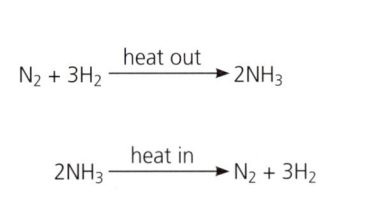

$$N_2 + 3H_2 \xrightarrow{\text{heat out}} 2NH_3$$

$$2NH_3 \xrightarrow{\text{heat in}} N_2 + 3H_2$$

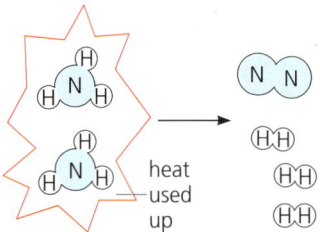

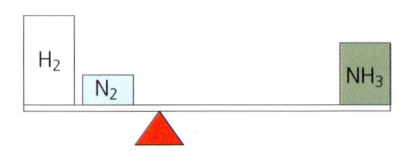

You know that heat speeds up *any* reaction. But here the forward reaction is exothermic – it gives out heat. The back reaction is endothermic – it takes it in.

If you heat the equilibrium mixture it will act to oppose the change. More ammonia will break down in order to use up the heat you've added.

The result is that the reaction reaches equilibrium faster, which is good, but the level of ammonia *decreases*. So you're worse off than before.

But if you decide to improve the yield by running at a very *low* temperature, the reaction will take too long to reach equilibrium!

138

2 Increasing the pressure

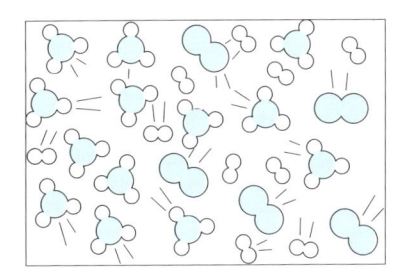

Pressure is caused by collisions between gas molecules and the walls of the container. So the fewer the molecules present, the lower the pressure.

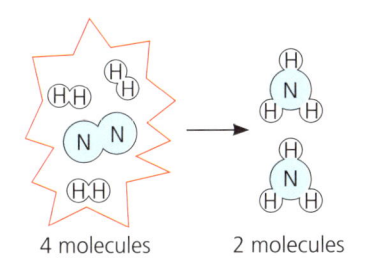

If you apply more pressure, the equilibrium mixture will act to oppose the change. More ammonia will form in order to reduce the number of molecules.

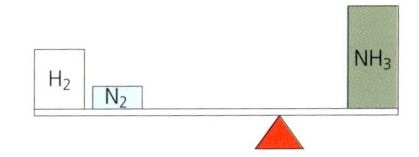

The result is that the level of ammonia *increases*. Equilibrium shifts to the right. Well done. You're on the right track.

3 Removing the ammonia

The equilibrium mixture is a balance between the levels of nitrogen, hydrogen, and ammonia present. Suppose you cool the mixture. Ammonia condenses first so you can run it off as a liquid. Then warm the remaining nitrogen and hydrogen again. *Et voilà*, more ammonia!

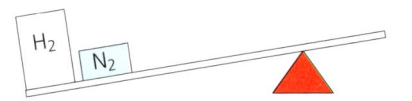

When you remove ammonia, nitrogen and hydrogen react to restore equilibrium.

4 Adding a catalyst

Iron acts as a catalyst for this reaction. But note that it speeds up the forward and back reactions *equally*. So although the reaction reaches equilibrium faster, the equilibrium position doesn't change. Still, the catalyst is worth using because it saves time and therefore money.

Choosing the optimum conditions

To make production economical, you have to balance all the above factors. You have to:
- use high pressure, and remove ammonia, to shift equilibrium to the right
- use a catalyst to reach equilibrium quickly
- use moderate heat as a compromise

These are the conditions that are applied, as you will see on the next spread.

Making ammonia: a summary

$$N_2 + 3H_2 \underset{}{\overset{exothermic}{\rightleftharpoons}} 2NH_3$$

4 molecules   2 molecules

To improve the yield:
- high pressure
- low temperature
- remove ammonia

To get a decent reaction rate:
- raise temperature (compromise!)
- use a catalyst

Remember, for reversible reactions between gases …

If the forward reaction is:
- exothermic, temperature ↑ means yield ↓
- endothermic, temperature ↑ means yield ↑

If there are fewer molecules of product than reactant, in the equation:
pressure ↑ means yield ↑

Questions

1. a Explain what goes on in an equilibrium mixture of nitrogen, hydrogen, and ammonia.
   b Why does this cause a problem in industry?
2. What is Le Chatelier's principle? Write it down.
3. In manufacturing ammonia, explain why:
   a high pressure is used   b ammonia is removed
   c only moderate heat is used
4. Sulphur dioxide (SO_2) and oxygen react to form sulphur trioxide (SO_3). The reaction is exothermic and reversible.
   a Write a balanced equation for it.
   b What happens to the yield of sulphur trioxide:
      i if you increase the pressure?
      ii if you increase the temperature?

7.14 Making ammonia in industry

It's a key chemical

Ammonia is one of the world's top chemicals, because it is needed to make fertilizers. Around 100 years ago, Europe was running short of natural fertilizer. If only ammonia could be made from the nitrogen in air, it would solve the problem. Not easy! Nitrogen is very unreactive and its reaction with hydrogen is reversible.

Then in 1908 a German chemist called Fritz Haber developed a process that worked. It is named after him, and is still used all over the world.

The Haber process for making ammonia

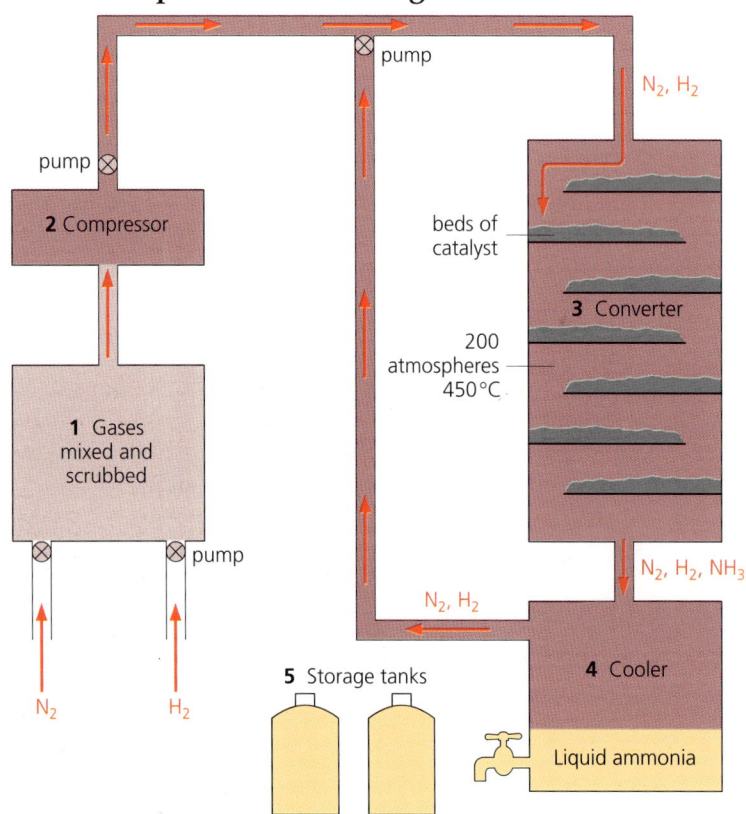

The Terra ammonia plant at Billingham in Cleveland. It runs 24 hours a day year after year, except for breaks for safety checks and maintenance.

The raw materials

Nitrogen
Air is about $\frac{4}{5}$ nitrogen and $\frac{1}{5}$ oxygen, with just small amounts of other gases.

The oxygen is removed by burning hydrogen in the air.

$$2H_2(g) + O_2(g) \longrightarrow 2H_2O(l)$$

That leaves mainly nitrogen.

Hydrogen
It is usually made from natural gas or **methane**, and water (as steam):

$$CH_4(g) + 2H_2O(g) \xrightarrow{catalyst} CO_2(g) + 4H_2(g)$$

Compounds in oil can also be **cracked** to give hydrogen. For example:

$$C_2H_6(g) \xrightarrow{catalyst} C_2H_4(g) + H_2(g)$$
ethane → ethene

1. The raw materials are hydrogen and nitrogen. (The box on the right shows how these are obtained.) They are mixed and **scrubbed** (cleaned) to remove impurities.
2. The gas mixture is **compressed** or pumped into a smaller space, until it reaches a pressure of 200 atmospheres.
3. Then it goes to the converter – a round tank containing beds of hot iron at around 450 °C. The iron catalyses the reaction:
   $$N_2(g) + 3H_2(g) \rightleftharpoons 2NH_3(g)$$
   Only about 15% of the mixture leaving the converter is ammonia.
4. The mixture is cooled until the ammonia condenses to a liquid. The nitrogen and hydrogen are **recycled** to the converter for another chance to react. Steps **3** and **4** are continually repeated.
5. The ammonia is run into tanks and stored as a liquid under pressure.

The economics of making ammonia

A typical ammonia plant (factory) makes around 1200 tonnes of liquid ammonia a day. Try to imagine that amount!

The company aims to make a profit. So it will reduce its costs where possible.

Choosing the temperature and pressure The graph on the right shows how the yield of ammonia changes with temperature and pressure. Note how the yield increases as the temperature falls, and as the pressure rises.

The yield of ammonia at different temperatures and pressures

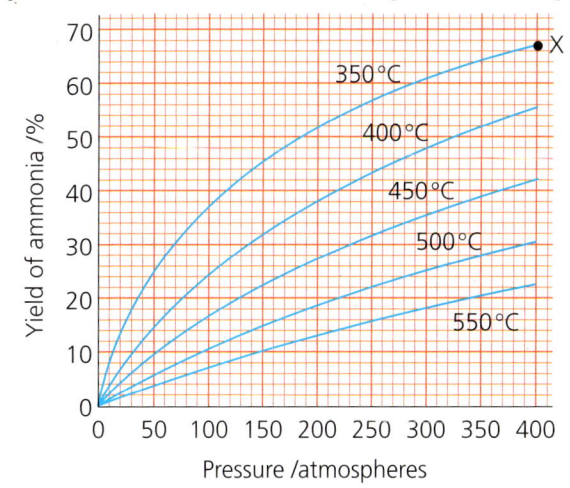

For the **optimum** (highest) yield you would choose a pressure of 400 atmospheres and a temperature of 350 °C. (See point X on the graph.) But instead the plant chooses 200 atmospheres and 450 °C. Why?

To raise the pressure of a gas you have to squeeze it into a smaller and smaller space. This is expensive – it needs powerful pumps and a lot of electricity to run them. So the choice of 200 atmospheres rather than 400 helps to save money. Note that the pipes, converter, and other tanks have to be really sturdy to withstand this pressure.

At 350 °C the reaction gives the best yields – but the rate is too slow. Time is money, when you aim to produce 1200 tonnes of ammonia a day. So the temperature is raised to 450 °C as a compromise.

So the conditions in the converter are *not* optimum. But that's where the process is so clever. Recycling the unreacted gases is the answer!

The catalyst All catalysts get 'poisoned' in the end, by impurities. The iron catalyst has other substances added to it, to slow down poisoning. Even so it must be replaced every few years, which means shutting the plant down. Chemists are still trying to improve the catalyst performance.

The raw materials Luckily air and water are easy to obtain. But you still need natural gas (methane) or compounds from oil, to make hydrogen. For this reason ammonia plants are usually within easy reach of natural gas terminals or oil refineries. In fact many oil and gas companies manufacture ammonia as a way to increase their profits.

Making ammonia: a summary

$$N_2 + 3H_2 \;\overset{\text{exothermic}}{\rightleftharpoons}\; 2NH_3$$

4 molecules → 2 molecules

To improve the yield:
- high pressure
- low temperature
- remove ammonia

To get a decent reaction rate:
- raise temperature (compromise!)
- use a catalyst

Fritz Haber (1868–1935), the brains behind the Haber process. Although his work helped Germany during World War I, he had to flee when Hitler came to power because he was Jewish.

Questions

1 Ammonia is made from nitrogen and hydrogen.
   a How are the nitrogen and hydrogen obtained?
   b What is the process for making ammonia called?
   c What catalyst is used? What does it do?
   d Write an equation for the reaction.

2 Look at the catalyst beds in the diagram for the Haber process. Why are they arranged that way?

3 A pressure of 400 atmospheres and a temperature of 350 °C would give a higher yield of ammonia. Explain why a *lower* pressure and *higher* temperature are used.

141

7.15 Fertilizers

What plants need

A plant needs carbon dioxide, light, and water. It also needs several different elements, mainly **nitrogen**, **potassium**, and **phosphorus**.

Nitrogen is needed for proteins for strong stems and healthy leaves.

Potassium helps a plant to survive frost and to resist disease.

Phosphorus is needed to help roots develop. It also helps crops ripen.

The plants obtain these from compounds in the soil, which they take in through their roots as solutions. The most important element is nitrogen, since every cell in the plant needs proteins. Plants take in nitrogen in the form of **nitrate** and **ammonium** ions.

Fertilizers

Every crop a farmer grows removes substances from the soil. Some get replaced naturally. For example:

Lightning makes oxygen and nitrogen in the air react, giving nitrogen oxides. These dissolve in rain giving acid rain – which reacts with compounds in the soil to give nitrates.

Animal manure contains some nitrates, and proteins that break down in the soil to ammonium compounds. It also returns potassium and phosphorus to the soil.

Some bacteria are **nitrogen-fixing**. They can convert nitrogen from the air into nitrates. They live in soil, and in **root nodules** like these on the roots of pea, bean, and clover plants.

But they can't replace compounds as fast as crops remove them, in these days of intensive farming. So soil would soon become useless without **chemical fertilizers** like these:

ammonium nitrate, NH_4NO_3 ammonium sulphate, $(NH_4)_2SO_4$
potassium sulphate, K_2SO_4 ammonium phosphate, $(NH_4)_3PO_4$

A fertilizer is any substance added to soil to make it more fertile.
The fertilizers above are made in factories and sold by the sackfull for sprinkling or spraying on fields. Three of those above provide nitrogen, so they are called **nitrogenous fertilizers**. And they are all salts. They are made by neutralizing acids, as you will see on the next page.

NPK fertilizer contains nitrogen (N), phosphorus (P) and potassium (K).

Ammonia: the key to nitrogenous fertilizers

The air is rich in nitrogen. But that's not much help to plants – they need it in the soil! And ammonia is the key. The Haber process gives us a way to 'fix' nitrogen from the air in factories and pass it on to plants.

Nitrogen usually reaches plants faster as nitrate ions than as ammonium ions. And here again ammonia is the key – it is the starting point for making nitric acid, which is then neutralized to give nitrates.

From ammonia to nitric acid

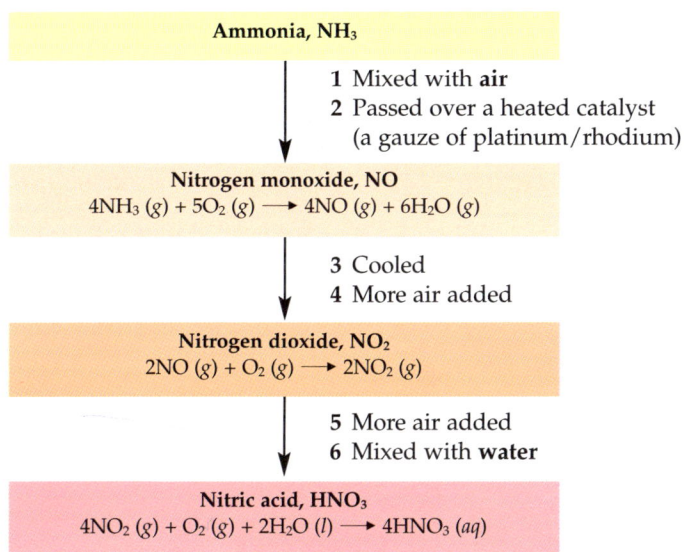

Ammonia, NH₃

↓ 1 Mixed with **air**
2 Passed over a heated catalyst
(a gauze of platinum/rhodium)

Nitrogen monoxide, NO
$4NH_3 (g) + 5O_2 (g) \rightarrow 4NO (g) + 6H_2O (g)$

↓ 3 Cooled
4 More air added

Nitrogen dioxide, NO₂
$2NO (g) + O_2 (g) \rightarrow 2NO_2 (g)$

↓ 5 More air added
6 Mixed with **water**

Nitric acid, HNO₃
$4NO_2 (g) + O_2 (g) + 2H_2O (l) \rightarrow 4HNO_3 (aq)$

The overall result is that ammonia is oxidized to nitric acid.

Next step: neutralization

Ammonium nitrate is one of the most common fertilizers. It is made by neutralizing nitric acid with ammonia solution:

$HNO_3 (aq)$ + $NH_3 (aq)$ → $NH_4NO_3 (aq)$
nitric acid   ammonia   ammonium nitrate

Ammonium sulphate is made by a similar neutralization:

$H_2SO_4 (aq)$ + $2NH_3 (aq)$ → $(NH_4)_2SO_4 (aq)$
sulphuric acid   ammonia   ammonium sulphate

Fertilizers are big business. UK farmers use around 2 million tonnes of chemical fertilizers a year – and the annual world consumption is around 130 million tonnes!

Fertilizers are made all year round, but sales are seasonal. A fertilizer pellet could lie buried in a 20-metre heap for several months. It must be able to survive storage without losing shape or sticking to other pellets.

Once ammonium nitrate is made, the water is evaporated off and the molten salt made into pellets by spraying it down a tall tower.

Questions

1 Name the three main elements that plants need from the soil, and say why they need them.
2 Describe two natural ways in which nitrogen from the air is converted into nitrates in the soil.
3 What are fertilizers and why are they needed?
4 Animal manure is a natural fertilizer. Explain why.
5 Name three nitrogenous fertilizers.
6 Why is the Haber process so important to farming?
7 a In the conversion of ammonia to nitric acid:
   i what are the three raw materials?
   ii what catalyst is used?
  b Overall, ammonia is *oxidized*. What does this mean?
8 Write a balanced equation for the reaction used to produce the important fertilizer ammonium nitrate.

7.16 The pros and cons of fertilizers

A green revolution?

A hundred years ago, most farms were small family farms. They used animal manure and kitchen waste as fertilizer. As machinery improved they could farm more efficiently, so food output increased – but slowly.

Then in 1913 the first ammonia plant was set up in Germany, using the Haber process – and a farming revolution had begun. It started slowly. World War I broke out, and all the ammonia was needed for making explosives. Next came a worldwide economic slump. Then World War II, and once again ammonia production was diverted to making war.

But after World War II the production of chemical fertilizers took off – and so did crop yields per hectare. Look at the graph below.

Tea break on a British farm in the 1930s. Farming then used natural fertilizers, and was much less intensive than it is today – but very hard work.

Some of this increase was due to improved farm machinery and plant breeds, but much was the result of using chemical fertilizers.

Feeding the world

Look at how the world's population has grown in the last 100 years. How has it been fed? The land available for farming has not changed a great deal. What has changed most is the use of fertilizers.

It is estimated that 40% of the world's 6 billion people are alive today thanks to the Haber process!

Today the world produces *more than enough food to feed everyone*. In 1998 world food production was enough to provide an adequate diet (2350 calories a day) for 6.9 billion people. The population was less than 6 billion.

But many people still go to bed hungry every night. Some poor countries produce too little food because farmers can't afford fertilizers, or better seeds, or machinery, and rains often fail. Other countries produce – and consume – more than they need. So food is not shared around the world equitably.

By 2025 there could be 8.3 billion people to feed. Without fertilizers this would be impossible.

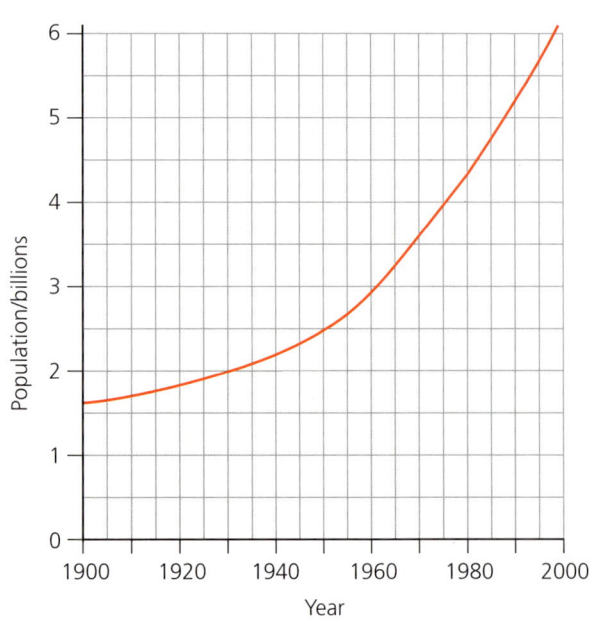

144

So – what are the problems?

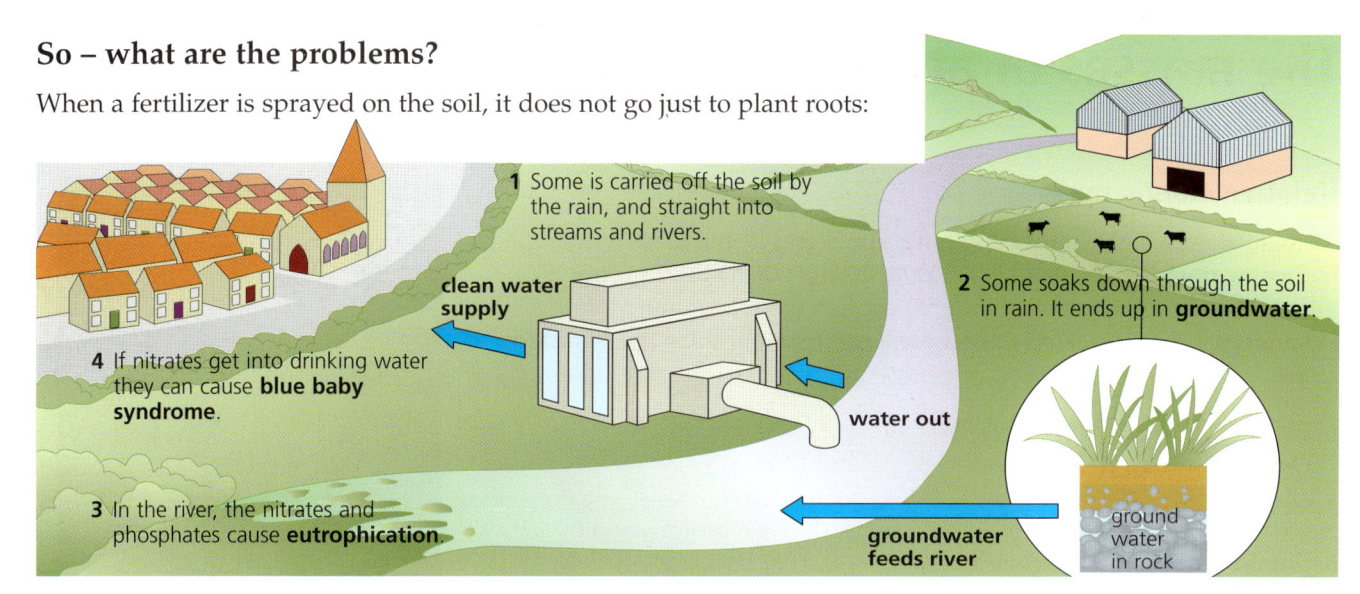

When a fertilizer is sprayed on the soil, it does not go just to plant roots:

1 Some is carried off the soil by the rain, and straight into streams and rivers.

2 Some soaks down through the soil in rain. It ends up in **groundwater**.

3 In the river, the nitrates and phosphates cause **eutrophication**.

4 If nitrates get into drinking water they can cause **blue baby syndrome**.

Now let's look more closely at these pollution problems.

Eutrophication Algae are tiny water plants. Fertilizer helps them grow – too well. They float on the river like a carpet, preventing photosynthesis in the plants below by blocking out the sunlight. When they die bacteria feed on them, causing them to decay – and use up oxygen from the water in the process. So the fish die from oxygen starvation.

Blue baby syndrome In your gut, nitrates are converted to nitrites. These combine with haemoglobin, so your blood can't carry so much oxygen around the body. This can cause oxygen starvation in young infants. Their skin takes on a blue tinge. The infant won't usually die, but its development may be affected.

A high concentration of nitrates in drinking water has also been linked to cancers, but this link has not been proven.

Tackling the pollution problems

Farmers can help by using fertilizers sparingly – and not on wet soil or in wet weather.

For drinking water, nitrates *can* be removed at water treatment works – but it is often cheaper to find a 'clean' source of water. However some plants in Europe use bacteria to remove nitrates, and this is likely to become more common. (The bacteria are then killed off by chlorine.)

Some people feel that **organic farming** is the best solution of all – where farmers use only natural fertilizers, and rotate the crops they grow or leave fields **fallow** (empty), to let the soil recover its goodness naturally.

A eutrophied river.

Questions

1 Give reasons why cereal yields per hectare almost trebled in the UK between 1946 and 1986.
2 It has been estimated that 40% of the world's population owe their lives to the Haber process. Explain why.
3 What would you expect to find in a eutrophied river?
4 Suggest steps you could take to help a eutrophied river recover.
5 Describe one health problem related to fertilizers.
6 What is *organic* farming? Would it be possible to feed the world with organic farming? Give reasons.

145

7.17 The masses of atoms

Relative atomic mass

A single atom weighs hardly anything. For example, the mass of a single hydrogen atom is only about 0.000 000 000 000 000 000 000 002 grams. So scientists had to find a simpler way to express the mass of an atom. Here is what they did. First they chose a carbon atom to be the standard atom.

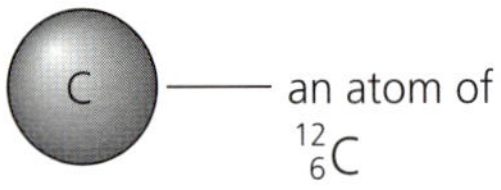

C —— an atom of $^{12}_{6}C$

Next, they fixed its mass as exactly 12 units. (It has 6 protons and 6 neutrons. They ignored the electrons.)

Then they compared all the other atoms with this standard atom, using a machine called a **mass spectrometer**, and found values for their masses, like this:

The mass spectrometer was invented by a British scientist called Aston, in 1919.

C	Mg	H
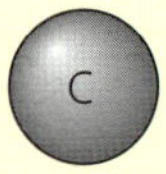 This is the standard atom, $^{12}_{6}C$. Its mass is exactly 12.	 This magnesium atom is twice as heavy as the standard atom, so its mass must be 24.	 This hydrogen atom is $\frac{1}{12}$ as heavy as the standard atom, so its mass must be 1.

Masses found in this way are called **relative**, since they are relative to the standard carbon atom.

Isotopes and A_r

But not all atoms of an element are exactly the same. For example, when scientists examined chlorine in the mass spectrometer, they found there were two types of chlorine atom:

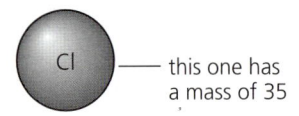

Cl —— this one has a mass of 35   Cl —— this one has a mass of 37

These atoms are the **isotopes** of chlorine. (They have different masses because one has two neutrons more than the other.)

It was found that 75% of chlorine atoms have a mass of 35 and 25% have a mass of 37. So the average mass of a chlorine atom was calculated to be 35.5. (Look at the box on the right.)

Most elements have isotopes, and these have to be taken into account when calculating relative atomic mass:
The relative atomic mass or A_r of an element is the average relative mass of its isotopes.

For most elements A_r is very close to a whole number, and is usually rounded off to a whole number to make calculations easier.

Calculating A_r for chlorine:
75% of chlorine atoms have a relative mass of 35.
25% have a relative mass of 37.

Average relative mass
$= 75\% \times 35 + 25\% \times 37$
$= \frac{75}{100} \times 35 + \frac{25}{100} \times 37$
$= 26.25 \quad + 9.25$
$= 35.5$

The relative atomic masses of some common elements

Element	Symbol	A_r
Hydrogen	H	1
Carbon	C	12
Nitrogen	N	14
Oxygen	O	16
Sodium	Na	23
Magnesium	Mg	24
Sulphur	S	32

Element	Symbol	A_r
Chlorine	Cl	35.5
Potassium	K	39
Calcium	Ca	40
Iron	Fe	56
Copper	Cu	64
Zinc	Zn	65
Iodine	I	127

Finding the mass of an ion:
Mass of sodium atom = 23, so
Mass of sodium ion = 23,
since a sodium ion is just a sodium atom minus an electron, and an electron has hardly any mass.
An ion has the same mass as the atom from which it is made.

The relative formula mass, M_r

Using a list of A_rs, it is easy to work out the mass of any molecule or group of ions. Check these examples using the information above:

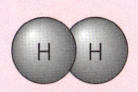

Hydrogen gas is made of molecules. Each molecule contains 2 hydrogen atoms, so its relative mass is 2.
($2 \times 1 = 2$)

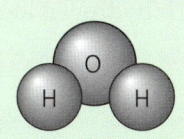

The formula of water is H_2O. A water molecule contains 2 hydrogen atoms and 1 oxygen atom, so its relative mass is 18.
($2 \times 1 + 16 = 18$)

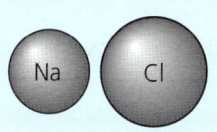

Sodium chloride (NaCl) contains a sodium ion for every chloride ion. The relative mass of a 'unit' of sodium chloride is 58.5.
($23 + 35.5 = 58.5$)

The mass of a substance found in this way is called its **relative formula mass** or M_r, since you work it out by adding up the relative atomic masses of the atoms in the formula.

So the M_r of hydrogen is 2, and of water is 18.
Here are some more:

Substance	Formula	Atoms in formula	Relative atomic mass (A_r)	Relative formula mass (M_r)
Nitrogen	N_2	2N	N = 14	$2 \times 14 =$ **28**
Ammonia	NH_3	1N 3H	N = 14 H = 1	$1 \times 14 =$ 14 $3 \times 1 =$ 3 Total = **17**
Magnesium nitrate	$Mg(NO_3)_2$	1Mg 2N 6O	Mg = 24 N = 14 O = 16	$1 \times 24 =$ 24 $2 \times 14 =$ 28 $6 \times 16 =$ 96 Total = **148**

Questions

1 What is the relative atomic mass of an element?
2 What is the A_r of the iodide ion, I^-?
3 Show that the relative formula mass for chlorine (Cl_2) is 71.
4 What is the M_r of butane, C_4H_{10}?
5 Work out the relative formula mass of:
  a oxygen, O_2
  b iodine, I_2
  c methane, CH_4
  d ethanol, C_2H_5OH
  e ammonium sulphate, $(NH_4)_2SO_4$

7.18 Percentage composition of a compound

Percentage composition

The percentage composition of a compound tells you *which* elements are in the compound and *how much* of each there is, as a percentage of the total mass.

Calculating the percentage of an element by mass

Relative atomic masses: C = 12, H = 1
Relative formula mass = 16

Methane is a compound of carbon and hydrogen, with the formula CH_4. The mass of a carbon atom is 12, and the mass of each hydrogen atom is 1, so the formula mass of methane is 16.

You can find what fraction of the total mass is carbon, and what fraction is hydrogen, like this:

mass of carbon as fraction of total mass = $\dfrac{\text{mass of carbon}}{\text{total mass}} = \dfrac{12}{16}$ or $\dfrac{3}{4}$

mass of hydrogen as fraction of total mass = $\dfrac{\text{mass of hydrogen}}{\text{total mass}}$
$= \dfrac{4}{16}$ or $\dfrac{1}{4}$

These fractions are usually written as percentages. To change a fraction to a percentage you just multiply it by 100:

$\dfrac{3}{4} \times 100 = \dfrac{300}{4} = 75$ per cent or 75% $\dfrac{1}{4} \times 100 = \dfrac{100}{4} = 25\%$

So 75% of the mass of methane is carbon and 25% is hydrogen. We say that the **percentage composition** of methane is 75% carbon, 25% hydrogen.

The famous scientist John Dalton and a pupil, collecting methane from a pond. Methane also occurs in oil wells and natural gas wells. No matter where it comes from, its composition is always the same.

The Law of Constant Composition

Methane is formed by the action of bacteria on organic material such as animal remains, dead plants, and human sewage. It occurs as natural gas (for example North Sea gas) formed by the decay of dead sea animals. You also find it in ponds and marshes, in sewage works, and in the compost heap at the bottom of the garden.

The methane you cook with, in North Sea gas, has exactly the same composition as John Dalton's.

But no matter where you find it, a methane molecule *always* contains one carbon atom and four hydrogen atoms. The percentage composition of methane is *always* 75% carbon, 25% hydrogen. This follows one of the basic laws of chemistry:
The Law of Constant Composition states that every pure sample of a given compound has exactly the same composition.

Calculating percentage composition

Here is how to calculate the percentage of an element in a compound:

1 Write down the formula of the compound.
2 Using a list of relative atomic masses (A_r), work out its formula mass.
3 Write the mass of the element you want, as a fraction of the total.
4 Multiply the fraction by 100, to give a percentage.

Example 1 The gas sulphur dioxide is a pollutant formed during the burning of fossil fuels. It dissolves in rainwater to make acid rain. It has the formula SO_2. Calculate the percentage of oxygen in the compound.

The formula mass of the compound is 64, as shown on the right.

Mass of oxygen in the formula = 32

Mass of oxygen as fraction of total = $\dfrac{32}{64}$

Mass of oxygen as percentage of total = $\dfrac{32}{64} \times 100 =$ **50%**

So the compound is **50% oxygen**.
This means it is also 50% sulphur (100% − 50% = 50%).

Sulphur dioxide, SO_2

A_r: S = 32, O = 16.
The formula contains 1 S and 2 O, so the relative formula mass is:

1 S = 32
2 O = 2 × 16 = 32
Total = 64

Example 2 Fertilizers contain nitrogen, which plants need to help them grow. One important fertilizer which is rich in nitrogen is ammonium nitrate, which has the formula NH_4NO_3. Calculate:
i the percentage of nitrogen in ammonium nitrate;
ii the mass of nitrogen in a 20 kg bag of the fertilizer.

i The formula mass of the compound is 80, as shown on the right.
The element we are interested in is nitrogen.

Mass of nitrogen in the formula = 28

Mass of nitrogen as fraction of total = $\dfrac{28}{80}$

Mass of nitrogen as percentage of total = $\dfrac{28}{80} \times 100 =$ **35%**

So the fertilizer is **35% nitrogen.**

ii The fertilizer is 35% nitrogen.

So the mass of nitrogen in a 20 kg bag = $\dfrac{35}{100} \times 20$ kg
= 7 kg

The bag contains **7 kg** of nitrogen.

Ammonium nitrate, NH_4NO_3

A_r: N = 14, H = 1, O = 16.
The formula contains 2 N, 4 H, and 3 O, so the relative formula mass is:

2 N = 2 × 14 = 28
4 H = 4 × 1 = 4
3 O = 3 × 16 = 48
Total = 80

Questions

1 A compound contains just oxygen and sulphur. It is 40% sulphur. What percentage is oxygen?
2 Find the percentage of:
 i hydrogen ii oxygen
 in ammonium nitrate, NH_4NO_3.
3 Calculate the percentage of copper in copper(II) oxide, CuO. (A_r: Cu = 64, O = 16.)
4 Calculate the percentage of nitrogen in the fertilizer ammonium sulphate, $(NH_4)_2SO_4$. (A_r: H = 1, N = 14, O = 16, S = 32.)

7.19 The formula of a compound

What a formula tells you

The formula of carbon dioxide is **CO₂**. Some molecules of it are shown on the right. You can see that:

| 1 carbon atom | combines with | 2 oxygen atoms |

The relative atomic mass of a carbon atom is 12.
The relative atomic mass of an oxygen atom is 16, so the total relative mass for two oxygen atoms is 32.
Therefore carbon and oxygen combine in a mass ratio of 12 to 32.

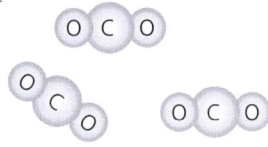

A_r: C = 12, O = 16

Knowing this, we can say:

| 12 g of carbon | combines with | 32 g of oxygen |

In the same way:
6 g of carbon combines with 16 g of oxygen
24 kg of carbon combines with 64 kg of oxygen, and so on.
The masses of substances that combine are *always in the same ratio*. You can work out the ratio from the formula, and also predict how many grams will combine.

How to find a formula from combining masses

If you know what masses combine, you can work out the formula. These are the steps:

| Find the masses that combine (in grams) by experiment. | → | Divide the mass of each substance by its relative atomic mass. | → | That tells you the **ratio** in which atoms combine. Put it in its simplest form. | → | Now you can write down a **formula** for the compound. |

A formula like this, using masses found by experiment, is called an **empirical formula**. (*Empirical* means *known by experiment*.)
The empirical formula shows the simplest ratio in which atoms combine.

Example 1 An experiment shows that 32 g of sulphur combine with 32 g of oxygen to form an oxide of sulphur. What is its empirical formula?

Draw up table like this:

Elements that combine	sulphur	oxygen
Masses that combine	32 g	32 g
Relative atomic masses (A_r)	32	16
Mass / A_r	32/32 = 1	32/16 = 2
Ratio in which atoms combine	1:2	
Empirical formula	SO₂	

So the oxide obtained in this case was **sulphur dioxide, SO₂**.

Sulphur combines with oxygen when it burns.

150

Example 2 Compound X is a **hydrocarbon**: it contains only carbon and hydrogen atoms. An experiment shows that X is 80% carbon and 20% hydrogen. What is its empirical formula?

X is 80% carbon and 20% hydrogen. So 100 g of X contains 80 g of carbon and 20 g of hydrogen. Draw up a table like this:

Elements that combine	carbon	hydrogen
Masses that combine	80 g	20 g
Relative atomic masses (A_r)	12	1
Mass / A_r	80/12 = 6.67	20/1 = 20
Ratio in which atoms combine	6.67 : 20 or 1:3 in its simplest form	
Empirical formula	**CH$_3$**	

But there's a problem. *No compound with the formula CH$_3$ exists.*
(A carbon atom bonds to 4 hydrogen atoms, not 3.)
A further experiment showed that the relative mass of a molecule of X was 30. The relative mass for CH$_3$ is 15.
So the *actual* formula for X is 2 × CH$_3$ or **C$_2$H$_6$**. (It is the gas ethane.)

An experiment to find the empirical formula

To work out the empirical formula, you need to know the masses of elements that combine. *The only way to do this is by experiment.*

For example, magnesium combines with oxygen to form magnesium oxide. The masses that combine can be found like this:

1 Weigh the crucible and lid, empty. Then add a coil of magnesium ribbon and weigh it again, to find the mass of magnesium.
2 Heat the crucible. Raise the lid carefully at intervals to let oxygen in. The magnesium burns brightly.
3 When burning is complete, let the crucible cool (still with its lid on). Then weigh it again. The increase in mass is due to oxygen.

Weighing showed that 2.4 g of magnesium combined with 1.6 g of oxygen. Draw up a table again:

Elements that combine	magnesium	oxygen
Masses that combine	2.4 g	1.6 g
Relative atomic masses (A_r)	24	16
Mass / A_r	2.4/24 = 0.1	1.6/16 = 0.1
Ratio in which atoms combine	1:1	
Empirical formula	**MgO**	

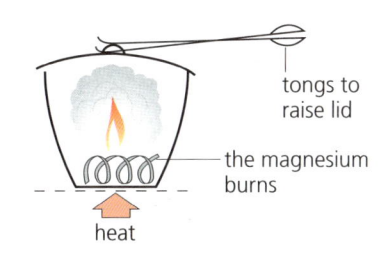

So the compound that formed is **magnesium oxide, MgO.**

Questions

1 **a** How many atoms of hydrogen combine with one carbon atom to form methane, CH$_4$?
  b How many grams of hydrogen combine with 12 grams of carbon to form methane?
2 What does the word *empirical* mean?
3 56 g of iron combine with 32 g of sulphur to form iron sulphide. Find its empirical formula.
  (A_r: Fe = 56, S = 32.)
4 An oxide of sulphur contains 40% sulphur and 60% oxygen. What is its empirical formula?

7.20 Equations for chemical reactions

What is a chemical equation?

When carbon is heated in oxygen they react together to form carbon dioxide. Carbon and oxygen are the **reactants**. Carbon dioxide is the **product**. You could show the reaction like this:

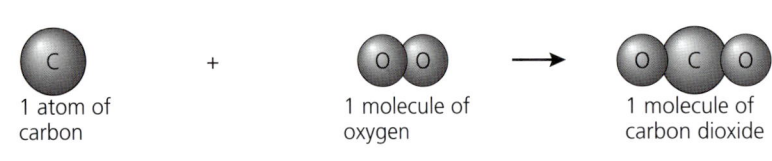

1 atom of carbon + 1 molecule of oxygen → 1 molecule of carbon dioxide

or in a shorter way, by using symbols, like this:

$$C + O_2 \rightarrow CO_2$$

This short way to describe the reaction is called a **chemical equation**.

In a coal fire, the main reaction is $C + O_2 \rightarrow CO_2$.

What is a balanced chemical equation?

When hydrogen and oxygen react together, the product is water:

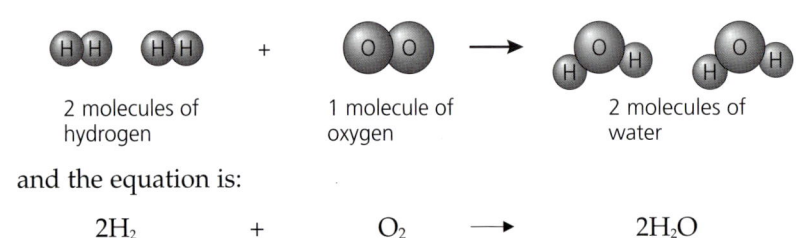

2 molecules of hydrogen + 1 molecule of oxygen → 2 molecules of water

and the equation is:

$$2H_2 + O_2 \rightarrow 2H_2O$$

Can you see why there is a 2 in front of H_2 and H_2O in the equation? Now look at the number of atoms on each side of the equation:

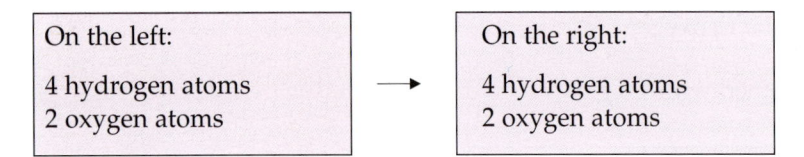

On the left:	On the right:
4 hydrogen atoms	4 hydrogen atoms
2 oxygen atoms	2 oxygen atoms

The numbers of hydrogen and oxygen atoms are the same on both sides of the equation. This is because atoms do not *disappear* during a reaction – they are just *rearranged*, as shown in the diagram. When the numbers of different atoms are the same on both sides, an equation is **balanced**. **All chemical equations should be balanced.**

Adding state symbols

Reactants and products may be solids, liquids, gases, or solutions. You can show their states by adding **state symbols** to the equations:

(s) for solid   (l) for liquid
(g) for gas   (aq) for aqueous solution (solution in water)

For the two reactions above, the equations with state symbols are:

$$C\,(s) + O_2\,(g) \rightarrow CO_2\,(g)$$
$$2H_2\,(g) + O_2\,(g) \rightarrow 2H_2O\,(l)$$

The reaction between hydrogen and oxygen gives out so much energy that it is used to power rockets. The hydrogen and oxygen are carried as liquids in the fuel tanks.

152

How to write the equation for a reaction

These are the steps to follow, when writing an equation:
1 Write the equation in words.
2 Now write the equation using symbols. Make sure all the formulae are correct.
3 Check that the equation is balanced, for each type of atom in turn. *Make sure you do not change any formulae.*
4 Add the state symbols.

Example 1 Calcium burns in chlorine to form calcium chloride, a solid. Write an equation for the reaction, using the steps above.

1 Calcium + chlorine ⟶ calcium chloride
2 $Ca + Cl_2 \longrightarrow CaCl_2$
3 Ca: 1 atom on the left and 1 atom on the right.
   Cl: 2 atoms on the left and 2 atoms on the right.
   The equation is balanced.
4 $Ca\,(s) + Cl_2\,(g) \longrightarrow CaCl_2\,(s)$

Example 2 In industry, hydrogen chloride is formed by burning hydrogen in chlorine. Write an equation for the reaction.

1 Hydrogen + chlorine ⟶ hydrogen chloride
2 $H_2 + Cl_2 \longrightarrow HCl$
3 H: 2 atoms on the left and 1 atom on the right.
   Cl: 2 atoms on the left and 1 atom on the right.
   The equation is *not* balanced. It needs another molecule of hydrogen chloride on the right. So a 2 is put *in front of* the HCl.
   $H_2 + Cl_2 \longrightarrow 2HCl$
   The equation is now balanced. Do you agree?
4 $H_2\,(g) + Cl_2\,(g) \longrightarrow 2HCl\,(g)$

Example 3 Magnesium burns in oxygen to form magnesium oxide, a white solid. Write an equation for the reaction.

1 Magnesium + oxygen ⟶ magnesium oxide
2 $Mg + O_2 \longrightarrow MgO$
3 Mg: 1 atom on the left and 1 atom on the right.
   O: 2 atoms on the left and 1 atom on the right.
   The equation is *not* balanced. Try this:
   $Mg + O_2 \longrightarrow 2MgO$
   (Note, the 2 goes *in front of* the MgO.)
   Another magnesium atom is now needed on the left:
   $2Mg + O_2 \longrightarrow 2MgO$
   The equation is balanced.
4 $2Mg\,(s) + O_2\,(g) \longrightarrow 2MgO\,(s)$

Magnesium burning in oxygen.

Questions

1 What do + and ⟶ mean, in an equation?
2 Copy these equations and balance them:
  a $Na\,(s) + Cl_2\,(g) \longrightarrow NaCl\,(s)$
  b $H_2\,(g) + I_2\,(g) \longrightarrow HI\,(g)$
  c $SO_2\,(g) + O_2\,(g) \longrightarrow SO_3\,(g)$
  d $NH_3\,(g) \longrightarrow N_2\,(g) + H_2\,(g)$
  e $C\,(s) + CO_2\,(g) \longrightarrow CO\,(g)$
  f $Al\,(s) + O_2\,(g) \longrightarrow Al_2O_3\,(s)$
3 Aluminium burns in chlorine to form aluminium chloride, $AlCl_3$ (s). Write an equation for the reaction.

7.21 Calculations from equations

What an equation tells you

When carbon burns in oxygen, the reaction can be shown as:

C	+	O O	→	O C O
1 atom of carbon		1 molecule of oxygen		1 molecule of carbon dioxide

or in shorthand using an equation:

$$C\,(s) + O_2\,(g) \longrightarrow CO_2\,(g)$$

This equation tells you that:

| 1 carbon atom | reacts with | 1 molecule of oxygen | to give | 1 molecule of carbon dioxide |

But the relative atomic mass of a carbon atom is 12.
The relative mass of an oxygen molecule is 32.
The relative mass of a carbon dioxide molecule is 44.

So the equation also tells us that the *masses* of carbon and oxygen reacting together, and carbon dioxide produced, are in the proportions 12 to 32 to 64. That means we can write:

| 12 g of carbon | react with | 32 g of oxygen | to give | 44 g of carbon dioxide |

and

| 6 g of carbon | react with | 16 g of oxygen | to give | 22 g of carbon dioxide |

and so on. *The masses are always in the same proportion.*

From a chemical equation you can tell how many grams of each substance react together and how many grams of product will form.

Does the mass change during a reaction?

Now look at what happens to the total mass, during the above reaction:
mass of carbon and oxygen at the start: 12 g + 32 g = **44 g**
mass of carbon dioxide at the end: **44 g**

The total mass has not changed during the reaction. This is because no atoms have disappeared. They have just been rearranged. This is one of the basic laws of chemistry.

The Law of Conservation of Mass states that the total mass remains unchanged during a chemical reaction.

Sometimes it appears that mass does change. When you burn coal in a fire only light ash is left. But carbon dioxide and other gases have escaped during the reaction. If you burn coal in a closed system where nothing can escape, you will find that mass is conserved.

Calculating masses from equations

These are the steps to follow:
1 Write the balanced equation for the reaction.
2 Write down the relative atomic mass or formula mass for each substance that takes part.
3 From the equation, write down reacting amounts in *grams*.
4 Once you know these, you can find any *actual* mass.

Example 1 Hydrogen burns in oxygen to form water.
What mass of oxygen is needed to burn 1 gram of hydrogen, and what mass of water is obtained?

1 The equation for the reaction is: $2H_2 (g) + O_2 (g) \longrightarrow 2H_2O (l)$
2 The relative atomic masses are: H = 1, O = 16.
 So the relative formula masses are: H_2 = 2, O_2 = 32, H_2O = 18.
3 So from the equation, the reacting amounts in grams are:
 $2H_2 (g)$ + $O_2 (g)$ $\longrightarrow$ $2H_2O (l)$
 2×2 g   32 g 2×18 g   or
 4 g   32 g   36 g
4 But you start with only 1 gram of hydrogen so the *actual* masses are:
 1 g   32/4 g   36/4 g   or
 1 g   8 g   9 g

So 1 g of hydrogen needs **8 g** of oxygen to burn and gives **9 g** of water.

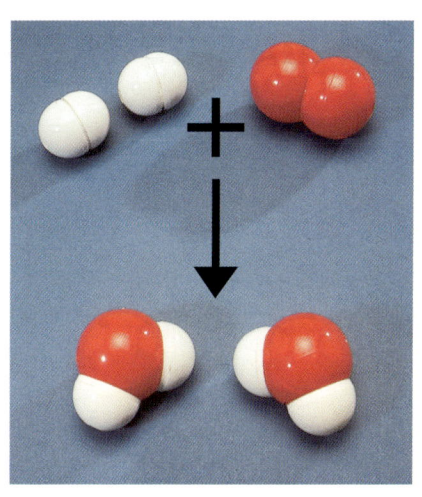

These models show how the atoms are rearranged during the reaction between hydrogen and oxygen. The equation is $2H_2 + O_2 \longrightarrow 2H_2O$.

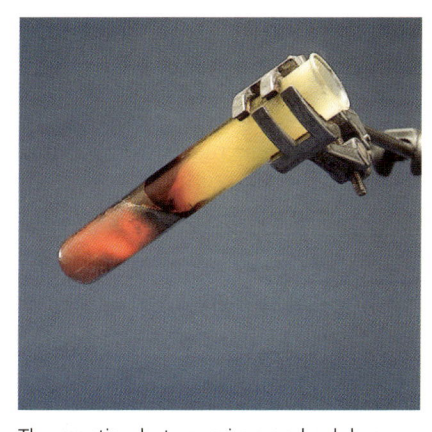

Example 2 When 7 g of iron are heated with excess sulphur, what mass of iron sulphide is formed and what mass of sulphur is used up?
(*Excess* sulphur means *more than enough* sulphur for the reaction.)

1 The equation for the reaction is: Fe *(s)* + S *(s)* $\longrightarrow$ FeS *(s)*
2 The relative atomic masses are: Fe = 56, S = 32.
 So the relative formula mass for FeS is 88.
3 From the equation, the reacting amounts are: Fe *(s)* + S *(s)* $\longrightarrow$ FeS *(s)*
                                                                 56 g   32 g   88 g
4 So for 7 g of sulphur (or $\frac{1}{8}$ of 56 g):   7 g   32/8 g   88/8 g
                                                   or 7 g   4 g   11 g

The reaction produces **11 g** of iron sulphide and uses up **4 g** of sulphur.

The reaction between iron and sulphur. It is an exothermic reaction, but heat is needed to get it started.

Questions

1 The total mass does not change during a reaction. Why?
2 The reaction between magnesium and oxygen is:
 $2Mg (s) + O_2 (g) \longrightarrow 2MgO (s)$
 a Write a word equation for the reaction.
 b Work out the reacting amounts for the equation above. (A_r: Mg = 24, O = 16.)
 c How many grams of oxygen react with:
  i 48 g of magnesium?   ii 12 g of magnesium?
3 Copper(II) carbonate breaks down on heating:
 $CuCO_3 (s) \longrightarrow CuO (s) + CO_2 (g)$
 a Write a word equation for the reaction.
 b Work out the reacting amounts for the equation above. (A_r: Mg = 24, O = 16.)
 c When 31 g of copper(II) carbonate are used:
  i what mass of carbon dioxide forms?
  ii what mass of solid remains after heating?

155

7.22 Calculating the volumes of gases

The volume of a gas

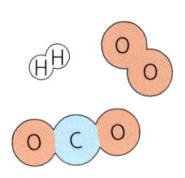

You might expect a given number of carbon dioxide molecules to take up *more* space than the same number of hydrogen or oxygen molecules. But no.

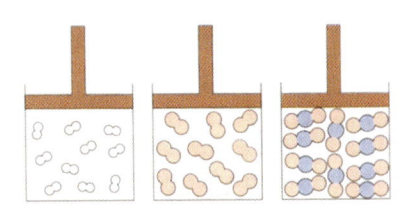

At a given temperature and pressure, the same number of particles of *any gas* take up the same volume, no matter what size the particles are.

Avogadro was an Italian chemist. In 1811 he stated that equal volumes of gases contain equal numbers of particles.

In other words at the same temperature and pressure, *each particle of any gas* occupies the same volume – strange as that sounds! Look again at the molecules, this time with their relative formula masses (M_r):

Each in effect occupies the same volume.

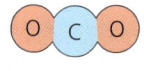

$M_r = 44$ $M_r = 2$ $M_r = 32$

Scaling up these masses to grams, it follows that 44 grams of carbon dioxide, 2 grams of hydrogen, and 32 grams of oxygen will each occupy the same volume at any given temperature and pressure.

At room temperature and pressure (rtp) this volume is **24 dm³**.
The same is true for any gas:
The M_r of a gas in grams occupies a volume of 24 dm³ at rtp.

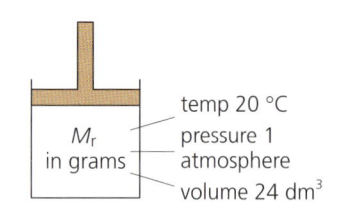

At room temperature and pressure (rtp) the M_r of a gas in grams occupies 24 dm³. (rtp is 20 °C and 1 atmosphere.)

Using M_r to find the volume of a gas

You can find the volume of a known mass of *any gas* at rtp like this:
1 Find the relative formula mass (M_r) of the gas.
2 Divide the given mass of the gas by the M_r to find what fraction of the M_r it is.
3 Multiply this number by 24 dm³.

Example 1 What volume do 22 g of carbon dioxide occupy at rtp?

Gas	carbon dioxide
Relative formula mass (M_r)	44
Mass / M_r	22/44 = 0.5
(Mass / M_r) × 24 dm³	0.5 × 24 = 12 dm³

22 g of carbon dioxide occupy a volume of **12 dm³** at rtp.

Example 2 What volume do 2 g of sulphur trioxide (SO_3) occupy at rtp?

Gas	sulphur trioxide
Relative formula mass (M_r)	80
Mass / M_r	2/80 = 0.025
(Mass / M_r) × 24 dm³	0.025 × 24 = 0.6 dm³

2 g of sulphur trioxide occupy **0.6 dm³** or **600 cm³** at rtp.

Calculating gas volumes in reactions

These are the steps to follow.
1 Write the balanced equation for the reaction.
2 Write down the relative atomic mass or formula mass for any solid or liquid in the equation, that you need for the calculation.
3 Use the equation to write down reacting amounts and volumes. (Use 24 dm^3 for each M_r of a gas.)
4 Once you know these you can find any *actual* quantity.

Example 1 Coal contains the impurity sulphur. When sulphur burns in air it forms sulphur dioxide. What volume of this polluting gas is produced when 8 g of sulphur burn?
1 The equation for the reaction is: $S(s) + O_2(g) \longrightarrow SO_2(g)$
2 The relative atomic mass of sulphur is: S = 32.
3 Use the equation to work out the reacting masses and volumes:
$$S(s) \longrightarrow SO_2(g)$$
$$32 \text{ g} \longrightarrow 24 \text{ dm}^3$$
4 But you start with only 8 g of sulphur, or a quarter of this amount, so:
$$8 \text{ g} \longrightarrow 24/4 \text{ dm}^3$$
$$\text{or } \mathbf{6 \text{ dm}^3}$$
8 g of sulphur produce **6 dm^3** of sulphur dioxide at rtp.

Example 2 What volume of carbon dioxide will be produced by reacting 1 g of calcium carbonate (CaCO$_3$) with excess hydrochloric acid?
1 The equation is: $CaCO_3(s) + 2HCl(aq) \longrightarrow CaCl_2(aq) + CO_2(g) + H_2O(l)$
2 Relative atomic masses: Ca = 40, C = 12, O = 16.
So the relative formula mass for calcium carbonate is 100.
3 Use the equation to work out reacting amounts and volumes:
$$CaCO_3(s) \longrightarrow CO_2(g)$$
$$100 \text{ g} \longrightarrow 24 \text{ dm}^3$$
4 But you start with 1g of calcium carbonate so divide both by 100:
$$1 \text{ g} \longrightarrow 24/100 \text{ dm}^3$$
$$\text{or } 0.24 \text{ dm}^3$$
So 1 g of calcium carbonate gives **0.24 dm^3** or **240 cm^3** of carbon dioxide at rtp.

Example 3 What volume of hydrogen will react with 12 dm^3 of oxygen to form water?
The equation for the reaction is: $2H_2(g) + O_2(g) \longrightarrow 2H_2O(l)$
The reacting volumes are: 48 dm^3 (2 × 24) + 24 dm^3
But you start with 12 dm^3 of oxygen, so divide both by 2:
24 dm^3 + 12 dm^3
So 12 dm^3 of oxygen react with **24 dm^3** of hydrogen at rtp to form water.

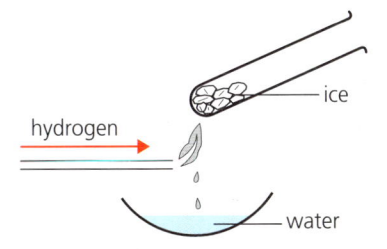

Smokeless fuels burn much more cleanly than ordinary coal. But they still produce some sulphur dioxide, because they contain sulphur compounds.

A jet of hydrogen burns in air or oxygen with a blue flame. It can be condensed to water on a cold surface.

Questions

(A_r: Ne = 20, O = 16, N = 14, H = 1, C = 12.)
1 What does rtp mean? What values does it have?
2 What is the volume of:
 a 5 g **b** 400 g of neon (Ne) at rtp?
3 Calculate the volume at rtp of:
 a 16 g of oxygen (O$_2$) **b** 1.7 g of ammonia (NH$_3$)
4 You burn 6 g of carbon in plenty of air:
$$C(s) + O_2(g) \longrightarrow CO_2(g)$$
What volume of oxygen will be used up?
5 If you burn the 6 g of carbon in limited air, the reaction is different: $2C(s) + O_2(g) \longrightarrow 2CO(g)$
What volume of oxygen will be used up this time?

7.23 Calculations on electrolysis

How much of each substance is produced?

If you electrolyse copper(II) chloride solution using carbon electrodes, copper and chlorine are produced:

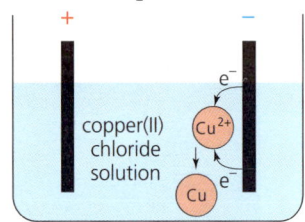

At the negative electrode	At the positive electrode
Copper ions grab electrons and deposit as atoms. Each copper ion needs 2 electrons to become an atom: $$Cu^{2+} + 2e^- \longrightarrow Cu$$ An equation like this for the reaction at an electrode is called a **half equation**.	Each chloride ion gives up 1 electron to form an atom. Then two atoms join to form a molecule: $$Cl^- - e^- \longrightarrow Cl \text{ then } Cl + Cl \longrightarrow Cl_2$$ so the half equation can be written as: $$2Cl^- - 2e^- \longrightarrow Cl_2$$

The two electrons given up by the chloride ions move along the wire in the current, and are taken up by one copper ion. So there is a direct link between the amounts of copper and chlorine produced:
For each atom of copper that forms, 1 molecule of chlorine is produced.

Using relative atomic masses we can say:
For every 64 g of copper produced, 71 g of chlorine form.

And changing M_r for chlorine to gas volume we can say:
For every 64 g of copper produced, 24 dm³ of chlorine form (at rtp).

Doing the calculations

So if you know the mass of one element produced in an electrolysis, you can find the mass of another from the equation. These are the steps:
1 Write a balanced **half equation** for the reaction at each electrode.
2 Now balance the two half equations *with each other* so that the number of electrons is the same in each.
3 Write down relative atomic and formula masses for the substances.
4 Add *grams* to these numbers to show the relative masses of each element produced.
5 Using the same **ratio** of masses, find the *actual* mass obtained.

Example 1 In the electrolysis of lead bromide, 2.07 g of lead were obtained. How much bromine was produced?
1 The half equations are:
   lead $Pb^{2+} + 2e^- \longrightarrow Pb$; bromine $2Br^- - 2e^- \longrightarrow Br_2$
2 These half equations are balanced with each other (2 electrons each).
3 A_r: Pb = 207, Br = 80. So M_r for Br_2 = 160.
4 So relative amounts obtained: 207 g of lead and 160 g of bromine.
5 But 2.07 g of lead were in fact produced, which means **1.6 g of bromine**.

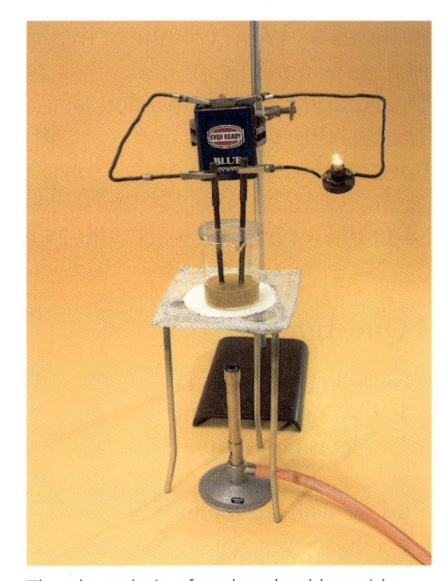

The electrolysis of molten lead bromide.

Example 2 Aluminium is extracted from aluminium oxide by electrolysis. In the extraction of 108 g of aluminium, what mass of oxygen is also produced?
1. The half equations are:
   aluminium $Al^{3+} + 3e^- \longrightarrow Al$; oxygen $2O^{2-} - 4e^- \longrightarrow O_2$
2. The number of electrons is not the same in each half equation. So they need to be balanced with each other:
   $4Al^{3+} + 12e^- \longrightarrow 4Al$ $6O^{2-} - 12e^- \longrightarrow 3O_2$
3. A_r: Al = 27, O = 16. Using these:
   $4Al = 4 \times 27 = 108$ $3O_2 = 3 \times (2 \times 16) = 96$
4. So the relative masses obtained are 108 g of aluminium and 96 g of oxygen.
5. So 96 g of oxygen would be produced.

So **96 g of oxygen** were produced in the electrolysis.

Aluminium is expensive to extract, because of the electricity needed for electrolysis. Recycling is much cheaper.

Calculating gas volumes in electrolysis

Follow the same steps but use 24 dm³ for every M_r of gas.

Example 3 In the electrolysis of calcium chloride, 10 g of calcium were obtained. What volume of chlorine would be collected at rtp? (The M_r of a gas in grams occupies 24 dm³ at rtp.)
1. The half equations are:
   calcium $Ca^{2+} + 2e^- \longrightarrow Ca$; chlorine $2Cl^- - 2e^- \longrightarrow Cl_2$
2. Both half equations have the same number of electrons.
3. A_r for Ca = 40, and the volume of chlorine produced is 24 dm³.
4. So the relative amounts obtained are 40 g of calcium and 24 dm³ of chlorine.
5. So for 10 g of calcium, the volume of chlorine produced would be 24/4 dm³ or 6 dm³.

So **6 dm³ of chlorine** would be collected at rtp.

Example 4 When water containing a little sulphuric acid is electrolysed, hydrogen and oxygen are produced. In effect the water is decomposed. In one experiment 50 cm³ of hydrogen were obtained. How much oxygen was produced?
1. The half equations are:
   hydrogen $2H^+ + 2e^- \longrightarrow H_2$; oxygen $2O^{2-} - 4e^- \longrightarrow O_2$
2. The number of electrons is not the same in each half equation. So they need to be balanced with each other:
   $4H^+ + 4e^- \longrightarrow 2H_2$ $2O^{2-} - 4e^- \longrightarrow O_2$
3. So the relative volumes are:
   2×24 dm³ = 48 dm³ of hydrogen to 24 dm³ of oxygen.
4. So for 50 cm³ of hydrogen, 25 cm³ of oxygen would be produced.

So **25 cm³ of oxygen** were produced.

Balancing half equations

$Cl^- - e^- \longrightarrow Cl_2$

One chlorine ion on the left but two chlorine atoms on the right – so this is not balanced.

1. First balance for chlorine:

   $2Cl^- - e^- \longrightarrow Cl_2$

2. Now balance the charges to give a final charge of zero:

   $2Cl^- - 2e^- \longrightarrow Cl_2$

Questions

1. At which electrode are electrons lost?
2. Copy, complete and balance these half equations:
   a $Na^+ + \ldots \longrightarrow Na$
   b $Mg^{2+} + \ldots \longrightarrow$
   c $I^- - \ldots \longrightarrow I_2$
   d $\ldots - \ldots \longrightarrow O_2$
3. If in the electrolysis of lead bromide 207 g of lead are obtained, what mass of bromine would also be produced?
4. In the electrolysis of aluminium oxide 108 g of aluminium are obtained. What **volume** of oxygen would also be obtained, at rtp?

159

Questions for Module 7

1. The rate of the reaction between magnesium and dilute hydrochloric acid could be measured using this apparatus:

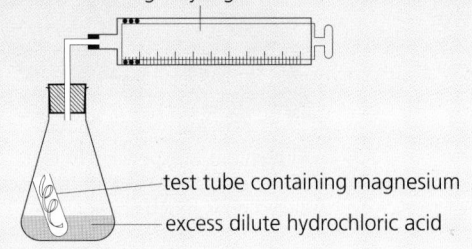

   gas syringe
   test tube containing magnesium
   excess dilute hydrochloric acid

   a. What is the purpose of:
      i. the test tube?
      ii. the gas syringe?
   b. How would you get the reaction to start?

2. Some magnesium and an *excess* of dilute hydrochloric acid were reacted together. The volume of hydrogen produced was recorded every minute, as shown in the table:

Time/min	0	1	2	3	4	5	6	7
Volume of hydrogen/cm^3	0	14	23	31	38	40	40	40

   a. What does an *excess* of acid mean?
   b. Plot a graph of the results.
   c. What is the *rate of reaction* (cm^3 of hydrogen per minute) during the:
      i. first minute?
      ii. second minute?
      iii. third minute?
   d. Explain why the rate of reaction changes during the reaction.
   e. What is the total volume of hydrogen produced in the reaction?
   f. How many minutes pass before the reaction finishes?
   g. What is the *average rate* of the reaction?
   h. How could you make the reaction go slower, while still using the same quantities of metal and acid?

3. You will need your graph from question 2. The experiment with magnesium and an excess of dilute hydrochloric acid was repeated. This time a different concentration of hydrochloric acid was used. The results were:

Time/min	0	1	2	3	4	5	6
Volume of hydrogen/cm^3	0	22	34	39	40	40	40

   a. Plot these results on the graph you drew for question 2.
   b. Which reaction was faster? How can you tell?
   c. In which experiment was the acid more concentrated? Give a reason for your answer.
   d. The same volume of hydrogen was produced in each experiment. What does that tell you about the mass of magnesium used?

4. Suggest a reason for each of the following observations:
   a. Magnesium powder reacts faster than magnesium ribbon, with dilute sulphuric acid.
   b. Hydrogen peroxide decomposes much faster in the presence of the enzyme catalase.
   c. The reaction between manganese carbonate and dilute hydrochloric acid speeds up when some concentrated hydrochloric acid is added.
   d. Zinc powder burns much more vigorously in oxygen than zinc foil does.
   e. The reaction between sodium thiosulphate and hydrochloric acid takes a very long time if carried out in an ice bath.
   f. Zinc and dilute sulphuric acid react much more quickly when a few drops of copper(II) sulphate solution are added.
   g. Drenching with water prevents too much damage from spilt acid.
   h. A car's exhaust pipe will rust faster if the car is used a lot.
   i. In fireworks, powdered magnesium is used rather than magnesium ribbon.
   j. In this country, dead animals decay quite quickly. But in Siberia, bodies of mammoths that died 30 000 years ago have been found fully preserved in ice.
   k. The more sweet things you eat, the faster your teeth decay.
   l. Food cooks much faster in a pressure cooker than in an ordinary saucepan.
   m. A concentrated solution of bleach is used to remove stains.
   n. Yeast bread that is not allowed to stand for a time before baking is not fit to be eaten.

5. a. Name four factors that affect the rate of a reaction, and describe the effect of changing each factor.
   b. Explain each factor using the idea of colliding particles.

6 When sodium thiosulphate reacts with hydrochloric acid, a precipitate forms. In an investigation, the time taken for the solution to become opaque was recorded. (*Opaque* means that you cannot see through it.) This apparatus was used for the experiment, in which only the concentration of thiosulphate was changed.

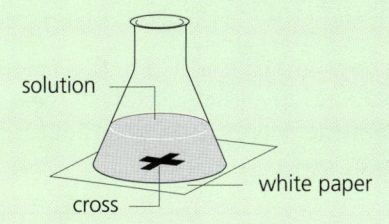

These are the results:

Experiment	A	B	C	D
Time taken/s	42	71	124	63

a How is the time measured?
b Name the precipitate that forms.
c What would be *observed* during the experiment?
d In which experiment was the reaction:
   i fastest? ii slowest?
e In which experiment was the sodium thiosulphate solution most concentrated? How can you tell?
f Suggest two other ways of speeding up this reaction.

7 Copper(II) oxide catalyses the decomposition of hydrogen peroxide. 0.5 g of the oxide was added to a flask containing 100 cm^3 of hydrogen peroxide solution. A gas was released. It was collected and its volume noted every 10 seconds. This table shows the results:

Time/s	0	10	20	30	40	50	60	70	80	90
Volume/cm^3	0	18	30	40	48	53	57	58	58	58

a What is a catalyst?
b Draw a diagram of suitable apparatus for this experiment.
c Name the gas that is formed.
d Write a balanced equation for the decomposition of hydrogen peroxide.
e Plot a graph of the volume of gas (vertical axis) against time (horizontal axis).
f Describe how the rate changes during the reaction.
g What happens to the concentration of hydrogen peroxide as the reaction proceeds?
h What chemicals are present in the flask after 90 seconds?
i What mass of copper(II) oxide would be left in the flask at the end of the reaction?
j Sketch on your graph the curve that might be obtained for 1.0 g of copper(II) oxide.
k Name one other chemical that catalyses this decomposition.

8 Ethanol can be made by the fermentation of sugar.

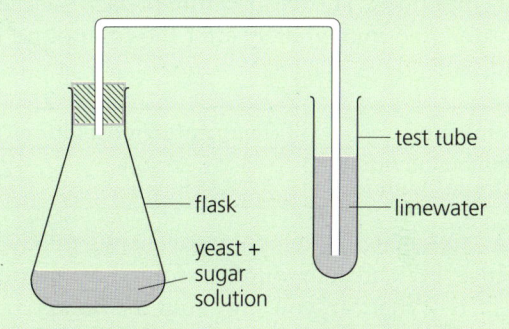

a What is present in the yeast that causes the fermentation of sugar?
b Which temperature is best for the reaction?
   i 0°C ii 10°C iii 25°C iv 75°C
   Explain your answer.
c i Which gas is released during fermentation?
   ii How could you prove this?
d i Name two drinks made by fermentation.
   ii What are the starting materials for each?
e Write an equation to represent the fermentation of glucose.

9 The **enzyme** *polyphenoxidase* is involved in the oxidation reaction that causes sliced apple to turn brown in air. Explain the following observations.
a When the apple is first cut open, the flesh is not brown.
b Browning is much slower when the sliced apple is placed in the fridge.
c No browning takes place if the apple is placed in boiling water for 30 seconds straight after slicing.
d Apple that has been pulped in a food mixer turns brown much faster than sliced apple does.
e The browning reaction does not take place if the sliced apple is dipped in lemon juice straight away.

10 Sodium hydroxide and hydrochloric acid react as shown in this equation:
   NaOH (aq) + HCl (aq) ⟶ NaCl (aq) + H₂O (l)
   The reaction is exothermic.
   a What type of chemical reaction is it?
   b Is heat given out or taken in during the reaction, or neither?
   c In that case, what happens to the temperature of the solutions as they react?
   d Draw an energy diagram for the reaction.

11 Water at 25 °C was used to dissolve two compounds. The temperature of each solution was measured immediately after the compound had dissolved.

Compound	Temperature of solution/°C
ammonium nitrate	21
calcium chloride	45

   a Calculate the temperature change for each.
   b i Which compound dissolved exothermically?
     ii What can you say about the bonds that are formed with water in the solution?
   c i Which dissolved endothermically?
     ii What can you say about the bonds that are formed with water in the solution?
   d For each solution, estimate the temperature of the solution if:
     i the amount of water was halved but the same mass of compound was used
     ii the mass of the compound was halved but the volume of water was unchanged
     iii both the mass of the compound and the volume of water were halved.

12 Methane is the main component of natural gas. The reaction between methane and oxygen is exothermic, giving out 890 kJ of energy.
   CH₄ (g) + 2O₂ (g) ⟶ CO₂ (g) + 2H₂O (g)
   a Explain in terms of bond breaking and bond making why this reaction is exothermic.
   b i Copy the diagram below and complete it to show the energy diagram for this reaction, including the activation energy.

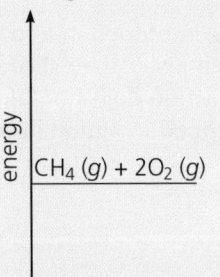

     ii Methane will not burn in air until a spark or flame is applied. Why not?
   c How much energy is given out when 1 g of methane burns? (A_r: C = 12, H = 1.)

13 Hydrogen peroxide decomposes very slowly to form water and oxygen:
   2H₂O₂ (aq) ⟶ 2H₂O (l) + O₂ (g)
   a A catalyst may be added to this reaction. Explain briefly what a catalyst does.
   b Here is the energy level diagram for the reaction shown above.

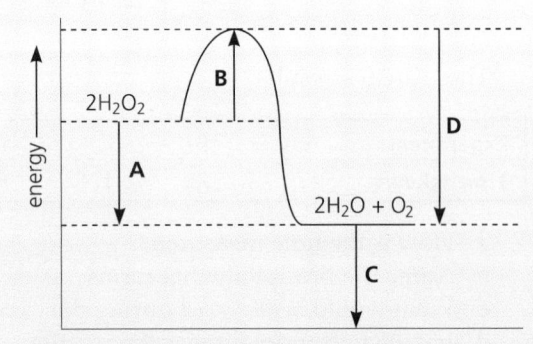

     i What is meant by the term activation energy?
     ii Which energy change, A, B, C, or D, is the activation energy?
     iii Explain, in terms of energy, the effect of a catalyst.
     iv Is the reaction exothermic or endothermic?
   c The following equation uses structural formulae to represent the decomposition of hydrogen peroxide:

   2 (H–O–O–H) ⟶ 2 (H–O–H) + O=O

   The table below shows the bond energies.

Bond	Energy required/kJ
O=O	498
O–O	146
H–O	464

     i Calculate the energy needed to break all the bonds in the reactants.
     ii Calculate the energy given out when new bonds are formed in the products.
     iii Calculate the energy change for this reaction.
     iv Is the reaction exothermic or endothermic? Explain your answer and compare it with the answer to b iv above.

14 In the following reaction involving methane (CH₄), 1664 kJ of energy are **taken in**.

H–C(H)(H)–H (g) ⟶ C (g) + 4H (g)

(Note: no bonds are **formed** in this reaction.)
a Why is this reaction **endothermic**?
b How many bonds are broken in this reaction?
c Calculate the bond energy of the C–H bond.
For a similar reaction involving ethane (C₂H₆), 2826 kJ of energy are taken in.
d List the bonds broken during the reaction.
e Calculate the bond energy of the C–C bond in ethane.

15 Hydrazine, N₂H₄, burns in oxygen as follows:

H₂N–NH₂ (g) + O=O (g) ⟶ N≡N (g) + 2 O(H)(H) (g)

a Count and list the bonds broken in this reaction.
b Count and list the new bonds formed.
c Calculate the total energy:
   i required to break the bonds
   ii released when the new bonds are made
(Bond energies (in kJ): N–H 391; N–N 158; N≡N 945; O–H 464; O=O 498.)
d Calculate the energy change in this reaction.
e Is the reaction exothermic or endothermic?
f Where is the energy transferred from and to?
g Would hydrazine be a suitable fuel? Why?

16 When chlorine gas is passed over iodine, the following reactions take place.
   Reaction 1: I₂ (s) + Cl₂ (g) ⟶ 2 ICl (l)
                                          brown
   Reaction 2: ICl (l) + Cl₂ (g) ⇌ ICl₃ (s)
                   brown   yellow
a What is the first change you would see as the chlorine is passed over the iodine?
b As more chlorine is passed, what further change would you see?
c Which of the reactions can be *reversed*?
d What change would you see as the chlorine gas supply is turned off? Explain your answer.

17 Hydrogen and bromine react together to form hydrogen bromide:
   H₂ (g) + Br₂ (g) ⇌ 2HBr (g)
a Which of the following will favour the formation of more hydrogen bromide?
   i adding more hydrogen
   ii removing bromine
   iii removing the product as it is formed
b Explain why increasing the pressure has no effect on the amount of product formed.

18 Ammonia is manufactured from nitrogen and hydrogen. 92 kJ of energy are given out.
a Is the forward reaction endothermic or exothermic?
b Explain why the yield of ammonia:
   i rises if you increase the pressure
   ii falls if you increase the temperature
c Why is the reaction carried out at 450 °C rather than a lower temperature?

19 The Haber process converts nitrogen and hydrogen to ammonia:
   N₂ (g) + 3H₂ (g) ⇌ 2NH₃ (g)
a The energy level diagram for this reaction is shown below. What does this diagram tell you about the reaction?

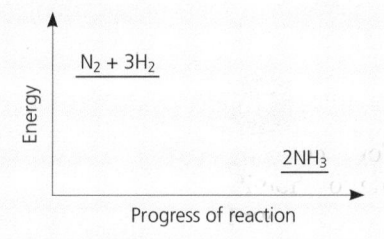

[Energy level diagram: N₂ + 3H₂ higher level, 2NH₃ lower level; axes: Energy vs Progress of reaction]

b The reaction is reversible. It reaches equilibrium.
   i What effect does a catalyst have on an equilibrium reaction?
   ii What catalyst is used in the Haber process?
c The following three reactions show how nitric acid is produced from ammonia.

                           platinum
ammonia + oxygen ⟶ nitrogen monoxide + steam

nitrogen monoxide + oxygen ⟶ nitrogen dioxide

nitrogen dioxide + oxygen + water ⟶ nitric acid

   i Ammonia is a raw material in this process. What are the other two raw materials?
   ii One use of nitric acid is to produce ammonium nitrate. Write a word equation for this reaction.
   iii What is ammonium nitrate used for?

163

20 Calculate the formula mass of the following compounds.
   a dinitrogen tetroxide, N_2O_4
   b lead oxide, Pb_3O_4
   c nitroglycerine, $C_3H_5(NO_3)_3$
   d hydrated iron(II) sulphate, $FeSO_4.7H_2O$
   e ethanoic acid, CH_3COOH
   (A_r: H = 1, C = 12, N = 14, O = 16, S = 32, Fe = 56, Pb = 207.)

21 The formula of calcium oxide is CaO. The relative atomic masses are: Ca = 40, O = 16. Copy and complete the following statements:
   a 40 g of calcium combine with ____ g of oxygen to form ____ g of calcium oxide.
   b 4.0 g of calcium combine with ____ g of oxygen to form ____ g of calcium oxide.
   c When 0.4 g of calcium reacts in oxygen, the increase in mass is ____ g.
   d If 112 g of calcium oxide were decomposed to calcium and oxygen, ____ g of calcium and ____ g of oxygen would be obtained.
   e The percentage mass of calcium in calcium oxide is ____ %.

22 Two samples of copper oxide were made by different methods. The oxides were then converted to copper. The following results were obtained:

	Sample 1	Sample 2
Mass of copper oxide	16.0 g	32.0 g
Mass of copper obtained	12.8 g	25.6 g

   a Calculate the percentage composition of copper in each sample.
   b Were the two oxides of the same or different composition?

23 In a reaction to make manganese from manganese oxide, the following results were obtained: 174 g of manganese oxide produced 110 g of manganese. (A_r: Mn = 55, O = 16.)
   a What mass of oxygen is there in 174 g of manganese oxide?
   b What is the empirical formula of manganese oxide?

24 The following results were obtained in an experiment to find the formula of magnesium oxide:

Mass of crucible = 12.5 g
Mass of crucible + magnesium = 14.9 g
Mass of crucible + magnesium oxide = 16.5 g
(A_r: Mg = 24, O = 16.)
   a What mass of magnesium was used in the experiment?
   b What mass of oxygen combined with the magnesium?
   c Use your answers to parts a and b to find the formula of magnesium oxide.

25 Zinc phosphide is made by heating zinc and phosphorus together. It is found that 8.4 g of zinc combines with 3.1 g of phosphorus.
   a Find the empirical formula for the compound. (A_r: Zn = 56, P = 31.)
   b Calculate the percentage of phosphorus in it.

26 Write a chemical equation for each of the following. You do not need to add state symbols. Example:

$2H_2 + O_2 \rightarrow 2H_2O$

   a $H_2 + Cl_2 \rightarrow HCl + HCl$
   b $N_2 + H_2 + H_2 + H_2 \rightarrow NH_3 + NH_3$
   c $I_2 + Cl_2 \rightarrow ICl + ICl$
   d $P_2 + Cl_2 + Cl_2 + Cl_2 \rightarrow PCl_3 + PCl_3$

27 Copy these equations and balance them:
   a $H_2 (g) + Br_2 (g) \rightarrow HBr (g)$
   b $Cl_2 (g) + KBr (aq) \rightarrow KCl (aq) + Br_2 (aq)$
   c $C_2H_4 (g) + O_2 (g) \rightarrow CO_2 (g) + H_2O (l)$
   d $Zn (l) + Fe_2O_3 (l) \rightarrow Fe (l) + ZnO (s)$
   e $TiCl_4 (l) + Mg (l) \rightarrow Ti (l) + MgCl_2 (l)$
   f $NH_3 (g) + O_2 (g) + H_2O (l) \rightarrow HNO_3 (l)$
   g $Pb(NO_3)_2 (s) \rightarrow PbO_2 (s) + NO_2 (g) + O_2 (g)$
   h $Al (s) + HCl (aq) \rightarrow AlCl_3 (aq) + H_2 (g)$
   i $C_2H_5OH (l) + O_2 (g) \rightarrow CO_2 (g) + H_2O (l)$

28 Mercury(II) oxide breaks down into mercury and oxygen when heated.
$$2HgO\ (s) \longrightarrow 2Hg\ (l) + O_2\ (g)$$
How much mercury and oxygen are produced when 21.7 g of mercury(II) oxide is heated?
(A_r: Hg = 201, O = 16.)

29 Iron(II) sulphide is formed when iron and sulphur react together:
$$Fe\ (s) + S\ (s) \longrightarrow FeS\ (s)$$
a How many grams of sulphur will react with 56 g of iron? (A_r: Fe = 56, O = 16.)
b If 7 g of iron and 10 g of sulphur are used, which substance is in excess?
c If 7 g of iron and 10 g of sulphur are used, name the substances present when the reaction is complete, and find the mass of each.
d What mass of iron would react completely with 10 g of sulphur?

30 The following equation represents a reaction in which iron is obtained from iron(III) oxide:
$$Fe_2O_3\ (s) + 3CO\ (g) \longrightarrow 2Fe\ (s) + 3CO_2\ (g)$$
a Write a word equation for the reaction.
b i What is the formula mass of iron(III) oxide? (A_r: Fe = 56, O = 16.)
   ii How much iron would be obtained if this mass of iron(III) oxide was reacted completely?
   iii What mass of iron would be obtained from 320 g of iron(III) oxide?
   iv What mass of iron would be obtained from 320 tonnes of iron(III) oxide?

31 Calcium metal reacts with water. This is the symbol equation for the reaction:
$$Ca\ (s) + 2H_2O\ (l) \longrightarrow Ca(OH)_2\ (aq) + H_2\ (g)$$
a Write a word equation for the reaction.
b Calculate the formula mass of calcium hydroxide.
(A_r: Ca = 40, O = 16, H = 1.)
c What volume of hydrogen would be obtained if 40 g of calcium were reacted completely?
(The relative formula mass (M_r) in grams, of a gas, has a volume of 24 000 cm^3 at room temperature and pressure.)
d Calculate the mass of calcium that would be needed to produce 600 cm^3 of hydrogen.

32 Magnesium and sulphuric acid react to form magnesium sulphate and hydrogen. The equation follows:
$$Mg\ (s) + H_2SO_4\ (aq) \longrightarrow MgSO_4\ (aq) + H_2\ (g)$$
a Calculate the mass of magnesium sulphate that would be obtained from 2 g of magnesium. Show your working.
(A_r: Mg = 24, S = 32, O = 16.)
b What volume of hydrogen would also be produced?
(The relative formula mass (M_r) in grams, of a gas, has a volume of 24 000 cm^3 at room temperature and pressure.)

33 Nitrogen monoxide reacts with oxygen to form nitrogen dioxide. The equation is:
$$2NO\ (g) + O_2\ (g) \longrightarrow 2NO_2\ (g)$$
a How many molecules of nitrogen dioxide are obtained from each nitrogen monoxide molecule?
b What volume of oxygen will react with 50 cm^3 of nitrogen monoxide?
c What is the total volume of the two reactants?
d What volume of nitrogen dioxide will be obtained?
e Explain why there is a reduction in volume in the reaction.

34 Nitroglycerine, $C_3H_5(NO_3)_3$, is used as an explosive. The equation for the explosion reaction is:
$$4C_3H_5(NO_3)_3\ (l) \longrightarrow 12CO_2\ (g) + 10H_2O\ (l) + 6N_2\ (g) + O_2\ (g)$$
a How many molecules of gas are produced in the equation?
b What volume (in dm^3) would these molecules of gas occupy at room temperature and pressure?
(The relative formula mass (M_r) in grams, of a gas, has a volume of 24 dm^3 at room temperature and pressure.)
c Why is the volume of gas actually produced likely to be much larger than this?

35 Sodium hydrogencarbonate (baking powder) decomposed as follows when heated:
$$2NaHCO_3\ (s) \longrightarrow Na_2CO_3\ (s) + H_2O\ (l) + CO_2\ (g)$$
a What is the relative molecular mass of sodium hydrogencarbonate?
(A_r: Na = 23, C = 12, O = 16.)
b Which gas is released when baking powder is heated?
c What volume of gas would be obtained if:
   i 84 g of baking powder were decomposed?
   ii 8.4 g of baking powder were decomposed?
(The relative formula mass (M_r) in grams, of a gas, has a volume of 24 dm^3 at room temperature and pressure.)

165

Module 8
Structures and bonding

What happens when elements react? To understand that we need to look at the atoms, those unimaginably tiny units from which elements are made.

There are over a hundred elements, which means over a hundred different sorts of atom. When elements react, atoms bond with each other. The result is hundreds of thousands of different compounds.

In some elements, the drive for atoms to bond with other atoms gives vigorous and even explosive reactions. These are the reactive elements. In only a few rare elements – the noble gases – do atoms exist singly, with no need to bond.

There are different types of bonding, and this is where it gets exciting. The bonding dictates the structure of the material. Structure dictates properties. And properties dictate what we use a material for. So bonding is a subject of great interest and importance to chemists.

In this module you can find out more about atoms and what makes them react. You can see how elements can be arranged in groups whose atoms react in a similar way. And you can learn about the different types of bonding, and the structures of the solids that result. ∎

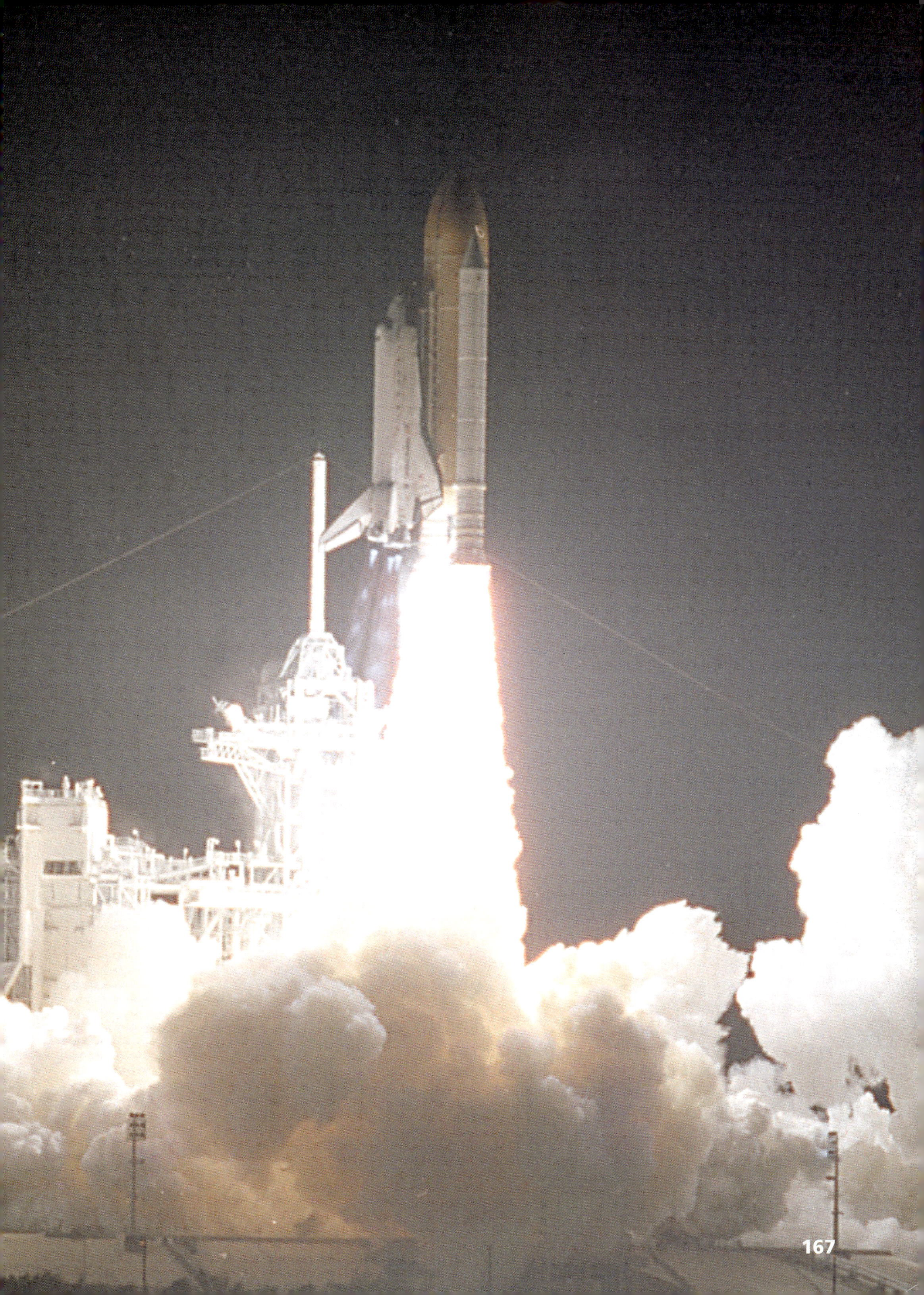

8.01 Atoms, elements, and compounds

Atoms

Sodium is made of tiny particles called sodium atoms.

Diamond is made of carbon atoms – different from sodium atoms.

Mercury is made of mercury atoms – different again!

Atoms are incredibly small. They are mostly empty space! Each atom consists of a **nucleus** and a cloud of particles called **electrons** that whizz around the nucleus. The drawing on the right shows what a sodium atom might look like, magnified about 100 million times.

The elements

Sodium is made of sodium atoms only, so it is an **element**.
An element contains only one kind of atom.

Scientists are still finding new elements. So far, 109 have been officially recognized and named. 90 of these are found in the Earth and atmosphere. The rest were made in the lab. Many of the 'artificial' elements last just a few seconds before decaying into other elements.

Symbols for the elements

To make life easy, each element has a symbol. These are usually made from the English or Latin name. For example the symbol for carbon is C. The symbol for potassium is K, from its Latin name *kalium*. Some elements are named after the people who discovered them.

The periodic table of the elements

Look at the table on the next page. It is called the **periodic table**. It is one of the most important and useful images in chemistry. It is like a map and address book for the elements, all rolled into one.

- It gives the names and symbols for the elements.
- Look at the numbered columns or **groups**. Groups are really families of elements. All the elements in a group have similar properties. So if you know how one of the elements in a group behaves, you can make a good guess about the others.
- The rows are called **periods**.
- Look at the zig-zag line. It separates the **metals** from the **non-metals**, with the non-metals on the right. You can see at a glance that most of the elements are metals.

Now look at the small numbers to the left of each symbol. These give you vital information about the element's atoms, as we will see later.

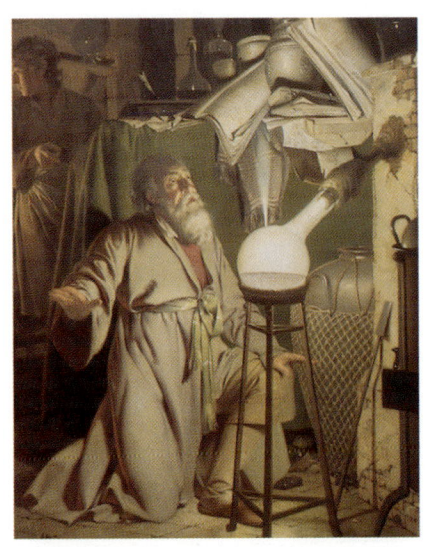

The elements have been 'discovered' one by one, over the centuries. This painting shows the alchemist who discovered phosphorus in the seventeenth century. To his amazement it glows in the dark.

The periodic table

Group 1	Group 2												Group 3	Group 4	Group 5	Group 6	Group 7	0
		1	$^1_1$H hydrogen															$^4_2$He helium
$^7_3$Li lithium	$^9_4$Be beryllium												$^{11}_5$B boron	$^{12}_6$C carbon	$^{14}_7$N nitrogen	$^{16}_8$O oxygen	$^{19}_9$F fluorine	$^{20}_{10}$Ne neon
$^{23}_{11}$Na sodium	$^{24}_{12}$Mg magnesium		The transition metals										$^{27}_{13}$Al aluminium	$^{28}_{14}$Si silicon	$^{31}_{15}$P phosphorus	$^{32}_{16}$S sulphur	$^{35.5}_{17}$Cl chlorine	$^{40}_{18}$Ar argon
$^{39}_{19}$K potassium	$^{40}_{20}$Ca calcium	$^{45}_{21}$Sc scandium	$^{48}_{22}$Ti titanium	$^{51}_{23}$V vanadium	$^{52}_{24}$Cr chromium	$^{55}_{25}$Mn manganese	$^{56}_{26}$Fe iron	$^{59}_{27}$Co cobalt	$^{59}_{28}$Ni nickel	$^{64}_{29}$Cu copper	$^{65}_{30}$Zn zinc	$^{70}_{31}$Ga gallium	$^{73}_{32}$Ge germanium	$^{75}_{33}$As arsenic	$^{79}_{34}$Se selenium	$^{80}_{35}$Br bromine	$^{84}_{36}$Kr krypton
$^{85}_{37}$Rb rubidium	$^{88}_{38}$Sr strontium	$^{89}_{39}$Y yttrium	$^{91}_{40}$Zr zirconium	$^{93}_{41}$Nb niobium	$^{96}_{42}$Mo molybdenum	$^{99}_{43}$Tc technetium	$^{101}_{44}$Ru ruthenium	$^{103}_{45}$Rh rhodium	$^{106}_{46}$Pd palladium	$^{108}_{47}$Ag silver	$^{112}_{48}$Cd cadmium	$^{115}_{49}$In indium	$^{119}_{50}$Sn tin	$^{122}_{51}$Sb antimony	$^{128}_{52}$Te tellurium	$^{127}_{53}$I iodine	$^{131}_{54}$Xe xenon
$^{133}_{55}$Cs caesium	$^{137}_{56}$Ba barium	$^{139}_{57}$La lanthanum	$^{178.5}_{72}$Hf hafnium	$^{181}_{73}$Ta tantalum	$^{184}_{74}$W tungsten	$^{186}_{75}$Re rhenium	$^{190}_{76}$Os osmium	$^{192}_{77}$Ir iridium	$^{195}_{78}$Pt platinum	$^{197}_{79}$Au gold	$^{201}_{80}$Hg mercury	$^{204}_{81}$Tl thallium	$^{207}_{82}$Pb lead	$^{209}_{83}$Bi bismuth	$^{210}_{84}$Po polonium	$^{210}_{85}$At astatine	$^{222}_{86}$Rn radon
$^{223}_{87}$Fr francium	$^{226}_{88}$Ra radium	$^{227}_{89}$Ac actinium																

$^{140}_{58}$Ce cerium	$^{141}_{59}$Pr praseodymium	$^{144}_{60}$Nd neodymium	$^{147}_{61}$Pm promethium	$^{150}_{62}$Sm samarium	$^{152}_{63}$Eu europium	$^{157}_{64}$Gd gadolinium	$^{159}_{65}$Tb terbium	$^{162}_{66}$Dy dysprosium	$^{165}_{67}$Ho holmium	$^{167}_{68}$Er erbium	$^{169}_{69}$Tm thulium	$^{173}_{70}$Yb ytterbium	$^{175}_{71}$Lu lutetium
$^{232}_{90}$Th thorium	$^{231}_{91}$Pa protactinium	$^{238}_{92}$U uranium	$^{237}_{93}$Np neptunium	$^{244}_{94}$Pu plutonium	$^{243}_{95}$Am americium	$^{247}_{96}$Cm curium	$^{247}_{97}$Bk berkelium	$^{251}_{98}$Cf californium	$^{252}_{99}$Es einsteinium	$^{257}_{100}$Fm fermium	$^{258}_{101}$Md mendelevium	$^{259}_{102}$No nobelium	$^{262}_{103}$Lw lawrencium

Compounds

Elements can combine with each other to form **compounds**.
A compound contains atoms of different elements joined together.
Although there are less than 120 elements, there are millions of compounds. This table shows three common ones:

Name of compound	Elements in it	How the atoms are joined up
Water	Hydrogen and oxygen	O with two H atoms attached
Carbon dioxide	Carbon and oxygen	O–C–O
Ethanol	Carbon, hydrogen, and oxygen	H-C-C-O-H with H atoms on carbons

Symbols for compounds

The symbol for a compound is called its **formula**. It is made from the symbols of the elements in it. So the formula for water is H_2O.
Note that the plural of **formula** is **formulae**.

Questions

1 What is an *atom*?
2 What is the centre part of an atom called?
3 Explain what an *element* is.
4 Which element has this symbol:
   a Ca?   b Mg?   c N?   d Ne?
5 Using the periodic table to help you, name three metals that you expect to behave in a similar way.
6 The ancient Romans gave Latin names to the elements they knew. Using the periodic table to help you, see if you can find the English names for these metals:
   a cuprum   b natrium   c ferrum   d plumbum
7 What is a *compound*?
8 Using the table above, give the formula for:
   a carbon dioxide   b ethanol

8.02 More about atoms

Protons, neutrons, and electrons

Atoms consist of a **nucleus** and a cloud of **electrons** that move round the nucleus. The nucleus is itself a cluster of two sorts of particle, **protons** and **neutrons**.

All the particles in an atom are very light. Their mass is measured in **atomic mass units**, rather than grams. Protons and electrons also have an **electric charge**:

Particle in atom	Mass	Charge
Proton ●	1 unit	Positive charge (1+)
Neutron ●	1 unit	None
Electron ●	Negligible	Negative charge (1−)

The nucleus is very tiny compared with the rest of the atom. If the atom was the size of a football stadium, the nucleus (sitting at the centre) would be the size of a pea!

How the particles are arranged

The sodium atom is a good one to start with. It has **11** protons, **11** electrons, and **12** neutrons. They are arranged like this:

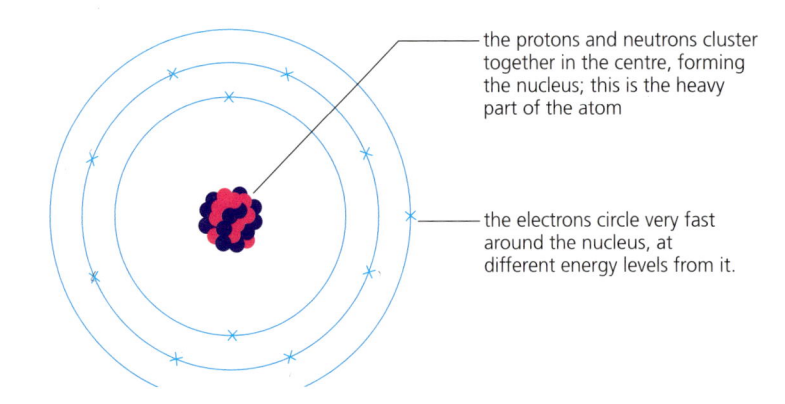

- the protons and neutrons cluster together in the centre, forming the nucleus; this is the heavy part of the atom
- the electrons circle very fast around the nucleus, at different energy levels from it.

Atomic number

A sodium atom has **11** protons. This fact could be used to identify it, because *only* a sodium atom has 11 protons. Every other atom has a different number of protons.
You can identify an atom by the number of protons in it.

The number of protons in an atom is called its **atomic number** or **proton number**. The atomic number of sodium is 11.

How many electrons?

The sodium atom also has 11 electrons. So it has an equal number of protons and electrons. The same is true for every sort of atom:
Every atom has an equal number of protons and electrons.

Because of this, atoms have no overall charge. The charge on the electrons cancels the charge on the protons.

The charge on a sodium atom:
●●●● 11 protons
●●●● Each has a charge of 1+
●●● Total charge 11+

××× × 11 electrons
××× × Each has a charge of 1−
××× Total charge 1−

Adding the charges: 11+
                     11−
                     ───
                      0

The answer is zero.
The atom has no overall charge.

170

Mass number

The mass of an atom is due to its protons and neutrons, since electrons have almost no mass. So the number of protons and neutrons in an atom is called its **mass number**:

mass number = number of protons + neutrons in an atom

The mass number for sodium atoms is 23. (11 + 12 = 23)

In the periodic table, sodium is shown as: $^{23}_{11}Na$.
This tells you that sodium atoms have a mass number of 23 and an atomic number of 11. So you know they have 12 neutrons too. (23 − 11 = 12)

The atoms of the first 20 elements

In the periodic table, the elements are arranged in order of increasing atomic number. Here are the first 20:

> **Try it yourself!**
> You can describe any element in shorthand if you know its symbol, atomic number, and mass number.
>
> Write them in this order:
>
> mass number
> atomic number **symbol**
>
> For example: $^{16}_{8}O$

Element	Symbol	Atomic number (protons)	Electrons	Neutrons	Mass number (protons + neutrons)
Hydrogen	H	1	1	0	1
Helium	He	2	2	2	4
Lithium	Li	3	3	4	7
Beryllium	Be	4	4	5	9
Boron	B	5	5	6	11
Carbon	C	6	6	6	12
Nitrogen	N	7	7	7	14
Oxygen	O	8	8	8	16
Fluorine	F	9	9	10	19
Neon	Ne	10	10	10	20
Sodium	Na	11	11	12	23
Magnesium	Mg	12	12	12	24
Aluminium	Al	13	13	14	27
Silicon	Si	14	14	14	28
Phosphorus	P	15	15	16	31
Sulphur	S	16	16	16	32
Chlorine	Cl	17	17	18	35
Argon	Ar	18	18	22	40
Potassium	K	19	19	20	39
Calcium	Ca	20	20	20	40

Questions

1. Name the particles that make up the atom.
2. Which particle has:
   a a positive charge? b no charge? c almost no mass?
3. An atom has 9 protons. Which element is it?
4. Why do atoms have no overall charge?
5. What does the term mean?
   a atomic number b mass number
6. Name each of these atoms, and say how many protons, electrons, and neutrons it has:
   $^{12}_{6}C$ $^{16}_{8}O$ $^{24}_{12}Mg$ $^{27}_{13}Al$ $^{64}_{29}Cu$

171

8.03 Isotopes and A_r

A reminder: how to identify an atom

Only sodium atoms have 11 protons.
You can identify an atom by the number of protons in it.

Isotopes

All hydrogen atoms have just 1 proton. But not all hydrogen atoms are identical. That's because some have more *neutrons* than others.

Most hydrogen atoms are like this, ^1_1H. It has one proton and one electron, but *no* neutrons.	But a few are like this. It has one neutron. It is called **deuterium**, ^2_1H.	And some are like this, with two neutrons. It is called **tritium**, ^3_1H.

The three atoms are called **isotopes** of hydrogen.
Isotopes are atoms of the same element with different numbers of neutrons.

Most elements have isotopes. For example calcium has six, magnesium three, silver two, and carbon three. This table shows the isotopes of carbon:

Name	Symbol	How many protons?	How many electrons?	How many neutrons?	Mass number
Carbon-12	$^{12}_6C$	6	6	6	12
Carbon-13	$^{13}_6C$	6	6	7	13
Carbon-14	$^{14}_6C$	6	6	8	14

All three occur naturally – for example in the carbon dioxide in air. But only about 1% of carbon atoms are carbon-13. Only a very tiny % are carbon-14.

Isotopes and radioactivity

Carbon-14 is **radioactive**. That means the atom is unstable. Sooner or later it breaks down or **decays**, giving out **radiation**. Many elements have radioactive isotopes.

Radioactive isotopes, or **radioisotopes,** decay randomly. We can't say whether a given atom of carbon-14 will decay in the next few seconds, or in a thousand years. But we do know how long it takes for half the radioisotopes in a sample to decay. This is called the **half-life**.

The half-life for carbon-14 is 5730 years. So if you have 100 carbon-14 atoms, 50 of them will have decayed 5730 years from now.

This helps us work out the age of ancient fragments of wood, bones, and cloth. All contain some carbon-14, so we can date them by measuring the very faint radiation they give out. This is called **carbon dating**.

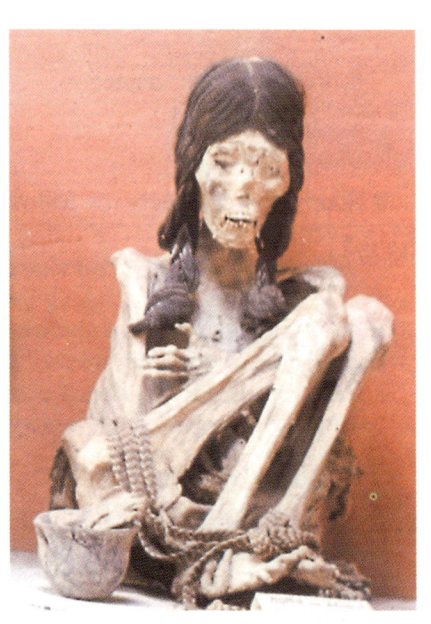

Carbon dating showed that this mummy from Chile was about 1100 years old.

Isotopes and relative atomic mass

As you saw, some atoms of an element are heavier than others.
The relative atomic mass of an element is the *average* mass of its atoms.
It is written in short form as A_r.

For example bromine has two natural isotopes:

Name	Symbol	Protons	Neutrons	Mass number	% of bromine atoms like this
Bromine-79	$^{79}_{35}Br$	35	(79 − 35) = 44	79	50%
Bromine-81	$^{81}_{35}Br$	35	(81 − 35) = 46	81	50%

Half the bromine atoms in the world have 44 neutrons and a mass number of 79. The other half have 46 neutrons and a mass number of 81. So it is easy to see that the average mass number for bromine is **80**. (It is unusual to have equal percentages of isotopes like this.)

How to work out relative atomic mass

You can use this formula to work out relative atomic mass:

relative atomic mass of an element =
  (% × mass number for the first isotope)
+ **(% × mass number for the second isotope)**
  and so on, for all its natural isotopes

For example chlorine also has two natural isotopes:

Name	Symbol	Protons	Neutrons	Mass number	% of chlorine atoms like this
Chlorine-35	$^{35}_{17}Cl$	17	18	35	75%
Chlorine-37	$^{37}_{17}Cl$	17	20	37	25%

Using the formula above, the relative atomic mass of chlorine

$= \frac{75}{100} \times 35 + \frac{25}{100} \times 37$ (Convert the percentage to a fraction by dividing it by 100!)

$= 26.25 + 9.25$

$= 35.5$

The relative atomic mass or A_r for chlorine is 35.5.

Questions

1 What is an isotope? Name the isotopes of hydrogen and write symbols for them.
2 How many isotopes does carbon have?
3 Carbon-14 is *radioactive*. What does that mean?
4 The radiation from radioactive material can be dangerous. But we take in some carbon-14 when we breathe. Suggest a reason why it does not harm us.
5 a What is the isotopic composition of bromine?
  b What is unusual about this?
6 Copper has two isotopes: $^{63}_{29}Cu$ and $^{65}_{29}Cu$.
  a How many neutrons does each isotope have?
  b 69% of copper atoms are copper-63. Calculate the relative atomic mass of copper.

8.04 How electrons are arranged

Electron shells

The electrons in an atom circle very fast around the nucleus, at different **energy levels** from it. These energy levels are called **electron shells**.

The first shell, closest to the nucleus, is the lowest energy level. The further a shell is from the nucleus, the higher the energy level.

Electrons occupy the lowest available energy level. But they can't all crowd into the first shell, because a shell can hold only a limited number of electrons, like this:

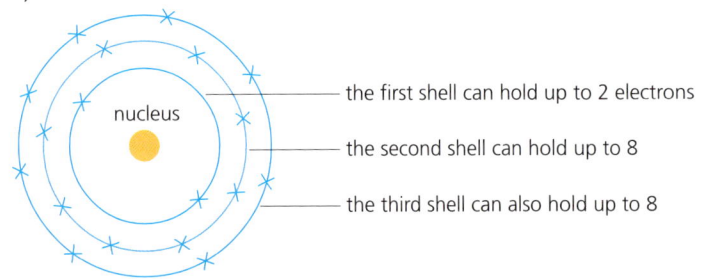

- the first shell can hold up to 2 electrons
- the second shell can hold up to 8
- the third shell can also hold up to 8

Niels Bohr, a Danish scientist, was the first person to put forward the idea of electron shells. He died in 1962.

So the electrons fill up the shells one by one, starting with the first shell. When a shell is full they start a new one.

The electron shells for the first 20 elements

This is how the shells fill up for the first 20 elements of the periodic table:

Element	Symbol	Atomic number	1st shell	2nd shell	3rd shell	4th shell
Hydrogen	H	1	1			
Helium	**He**	**2**	**2**			
Lithium	Li	3	2	1		
Beryllium	Be	4	2	2		
Boron	B	5	2	3		
Carbon	C	6	2	4		
Nitrogen	N	7	2	5		
Oxygen	O	8	2	6		
Fluorine	F	9	2	7		
Neon	**Ne**	**10**	**2**	**8**		
Sodium	Na	11	2	8	1	
Magnesium	Mg	12	2	8	2	
Aluminium	Al	13	2	8	3	
Silicon	Si	14	2	8	4	
Phosphorus	P	15	2	8	5	
Sulphur	S	16	2	8	6	
Chlorine	Cl	17	2	8	7	
Argon	**Ar**	**18**	**2**	**8**	**8**	
Potassium	K	19	2	8	8	1
Calcium	Ca	20	2	8	8	2

Only three of the elements, helium, neon, and argon, have atoms with full outer shells. As you'll see later, this makes them unreactive.

Electronic configuration and the periodic table

Electronic configuration means the arrangement of electrons in an atom. The electronic configurations for the first 20 elements are shown below, this time in rows and columns, to match the periodic table:

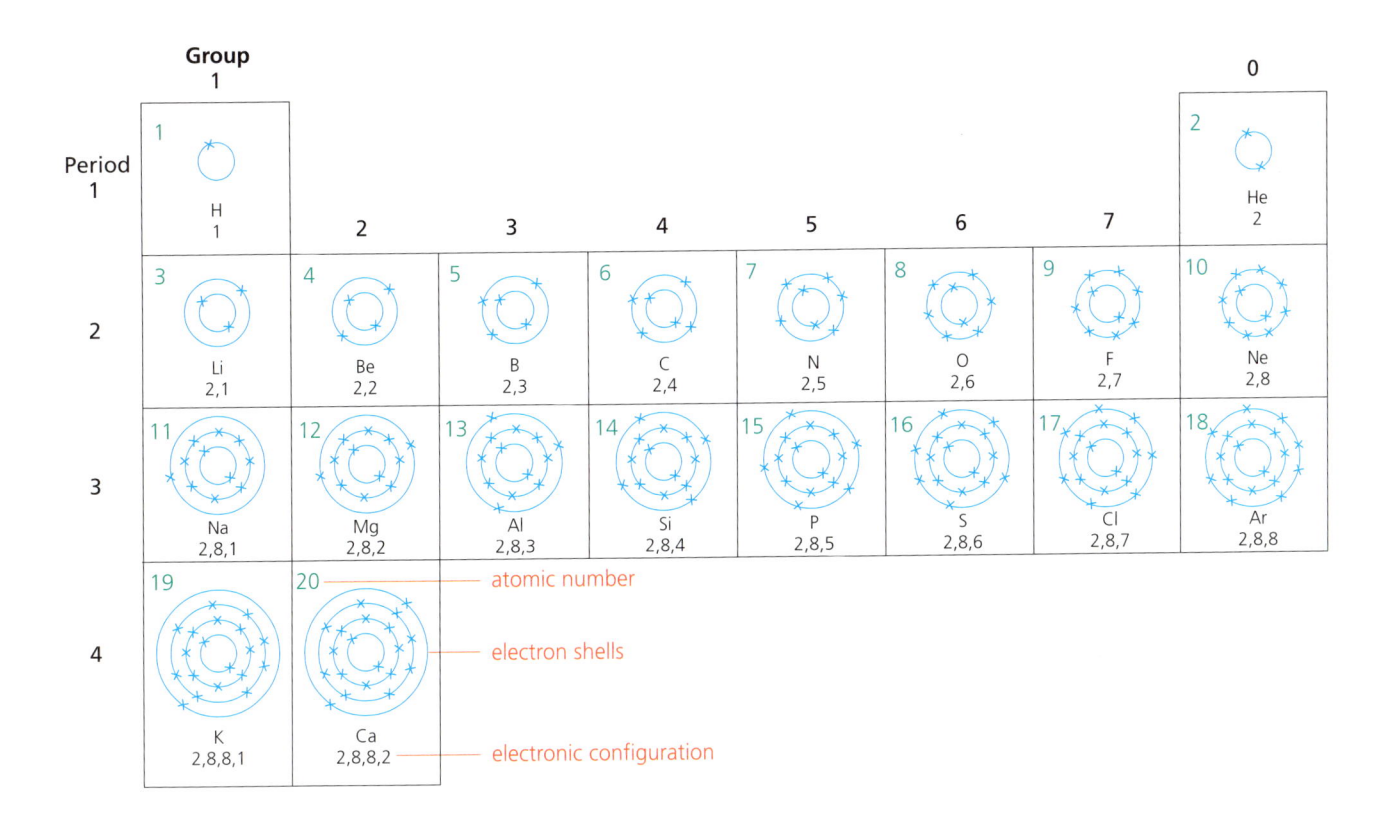

Period \ Group	1	2	3	4	5	6	7	0
1	1 H 1							2 He 2
2	3 Li 2,1	4 Be 2,2	5 B 2,3	6 C 2,4	7 N 2,5	8 O 2,6	9 F 2,7	10 Ne 2,8
3	11 Na 2,8,1	12 Mg 2,8,2	13 Al 2,8,3	14 Si 2,8,4	15 P 2,8,5	16 S 2,8,6	17 Cl 2,8,7	18 Ar 2,8,8
4	19 K 2,8,8,1	20 Ca 2,8,8,2						

- atomic number
- electron shells
- electronic configuration

Note these important points:

- The shells fill in order, starting with the inner one (lowest energy level).
- All the elements in a group have the same number of electrons in their outer shells.
- The group number is the same as the number of outer electrons.
- The elements in Group 0 (the noble gases) have full shells.
- The period number shows how many shells of electrons there are.

Questions

1 An atom has 13 electrons. Draw a diagram to show how its electrons are arranged. Which element is it?
2 What does *electronic configuration* mean?
3 You can show the electronic configuration for sodium as 2,8,1. Do the same for:
   a lithium   b magnesium   c hydrogen
4 An atom has 5 electrons in its outer shell. To which group of the periodic table does it belong?
5 How many shells of electrons are there in the atoms of the elements of Period 3?
6 Which of the 20 elements have full outer shells of electrons? Write the electronic configuration for each of them.

175

8.05 How ideas of the atom developed

That big idea again

All chemistry depends on one big idea: that everything is made of atoms. So when you make something in the school lab, you are really making billions of tiny atoms react together. Amazing!
But how did chemists find out about atoms? It's a long story.

It began with the Ancient Greeks

The Greek philosophers thought hard about the world around them. Is water continuous matter, or lots of separate bits? Is air just empty space? If you crush a stone to dust, and then crush the dust, will you end up with bits that you can't break up further?

Around 450 BC, the philosopher Democritus came up with an answer: everything is made of tiny particles that can't be divided. He called them **atoms**, which means *indivisible*. He guessed they came in four colours: white, black, red, and green. He suggested different shapes and sizes: large round atoms that taste sweet, and small sharp ones that taste sour. White atoms are smooth, and black ones jagged.

Everything in the world is made up of these atoms mixed in different quantities.

Other philosophers thought this was nonsense. Aristotle (384–270 BC) believed that everything was made of four elements – earth, air, fire, and water – mixed in different amounts. A stone has a lot of earth but not much water. No matter how small you break it up, each tiny bit will still have the properties of stone.

The Greek philosopher Aristotle (384–270 BC). A lot of thinking – but no experiments!

On to the alchemists

The Greek philosophers did a lot of heavy thinking – but no experiments. The **alchemists** (AD 1400–1650) were different. They experimented day and night, mixing this with that, in the hopes of finding the **elixir of life** (to keep us young) and of making **gold** from cheaper metals.

The alchemists believed that everything was made of three elements: mercury, sulphur, and salt. You could turn a 'base' metal like lead into gold by adding just the right amount of mercury.

Not surprisingly, they did not suceed in making gold. But they made many substances look like gold, by using secret recipes to deposit compounds on their surfaces.

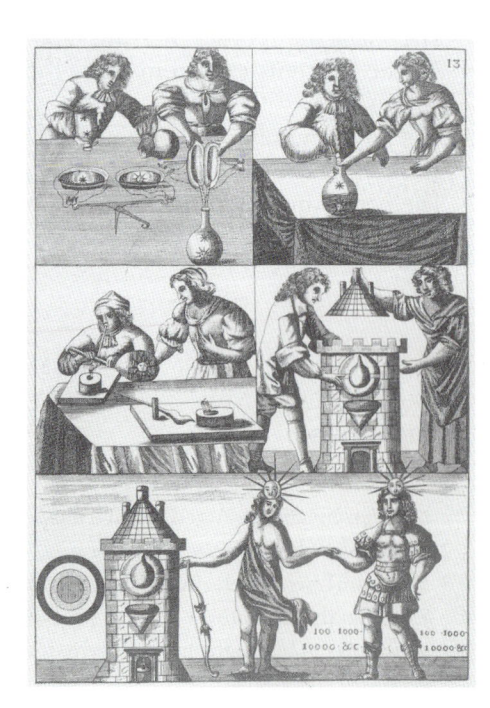

Alchemists at work. They perfected techniques we still use today, such as distillation and crystallization. They used many symbols from astrology and myths in their recipes.

176

Make way for us chemists

In the end, the alchemists got a reputation as cheats and liars, who swindled 'grants' from rich men with the promise of gold. They eventually gave way to a new breed of honest **chemists**.

By now the idea of atoms was almost forgotten. But in 1661 the scientist Robert Boyle showed that a gas can be compressed into a smaller space. He concluded that gas is made of particles with empty space between them. (He also came up with the definition of an element that still works today: **an element is any substance that can't be broken down into simpler substances.**)

In 1799, over 130 years later, the French chemist Joseph Louis Proust showed that copper(II) carbonate always contained the same proportions by weight of copper, carbon, and oxygen, no matter how you made it: 5 parts of copper to 2 of carbon to 1 of oxygen. This suggested that copper, carbon, and oxygen were made of particles, and these always combined in the same ratios.

Dalton's dilemma

The English chemist John Dalton puzzled over these discoveries. In 1803 he concluded: if elements *really were made of indivisible particles* then everything made sense. He called the particles **atoms**, as a tribute to the Greeks. Atoms of one element could combine with atoms of another element in a fixed ratio.

This time the idea of atoms caught on really fast, because it fitted in with the results from so many experiments.

Jiggling pollen grains

No one could *prove* that matter was made of separate particles, since they were too small to see. But in 1827 a Scottish botanist called Robert Brown did some clever detective work. He was studying pollen grains in water under a microscope. They jiggled around. He thought they were 'alive'. But then he tried dust. It did the same! He deduced that the pollen grains were being struck and pushed about by water particles. So tiny separate particles really did exist. They were not just theory.

And then …

In 1955 Erwin Müller, an American, developed a machine called a **field-ion microscope**. It could 'picture' the tip of a needle, magnified 5 million times! The atoms in the needle showed up as little dots.

Today the **scanning electron microscope** gives us great images of individual atoms. Meanwhile of course, for many decades, scientists have wondered what is *inside* atoms. And that's another story.

Robert Boyle (1627–91). He was born in Ireland but did most of his work in England. He put forward Boyle's Law for gases. And yes, it is a wig.

Dust particles in a shaft of sunlight jiggle around because they are struck by air particles.

Questions

1 Using the ideas of Democritus, describe the atoms you'd expect to find in a can of cola.
2 What did Democritus get right about atoms?
3 How many years elapsed between the statements of Democritus and Dalton about the atom?
4 Unlike Democritus, Dalton found that his ideas about the atom were readily accepted. Why?
5 The discovery that pollen grains jiggled in water was an important step in science. Why?
6 Dust particles jiggle about in air too. Why?

8.06 The atom: the inside story

Bring on the physicists

200 years ago, chemists had finally begun to accept that everything was indeed made of tiny indivisible particles: atoms. Today we know that atoms are not indivisible. In the last 100 years or so we have learned a great deal about the particles inside them, thanks mainly to the physicists.

First, the electrons

In 1897, over 100 years ago, the English physicist J.J. Thomson was investigating **cathode rays**: glowing mystery rays that appeared in an empty glass tube plugged into an electric circuit.

Thomson deduced that these rays were in fact streams of charged particles, *much smaller than atoms* – in fact they were bits from atoms. He called them *corpuscles* but soon the name got changed to **electrons**.

It was a shock to find that atoms were not the smallest particle after all! Thomson suggested that the electrons were stuck on the atoms like raisins on a bun. The rest of the atom (the bun) had a positive charge.

More strange rays

A year earlier, a French physicist called Becquerel had noticed something very odd. Some photographic plates (like our photo film) left in a drawer had gone foggy – struck by strange rays. These could only have come from the sample of uranium that was also in the drawer. He had discovered **radioactivity**.

Uranium gives out **alpha particles**. (We know now that these are clusters of two protons and two neutrons.) They turned out to be very useful. They have a positive charge and are over 7000 times heavier than electrons. You can speed them up and shoot them like tiny bullets.

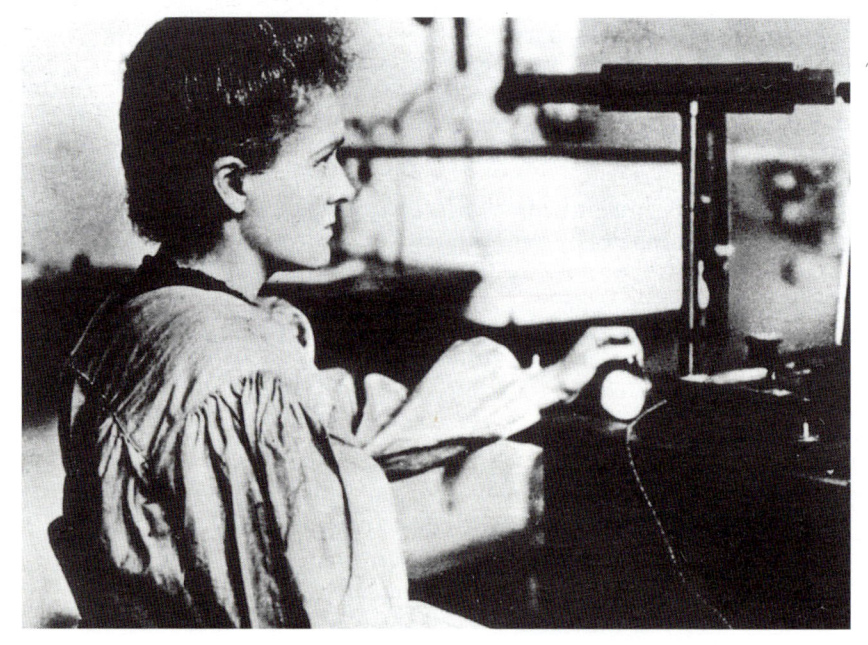

Dalton's atom

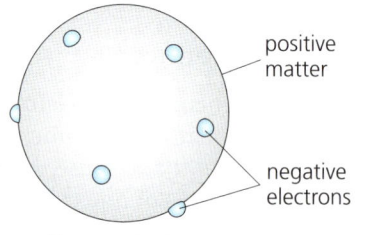

Thomson's atom (positive matter, negative electrons)

The Polish scientist Marie Curie (1867–1934) heard about Becquerel's discovery. She began to look for other radioactive substances – and soon discovered the elements polonium and radium.

Marie Curie spent much of her life studying ways to use radioactivity in medicine. Sadly she herself died from leukaemia caused by exposure to radium.

The nucleus and protons

In 1911, in England, the physicist Ernest Rutherford was experimenting with alpha particles. He shot a stream of them at some gold foil. Most of the particles went right through it. But some bounced back!

Rutherford deduced that an atom is mostly empty space, which the alpha particles pass through unobstructed. But there is something small and dense at the centre that can bounce them back. He had discovered the **nucleus**. He assumed it was made up of particles of positive charge, and called them **protons**.

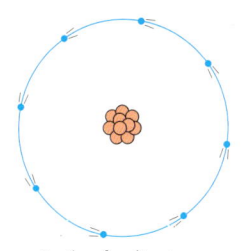

Those electron shells

If the nucleus is positive, why don't the electrons rush straight into it? In 1913 Niels Bohr came up with the theory of 'electron shells'. It fitted all the experiments.

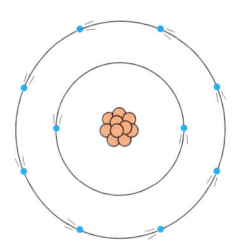

At last, the neutrons

In 1930, two German physicists, Bothe and Becker, shot alpha particles at beryllium – and knocked a stream of new particles from it. In 1932 the English physicist James Chadwick found that these particles had the same mass as protons, but no charge. He named them **neutrons**.

So finally, 129 years after Dalton proposed the atom, the chemist's model of it was complete.

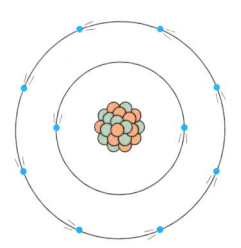

Rutherford's atom

Bohr's atom

Chadwick's atom

The whole truth

But now for the whole truth – are you ready? The model of the atom that you use works well for chemists. It explains how elements behave. But it is only a **model** – a simplified picture.

In fact atoms are far more complex than our model. Physicists have discovered around 50 different **elementary particles** within atoms. They include the **up** and **anti-up**, the **charm** and **anti-charm**, and the **strange** and **anti-strange**. There may be more to discover.

It turns out that those tiny atoms, far far too small to see, are each a throbbing universe, with bullets of energy zapping to and fro. And they are what you're made of. It's enough to make you want to lie down.

Sir James Chadwick, who gave the neutron its name.

Questions

1 Dalton put forward his ideas about atoms in 1803. How many years later were electrons discovered?
2 Often discoveries in science are not just the work of one person. They build on work by many others. Who is credited with the discovery of:
   a the electron?   **b** the nucleus?   **c** the neutron?
3 Suggest a reason why discoveries of atomic particles (particles inside the atom) were made by physicists rather than chemists.
4 **a** What you learn in chemistry about atoms is a simplified model. Explain.
   b Why do chemists continue to use this model?

179

8.07 Why compounds form

Most elements form compounds

When sodium is heated and placed in a jar of chlorine, it burns with a bright flame.

The result is a white solid, which has to be scraped from the sides of the jar.

The white solid is called **sodium chloride**. It is formed by atoms of sodium and chlorine joining together, so it is a **compound**. The reaction can be described like this:

    sodium + chlorine ⟶ sodium chloride

The + means *reacts with*, and the ⟶ means *to form*.
Most elements react to form compounds. For example:

    lithium + chlorine ⟶ lithium chloride
    hydrogen + chlorine ⟶ hydrogen chloride

The noble gases do not usually form compounds

The noble gases are different from other elements, in that they do not usually form compounds. So their atoms are described as **unreactive** or **stable**. They are stable because their outer electron shells are *full*:

A full outer shell makes an atom stable.

Helium atom, full outer shell: *stable*

Neon atom, full outer shell: *stable*

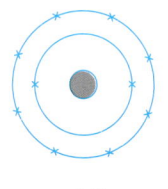

Argon atom, full outer shell: *stable*

2

2,8

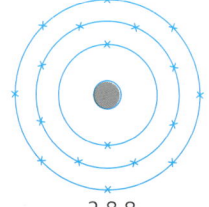

2,8,8

Only the noble gas atoms have full outer shells. The atoms of all other elements have incomplete outer shells. That is why they react.
By reacting with each other, atoms can obtain full outer shells and so become stable.

Several of the noble gases are used in lighting. For example, xenon is used in lighthouse lamps, like this one. It gives a beautiful blue light.

180

How sodium atoms get a full outer shell

The sodium atom has just 1 electron in its outer shell. It can obtain a full outer shell by losing this electron to another atom. It becomes a **sodium ion**:

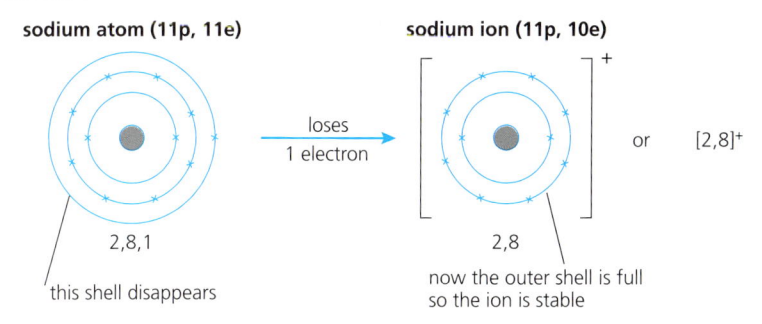

sodium atom (11p, 11e) → loses 1 electron → sodium ion (11p, 10e)

2,8,1 (this shell disappears) → 2,8 (now the outer shell is full so the ion is stable) or $[2,8]^+$

The sodium ion has 11 protons but only 10 electrons, so it has a charge of +1, as you can see from the box on the right.
The symbol for sodium is Na, so the sodium ion is called **Na⁺**.
The + means *1 positive charge*. Na⁺ is a **positive ion**.

The charge on a sodium ion:	
charge on 11 protons	11+
charge on 10 electrons	10–
total charge	1+

How chlorine atoms get a full outer shell

A chlorine atom has 7 electrons in its outer shell. It can reach a full shell by accepting just 1 electron from another atom. It becomes a **chloride ion**:

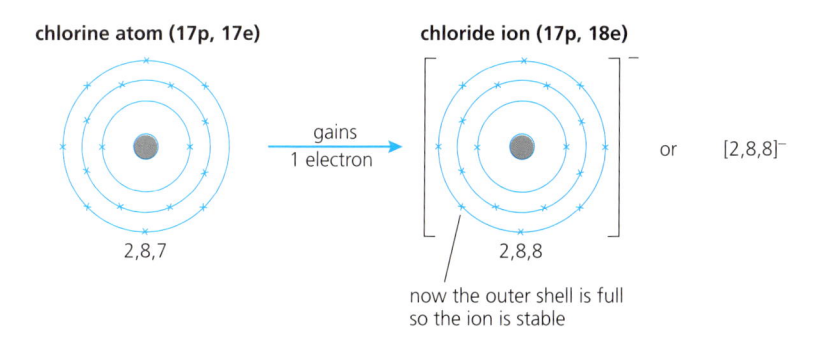

chlorine atom (17p, 17e) → gains 1 electron → chloride ion (17p, 18e)

2,8,7 → 2,8,8 (now the outer shell is full so the ion is stable) or $[2,8,8]^-$

The chloride ion has a charge of 1–, so it is a **negative ion**.
Its symbol is **Cl⁻**.

The charge on a chloride ion:	
charge on 17 protons	17+
charge on 18 electrons	18–
total charge	1–

Ions

Any atom becomes an ion if it loses or gains electrons.
An ion is a charged particle. It is charged because it contains an unequal number of protons and electrons.

Questions

1. What is another word for *unreactive*?
2. Why are the noble gas atoms unreactive?
3. Explain why all other atoms are reactive.
4. Draw a diagram to show how a sodium atom obtains a full outer shell.
5. Explain why the sodium ion has a charge of 1+.
6. Draw a diagram to show how a chlorine atom can obtain a full outer shell.
7. Write down the symbol for a chloride ion. Why does this ion have a charge of 1–?
8. Explain what an ion is, in your own words.
9. Why do noble gas atoms *not* form ions?

8.08 The ionic bond

How sodium and chlorine atoms bond

A sodium atom can lose one electron, and a chlorine atom can gain one, to obtain full outer shells. So when a sodium atom and a chlorine atom react together, the sodium atom loses its electron *to the chlorine atom*, and two ions are formed. Here, sodium electrons are shown as ● and chlorine electrons as ×, but remember that all electrons are exactly the same:

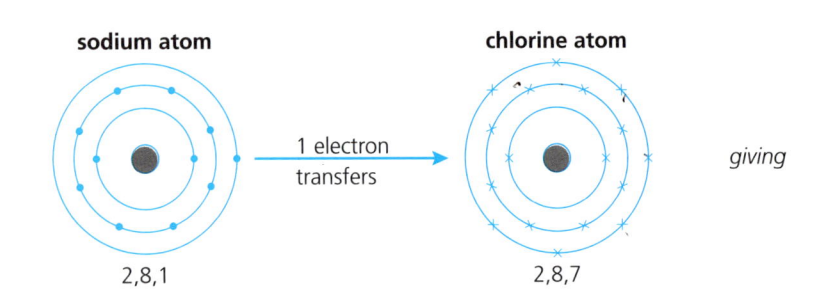

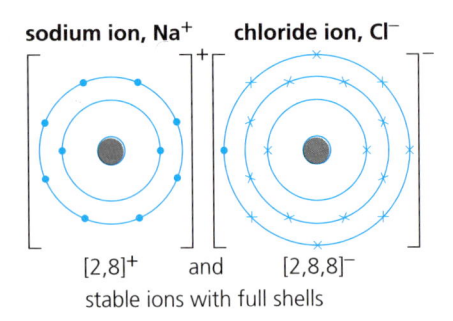

sodium atom — 1 electron transfers → chlorine atom — *giving* → sodium ion, Na^+ and chloride ion, Cl^-

2,8,1   2,8,7 $[2,8]^+$ and $[2,8,8]^-$
                              stable ions with full shells

The two ions have opposite charges so they attract each other. The force of attraction between them is strong. It is called an **ionic bond**.

How solid sodium chloride is formed

When sodium reacts with chlorine, billions of sodium and chloride ions form, and are attracted to each other. But the ions do not stay in pairs. They cluster together so that each ion is surrounded by six ions of opposite charge. They are held together by strong ionic bonds.

The pattern grows until a giant structure of ions is formed. It is called 'giant' because it may contain many billions of ions. This giant structure is the compound **sodium chloride**, or **common salt**.

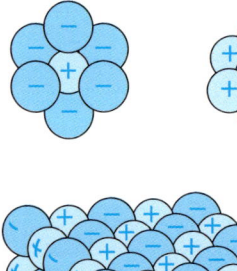

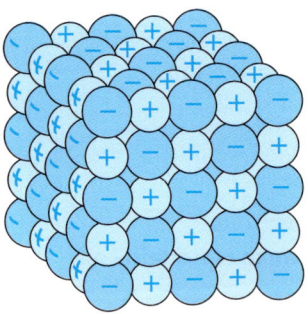

Because sodium chloride is made of ions, it is called an **ionic compound**. It contains one Na^+ ion for each Cl^- ion, so its formula is **NaCl**. The charges in the structure add up to zero:

the charge on each sodium ion is	1+
the charge on each chloride ion is	1−
total charge	0

The compound therefore has no overall charge.

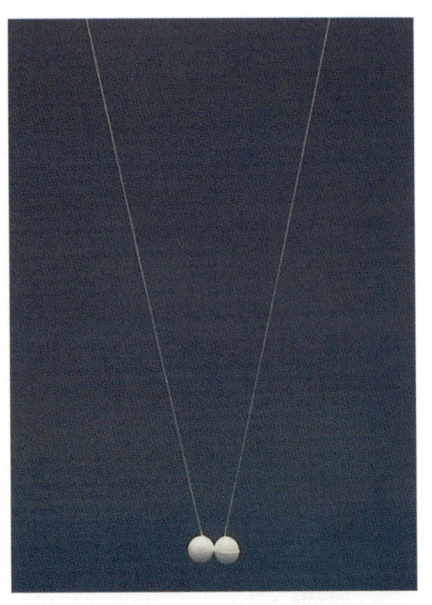

These polystyrene spheres have been given opposite charges, so they are attracted to each other. The same happens with ions of opposite charge.

Other ionic compounds

Sodium is a **metal**, and chlorine is a **non-metal**. They react together to form an **ionic compound**. Other metals can also react with non-metals to form ionic compounds. Below are two more examples.

Magnesium and oxygen A magnesium atom has 2 outer electrons and an oxygen atom has 6. Magnesium burns fiercely in oxygen. During the reaction, each magnesium atom loses its 2 outer electrons to an oxygen atom. Magnesium ions and oxide ions are formed:

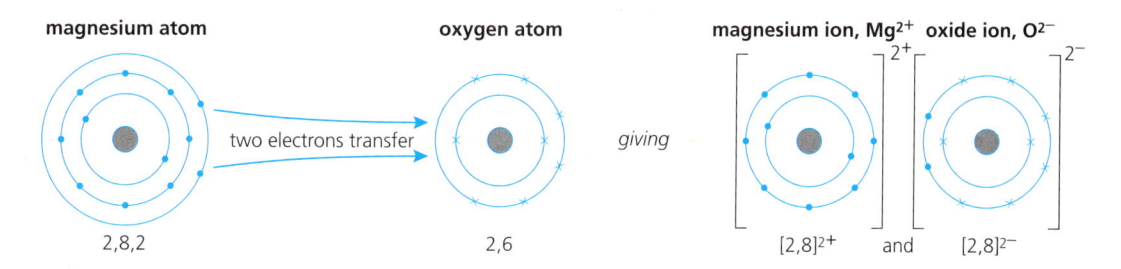

magnesium atom → two electrons transfer → oxygen atom → giving → magnesium ion, Mg^{2+} and oxide ion, O^{2-}

2,8,2   2,6 $[2,8]^{2+}$ and $[2,8]^{2-}$

The ions attract each other because of their opposite charges. Like the ions on the last page, they group together into a giant ionic structure.

The resulting compound is called **magnesium oxide**. Magnesium oxide contains one magnesium ion for each oxide ion, so its formula is **MgO**. The compound has no overall charge.

The charge on magnesium oxide:
charge on each magnesium ion   2+
charge on each oxide ion   2−
                       total charge   0

Magnesium and chlorine To obtain full outer shells a magnesium atom must lose 2 electrons, and a chlorine atom must gain 1 electron. So when magnesium burns in chlorine, each magnesium atom reacts with *two* chlorine atoms, to form **magnesium chloride**:

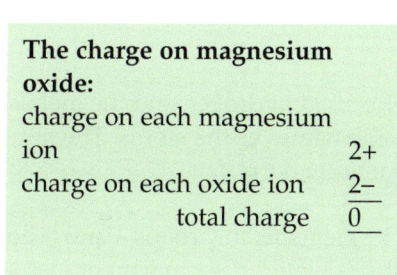

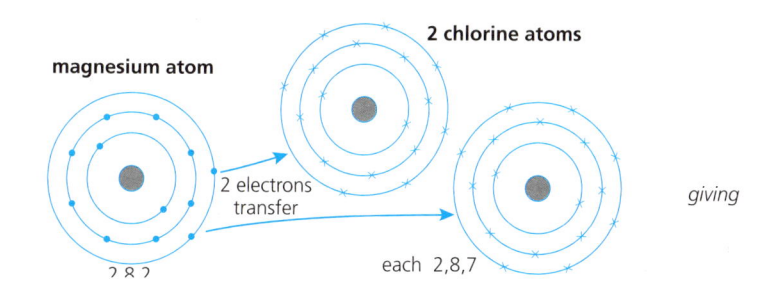

magnesium atom + 2 chlorine atoms → 2 electrons transfer → giving → magnesium ion, Mg^{2+} and two chloride ions, Cl^-

2,8,2   each 2,8,7 $[2,8]^{2+}$ each $[2,8]^-$

The ions form a giant ionic structure, with *two* chloride ions for each magnesium ion. The formula of magnesium chloride is therefore **MgCl$_2$**. The compound has no overall charge. Can you explain why?

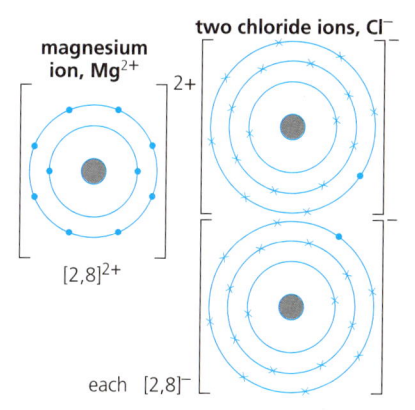

Questions

1. Draw a diagram to show what happens when a sodium atom reacts with a chlorine atom.
2. What is an ionic bond?
3. Sketch the structure of sodium chloride, and explain why its formula is NaCl.
4. Explain why:
    a. a magnesium ion has a charge of 2+
    b. the ions in magnesium oxide stay together
    c. magnesium chloride has no overall charge
    d. the formula of magnesium chloride is MgCl$_2$

183

8.09 Some other ions

Ions of the first twenty elements

Not every element forms ions during reactions. In fact, out of the first twenty elements in the periodic table, only twelve easily form ions. These ions are given below, with their names.

Group 1	2		3	4	5	6	7	0
		H^+ hydrogen						none
Li^+ lithium	Be^{2+} beryllium					O^{2-} oxide	F^- fluoride	none
Na^+ sodium	Mg^{2+} magnesium		Al^{3+} aluminium			S^{2-} sulphide	Cl^- chloride	none
K^+ potassium	Ca^{2+} calcium		transition metals					

Note that hydrogen and the metals form **positive ions**, which have the same names as the atoms. The non-metals form **negative ions** and their names end in *-ide*.

The elements in Groups 4 and 5 do not usually form ions, because their atoms would have to gain or lose several electrons, and that takes too much energy. The elements in Group 0 do not form ions because their atoms already have full shells.

The names and formulae of their compounds

The names   To name an ionic compound, you just put the names of the ions together, with the positive one first:

Ions in compound	Name of compound
K^+ and F^-	Potassium fluoride
Ca^{2+} and S^{2-}	Calcium sulphide

Bath salts contain Na^+ and CO_3^{2-} ions; Epsom salts contain Mg^{2+} and SO_4^{2-} ions. Can you give the names and formulae of the three main compounds illustrated here?

The formulae   The formulae of ionic compounds can be worked out by the following steps. Two examples are given below.

1 Write down the name of the ionic compound.
2 Write down the symbols for its ions.
3 The compound must have no overall charge, so **balance** the ions, until the positive and negative charges add up to zero.
4 Write down the formula without the charges.

Example 1
1 Lithium fluoride.
2 The ions are Li^+ and F^-.
3 One Li^+ is needed for every F^-, to make the total charge zero.
4 The formula is LiF.

Example 2
1 Sodium sulphide.
2 The ions are Na^+ and S^{2-}.
3 Two Na^+ ions are needed for every S^{2-} ion, to make the total charge zero: $Na^+ Na^+ S^{2-}$.
4 The formula is Na_2S. (What does the ₂ show?)

Transition metal ions

Some transition metals form only one type of ion:

- silver forms only Ag^+ ions
- zinc forms only Zn^{2+} ions

but most of them can form more than one type. For example, copper and iron can each form two:

Ion	Name	Example of compound
Cu^+	copper(I) ion	copper(I) oxide, Cu_2O
Cu^{2+}	copper(II) ion	copper(II) oxide, CuO
Fe^{2+}	iron(II) ion	iron(II) chloride, $FeCl_2$
Fe^{3+}	iron(III) ion	iron(III) chloride, $FeCl_3$

The (II) in a name shows that the ion has a charge of 2+. What do the (I) and (III) show?

Compound ions

So far, all the ions have been formed from single atoms. But ions can also be formed from groups of joined atoms. These are called **compound ions**.

The most common ones are shown on the right. Remember, each is just one ion, even though it contains more than one atom.

NH_4^+, the ammonium ion

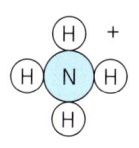

OH^-, the hydroxide ion

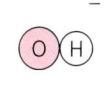

NO_3^-, the nitrate ion

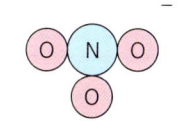

SO_4^{2-}, the sulphate ion

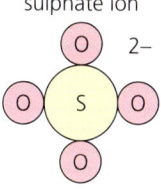

CO_3^{2-}, the carbonate ion

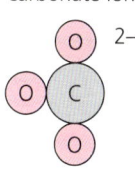

HCO_3^-, the hydrogencarbonate ion

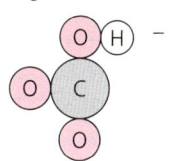

The formulae for their compounds can be worked out as before. Some examples are shown below.

Example 3
1 Sodium carbonate.
2 The ions are Na^+ and CO_3^{2-}.
3 Two Na^+ are needed to balance the charge on one CO_3^{2-}.
4 The formula is Na_2CO_3.

Example 4
1 Calcium nitrate.
2 The ions are Ca^{2+} and NO_3^-.
3 Two NO_3^- are needed to balance the charge on one Ca^{2+}.
4 The formula is $Ca(NO_3)_2$. Note that a bracket is put round the NO_3, before the $_2$ is put in.

Questions

1 Explain why a calcium ion has a charge of 2+.
2 Why is the charge on an aluminium ion 3+?
3 Write down the symbols for the ions in:
   a potassium chloride   b calcium sulphide
   c lithium sulphide   d magnesium fluoride
4 Now work out the formula for each compound in question 3.
5 Work out the formula for each compound:
   a copper(II) chloride   b iron(III) oxide
6 Write a name for each compound:
   $CuCl$, FeS, Na_2SO_4, $Mg(NO_3)_2$, NH_4NO_3, $Ca(HCO_3)_2$
7 Work out the formula for:
   a sodium sulphate   b potassium hydroxide
   c silver nitrate   d ammonium nitrate

8.10 Ionic compounds and their properties

Remember
Ionic compounds are formed when metals react with non-metals. Examples include sodium chloride (NaCl) and magnesium oxide (MgO).

The structure of ionic compounds
Sodium chloride is ordinary table salt. You already know quite a lot about its structure:

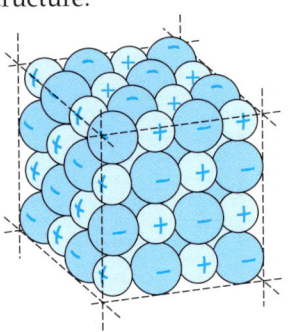

The sodium and chloride ions are packed in a regular arrangement or **lattice** like this. They are held together by strong ionic bonds.

The pattern repeats millions of times. The result is a crystal. This is a crystal of sodium chloride magnified 35 times.

The crystals look white and shiny. A box of table salt contains millions of them.

Sodium chloride is a typical ionic compound. In all ionic compounds the ions are held together in a lattice by strong ionic bonds, forming a **giant structure**. Since the ions are in a regular pattern, all ionic solids are crystalline.

Note that 'giant' does not mean the lattice is huge. (Most crystals don't grow that big.) It means it can contain many billions of particles.

Some ionic solids you can find in the laboratory:

sodium chloride
sodium hydroxide
copper(II) sulphate
magnesium oxide
silver nitrate

What happens when you heat an ionic compound?
It's hard to imagine salt melting – but it will if you heat it enough!

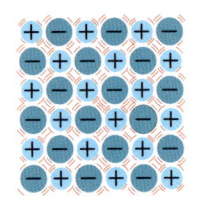

When you heat an ionic solid the ions take in energy and start to vibrate to and fro in the lattice. When the temperature reaches the **melting point** …

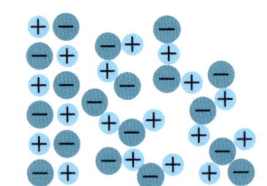

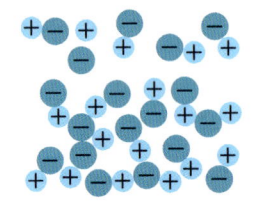

… they vibrate so much that they break away from the lattice. Now the substance is a **liquid**. The ions are free to move. If you keep on heating …

… some ions get enough energy to escape as a **gas**. This is called **evaporation**. When the temperature reaches the **boiling point** all the liquid boils into a gas.

If you cool sodium chloride gas the reverse happens. The ions lose more and more energy, and eventually settle back into a lattice.

It's not just ionic solids that behave like this. As you'll see, most solids are a lattice of particles and follow this pattern of melting and boiling.

Because of its high melting point, magnesium oxide is used to line furnaces.

The properties of ionic compounds

1 Ionic compounds have high melting and boiling points.
For example:

Compound	Melting point/°C	Boiling point/°C
Sodium chloride	801	1413
Magnesium oxide	2852	3600

This is because the ionic bonds are very strong, so it takes a lot of heat energy to break up the lattice. Room temperature is not nearly high enough to do this, so all ionic compounds are solid at room temperature.

Note that magnesium oxide has a far higher melting and boiling point than sodium chloride does. That's because its ions have double the charge (Mg^{2+} and O^{2-} compared with Na^+ and Cl^-) so its ionic bonds are stronger.

2 Ionic compounds are usually soluble in water.
The water molecules can attract the ions away from the lattice. The ions can now move freely, surrounded by water molecules.

3 Ionic compounds can conduct electricity when they are melted or dissolved.
A solid ionic compound will not conduct electricity. But when it melts the lattice breaks up and the ions are free to move. Since they are charged, this means they can conduct electricity.

The ions also move freely when the compound is dissolved in water. So the aqueous solution of an ionic compound conducts too.

The ionic solid sodium chloride is used to bring out the taste in food. Like all ionic solids it dissolves in moisture, and we taste the sodium ions.

Questions

1 What is a *lattice*?
2 Explain how a crystal of sodium chloride is formed.
3 A crystal of sodium chloride has flat faces. See if you can explain why.
4 Why do ionic solids have high melting points?
5 A solution of salt in water can conduct electricity. Why?
6 Why is the melting point of magnesium oxide higher than that of sodium chloride?

8.11 The covalent bond

Sharing electrons

When two non-metal atoms react together, *both of them need to gain electrons* to reach full shells. They can manage this only by sharing electrons between them. Atoms can share only their outer electrons, so just the outer electrons are shown in the diagrams below.

Hydrogen A hydrogen atom has only one electron. Its shell can hold two electrons, so is not full. When two hydrogen atoms get close enough, their shells overlap and then they can share electrons:

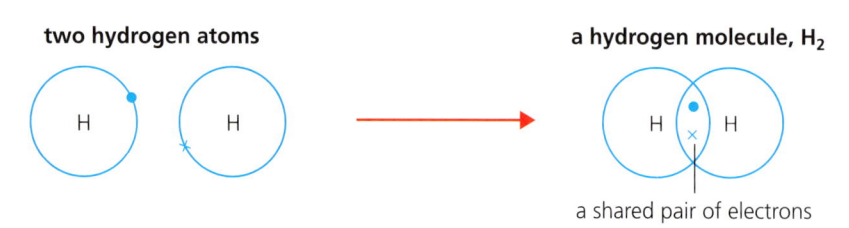

two hydrogen atoms → a hydrogen molecule, H_2

a shared pair of electrons

Because the atoms share electrons, there is a strong force of attraction between them, holding them together. This force is called a **covalent bond**. The bonded atoms form a **molecule**.

A molecule is a group of atoms which are held together by covalent bonds.

Hydrogen gas is made up of hydrogen molecules, and for this reason it is called a **molecular** substance. Its formula is H_2.
Several other non-metals are also molecular. For example:

chlorine, Cl_2 iodine, I_2 oxygen, O_2
nitrogen, N_2 sulphur, S_8 phosphorus, P_4

Chlorine A chlorine atom needs a share in one more electron, to obtain a full shell. So two chlorine atoms bond covalently like this:

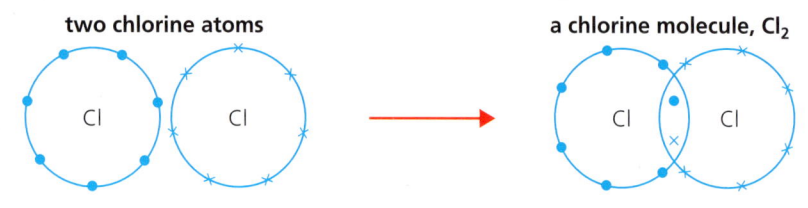

two chlorine atoms → a chlorine molecule, Cl_2

Oxygen The formula for oxygen is O_2, so each molecule must contain two atoms. Each oxygen atom has only six outer electrons; it needs a share in two more to reach a full shell:

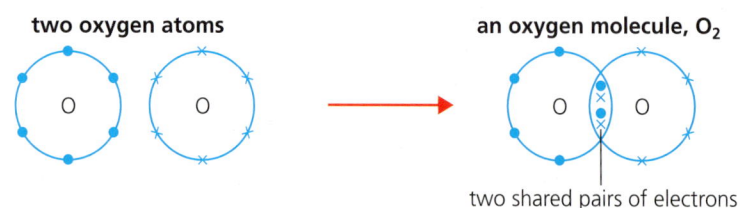

two oxygen atoms → an oxygen molecule, O_2

two shared pairs of electrons

Since the oxygen atoms share two pairs of electrons, the bond between them is called a **double covalent bond**, or just a **double bond**.

A model of the hydrogen molecule. The molecule can be shown as H–H. The line represents a single bond.

In H_2, the ₂ tells you the number of hydrogen atoms in each molecule.
Because it has two atoms in each molecule, hydrogen is described as **diatomic**.

A model of the chlorine molecule.

A model of the oxygen molecule. The molecule can be shown as O=O. The lines represent a double bond.

Covalent compounds

You have just seen that several non-metal elements exist as molecules. A huge number of compounds also exist as molecules. In a molecular compound, atoms of *different* elements share electrons with each other. These compounds are often called **covalent compounds** because of the covalent bonds in them. Water, ammonia, and methane are all covalent compounds.

Water   The formula of water is H_2O. In each molecule, an oxygen atom shares electrons with two hydrogen atoms, and they all reach full shells:

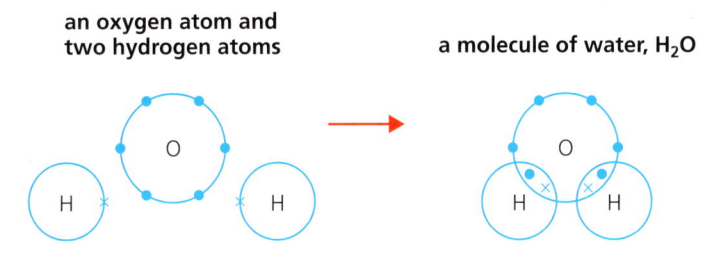

A model of the water molecule.

Ammonia   Its formula is NH_3. Each nitrogen atom shares electrons with three hydrogen atoms, and they all reach full shells:

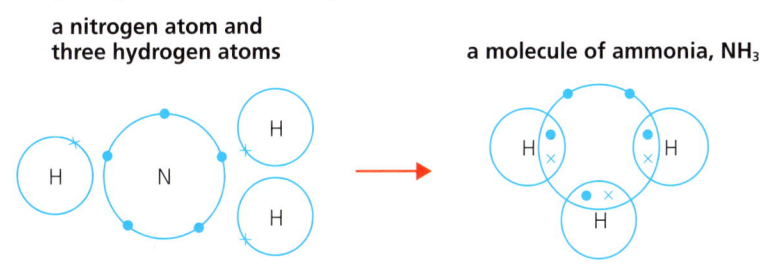

A model of the ammonia molecule.

Methane   Its formula is CH_4. Each carbon atom shares electrons with four hydrogen atoms, and they all obtain full shells:

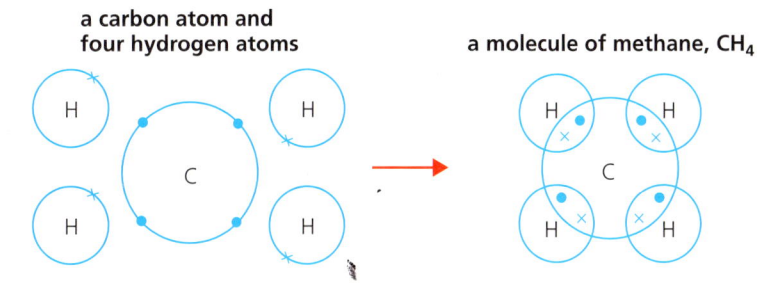

A model of the methane molecule.

Questions

1 What is the name of the bond between atoms that share electrons?
2 What is a molecule?
3 Give five examples of molecular elements.
4 Draw a diagram to show the bonding in:
   a chlorine   **b** oxygen
5 The bond between two oxygen atoms is called a double bond. Why?
6 Show the bonding in a molecule of:
   a water   **b** methane
7 Hydrogen chloride (HCl) is also molecular. Draw a diagram to show the bonding in it.

189

8.12 Covalent substances

Remember
Molecules are made of non-metal atoms joined by strong covalent bonds. Oxygen (O_2), water (H_2O), and ammonia (NH_3) exist as molecules.

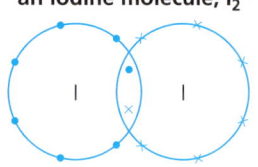

Molecular solids
Iodine is a good example of a molecular solid. Each iodine molecule contains two atoms held together by a strong covalent bond.

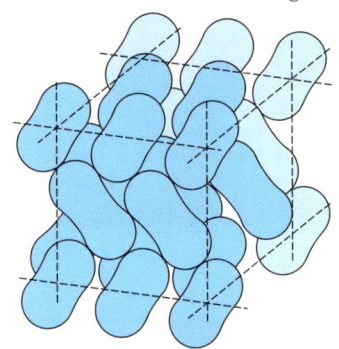

an iodine molecule, I_2

The iodine molecules are held in a **lattice** like this, in a regular pattern. But the forces between them are weak.

The pattern repeats millions of times. The result is a crystal. This is a single iodine crystal, magnified 15 times.

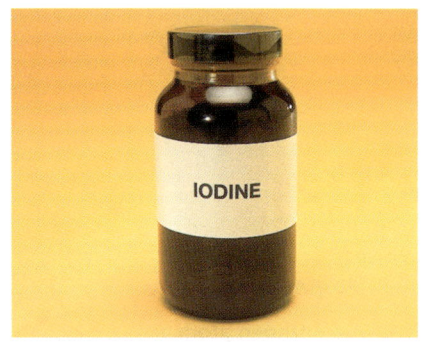

Iodine crystals are grey-black and shiny. This jar contains hundreds of thousands of them.

Iodine is typical of molecular solids. In all molecular solids the molecules are held in a regular pattern in a lattice. So the solids are crystalline. The forces *between* the molecules (the **intermolecular forces**) are weak.

Some properties of molecular solids
1 **Molecular solids have low melting and boiling points**. This is because the forces between the molecules are weak. In fact many molecular substances melt and even boil below room temperature, which is why they are liquids or gases at room temperature.
2 **Molecular solids do not conduct electricity.** Molecules are not charged, so molecular substances cannot conduct electricity even when melted.

Some molecular substances and their state at room temperature:

Solids   iodine
           sulphur

Liquids bromine
           water

Gases   nitrogen
           carbon dioxide

All have low melting points.

But not all covalent solids are molecular
All the substances in this table have covalent bonds. But just compare their melting points:

Substance	Melting point/°C
Hydrogen	–259
Oxygen	–219
Water	0
Silicon dioxide (silica)	1610
Carbon (as diamond)	>3550

Clearly diamond and silica are not molecular solids, with a weak lattice. Their melting points are far too high for that. In fact they form **giant covalent structures** rather than molecules.

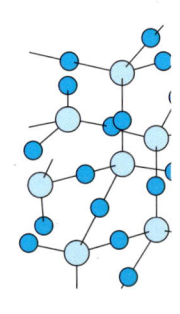

Silicon dioxide is made up of oxygen atoms and silicon atoms. Billions of them bond together like this to give a giant structure. So its melting point is high.

190

Diamond – a giant covalent structure

Diamond is made of carbon atoms held in a strong lattice.

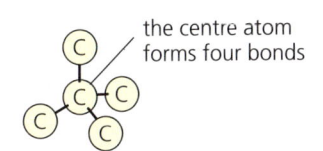

the centre atom forms four bonds

A carbon atom forms covalent bonds to *four* others, as shown above. Each outer atom then bonds to three more, and so on.

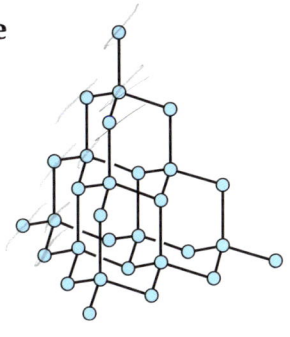

Eventually millions of carbon atoms are bonded together, in a giant covalent structure. This is part of it.

The result is a single crystal of diamond like this one. It has been cut, shaped, and polished to make it sparkle.

Diamond has these properties:
1 It is very hard, because each atom is held in place by four strong covalent bonds. In fact it's the hardest substance on Earth.
2 For the same reason it has a very high melting and boiling point.
3 It can't conduct electricity because there are no ions or free electrons to carry the charge.

Graphite – a different giant covalent structure

Graphite is made of carbon atoms too, but in a different giant structure.

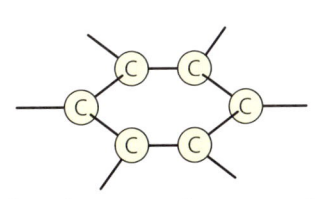

Each carbon atom forms covalent bonds to *three* others. This gives rings of six atoms.

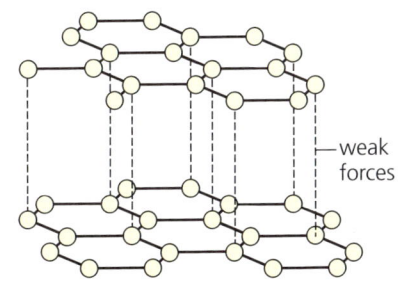

These join to make flat sheets that lie on top of each other, held together by weak forces.

—weak forces

Even at low magnification the layered structure of graphite shows up clearly.

Diamond and graphite are **allotropes** of carbon – two forms of the same element.

Graphite has these properties:
1 Unlike diamond it is soft and slippery. That's because the sheets can slide over each other easily.
2 Unlike diamond it is a good conductor of electricity. That's because each atom has four outer electrons but forms only three bonds. The fourth electron is free to move through a layer of the graphite, carrying charge.

Questions

1 A substance melts at 20 °C. What type of structure do you think it has? Give reasons.
2 Explain why many molecular substances are gases or liquids at room temperature, and give four examples.
3 Describe the structure and bonding in silicon dioxide. (It is similar to diamond.)
4 a Explain why silicon dioxide has a high melting point.
  b Suggest reasons why it is lower than diamond's.
5 You can buy solid air fresheners in shops. Are these substances giant structures, or molecular? Explain.
6 Explain why graphite:
  a conducts electricity   b is used as a lubricant

8.13 Metals: more giant structures

Clues from melting points

Compare these melting points:

Structure	Example	Melting point /°C
Molecular	Carbon dioxide	−56
	Water	0
Ionic	Sodium chloride	801
	Magnesium oxide	2852
Giant covalent	Diamond	>3550
	Silica	1610
Metal	Iron	1535
	Copper	1083

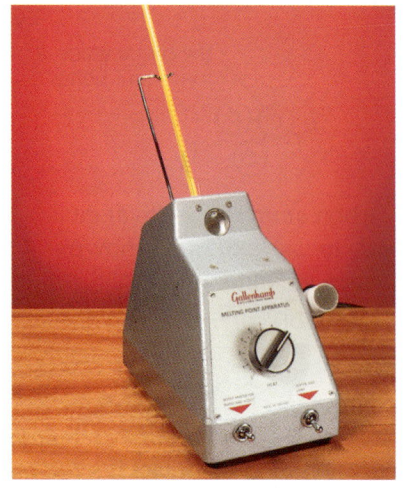

Equipment for measuring melting points in the school lab. It can heat substances up to 300 °C – so it's no good for sodium chloride!

The table shows clearly that:
- **molecular substances have low melting points.** That's because the forces between molecules in the lattice are weak.
- **giant structures such as sodium chloride and diamond have much higher melting points.** That's because the bonds between ions or atoms within giant structures are very strong.

Now look at the metals. They too have high melting points – not as high as for diamond, but much higher than for carbon dioxide or water. This gives us a clue that they too might be giant structures. And so they are, as you'll see below.

The structure of metals

In metals, the atoms are packed tightly together in a regular pattern. The tight packing allows outer electrons to separate from their atoms. The result is a lattice of ions in a sea of electrons. Look at copper:

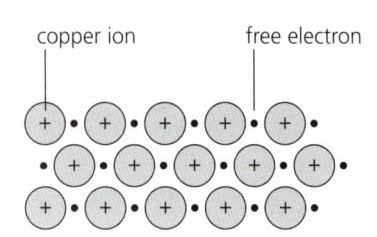

The copper ions are held together by their attraction to the electrons between them. The strong forces of attraction are called **metallic bonds**.

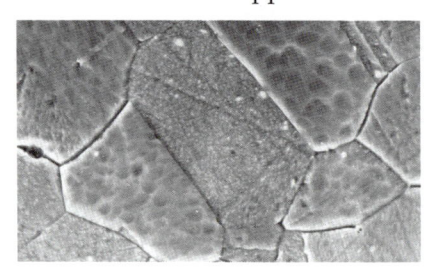

The regular arrangement of ions results in **crystals** of copper. This shows the crystals in a piece of copper magnified 1000 times.

The copper crystals are called **grains**. A lump of copper consists of millions of grains joined together. You need a microscope to see them.

It's the same with all metals. Because the ions are held in a regular lattice, metals are crystalline. (The crystals join at different angles as you can see above.) You might expect the positive metal ions to repel each other. But the free electrons hold them in the lattice with strong metallic bonds. That is why copper does not melt until 1083 °C.

Some properties of metals

1 Metals usually have high melting and boiling points.
That's because it takes a lot of heat energy to break up the lattice, with its strong metallic bonds. But there are some exceptions.
For example sodium melts at only 98°C. Mercury melts at −39°C so it is a liquid at room temperature.

2 Metals are malleable and ductile.
Malleable means they can be bent and pressed into shape. *Ductile* means they can be drawn out into wires. This is because the layers can slide over each other.

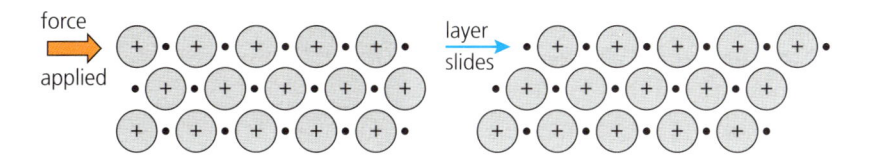

Metals: malleable, ductile, and sometimes very glamorous. This is a silver bracelet.

The layers can slide without the metallic bond breaking, because the electrons are free to move too.

3 Metals are good conductors of heat.
That's because the free electrons take in heat which makes them move faster. They quickly transfer the heat energy through the metal structure.

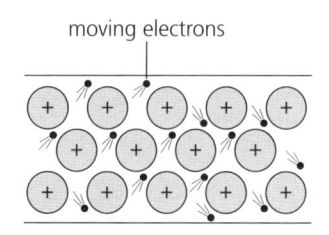

4 Metals are good conductors of electricity.
That's because the free electrons can move through the lattice carrying charge, when a voltage is applied across the metal structure.
Silver is the best conductor of all the metals. Copper is next – but used much more than silver since it is cheaper.

Questions

1 Describe in words the structure of a metal.
2 What is a metallic bond?
3 What does *malleable* mean?
4 Explain why metals can be drawn out into wires without breaking.
5 a Explain why metals can conduct electricity.
  b Would you expect molten metals to conduct? Give reasons for your answer.
6 We make use of their malleability to turn some metals into saucepans. Give two other examples of uses of metals that depend on:
   a their malleability   b their ductility
   c their ability to conduct electricity
7 Mercury forms ions with a charge of 2+. It hardens (freezes) at −39°C. Try drawing a diagram to show the structure of solid mercury.

193

8.14 How the periodic table developed

Life before the periodic table

Imagine you find a box of jigsaw pieces. You really want to build that jigsaw – but there is no picture on the box to help you, and many pieces are missing, and the image on some pieces is not complete. How frustrating!

That's how many chemists felt, around 150 years ago. They had found more and more new elements. For example 24 metals including lithium, sodium, potassium, calcium, and magnesium had been discovered in the 44 years from 1801 to 1844. Chemists could tell these fitted a pattern of some kind – they could see fragments of the pattern. But they could not work out what the overall pattern was.

And then the periodic table was published in 1869, and everything began to make sense.

A really clever summary

The periodic table is *the* summary of chemistry. It names the ingredients that make up our world – the elements. It shows the families they fit in, and how these relate to each other. It even tells you a great deal about their atoms – like the numbers of protons, electrons, and electron shells.

Today we take the periodic table for granted. But it took hundreds of years and the hard work of hundreds of chemists to develop. There were some good tries along the way, like these two 'laws'.

The Law of Triads

In 1817 the German scientist Johann Döbereiner noticed that calcium, strontium, and barium had similar properties, and that the atomic weight of strontium was halfway between the other two. (Instead of *atomic weight* we now use *relative atomic mass*.) He found the same pattern with chlorine, bromine, and iodine, and lithium, sodium, and potassium.

So he put forward the **Law of Triads**: *Elements exist in threesomes or triads with similar properties; the atomic weight of the middle one is the average of the other two.*

The Law of Octaves

By 1863, 56 elements were known. John Newlands, an English chemist, noted that there were many pairs of similar elements. In each pair, the atomic weights differed by a multiple of 8. So he produced a table with the elements in order of increasing atomic weight, and put forward the **Law of Octaves**: *an element behaves like the eighth one following it in the table.*

This was the first table to show a **periodic** or repeating pattern of properties. But it was not widely accepted because there were too many inconsistencies. For example he put copper and sodium in the same group even though they have very different properties.

Potassium burning. Potassium was just one of the exciting new discoveries in the early nineteenth century.

A triad

	atomic weight
calcium	40
strontium	**88**
barium	137

Average for calcium and barium = $\dfrac{40 + 137}{2}$ = **88.5**

Close!

An octave

	atomic weight
calcium	40
magnesium	−24
	$\overline{16}$ or 2×8

So there!

The periodic table arrives

Dmitri Ivanovich Mendeleev was born in Siberia in 1834, the youngest of 17 children. By the time he was 32 he was a Professor of Chemistry.

Mendeleev had collected a huge amount of data about the elements, from his own research and from scientists around the world. He was writing a textbook for his students and wanted to find a pattern to make sense of the data for them. So he made a card for each of the known elements (by then 63). He played around with the cards on a table, arranging the elements in order of increasing atomic weight, and then into groups with similar behaviour. The result was the periodic table. It was published in 1869.

Mendeleev's table was a big improvement on Newlands'. To avoid inconsistencies he took a big gamble, leaving gaps for elements that had not yet been found. He even named three of these: **eka-aluminium**, **eka-boron**, and **eka-silicon**, and predicted their properties. Some people thought this was crazy. But soon three new elements were discovered, matching his predictions. This helped to convince other chemists, and his table was finally accepted.

Professor Dmitri Mendeleev.

Atomic structure and the periodic table

Mendeleev had arranged the elements in order of atomic weight. But he was puzzled, because he then had to swop some to get them into the right families. For example potassium has a lower atomic weight than argon, so potassium should come first. But clearly a reactive metal like potassium belongs to Group 1. So he switched those two around.

Then in 1911 the proton was discovered. It soon became clear that the number of protons – the **atomic number** – was the key factor in deciding an element's position in the table, and not the atomic weight. So Mendeleev was right to swop those elements. But it was still not clear why.

However, over the next 20 years, ideas about the atom developed rapidly. The number of protons dictates the number of electrons. The electrons are arranged in shells. All the elements in a group have the same number of outer shell electrons. These dictate how the element will behave. The periodic table made sense!

The table has grown a lot since Mendeleev's day, and now shows 109 elements. But his table, over 130 years old, is still the blueprint for it.

Ernest Rutherford discovered the proton in 1911.

Questions

1 Look at the data below. In what way does it fit:
  i the Law of Triads?   ii the Law of Octaves?

Element	Atomic number	Mass number of main isotope
Li	3	7
Na	11	23
K	19	39
Rb	37	85

2 Rubidium appears to be the odd one out, in the data for question 1. Suggest a reason for this. (Check the periodic table.)
3 Our table has more elements than Mendeleev's. How many more?
4 Today, what do we know to be the key factor in deciding which group an element belongs to?
5 Which is the main factor in deciding how an element reacts?

195

8.15 The patterns within Group 1

The trends in physical properties

The Group 1 metals are a family. But like all families, each member of the group is a little different. Look at this table:

Metal	Symbol	This metal is silvery and …	Density/ g/cm³	Melts at … /°C	Boils at… /°C
Lithium	Li	Soft	0.53	181	1342
Sodium	Na	A little softer	0.97	98	883
Potassium	K	Softer still	0.86	63	760
Rubidium	Rb	Even softer	1.53	39	686
Caesium	Cs	The softest of the five	1.88	29	669

As you can see, each property shows an overall increase or decrease as you go down the table. This kind of pattern is called a **trend**.
You can summarize the trends in the table like this:

**lithium
sodium
potassium
rubidium
caesium**

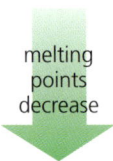

softness increases · density increases · melting points decrease · boiling points decrease

Why they have similar chemical properties

You know that the members of the Group 1 family react in a similar way with water, chlorine, and oxygen. Why is this? The answer is simple: *because they all have the same number of outer shell electrons.*

Li 2,1   Na 2,8,1   K 2,8,8,1

Atoms with the same number of outer shell electrons react in a similar way.

Strong family similarities – just like the alkali metals.

The trend in reactivity

Compare the reactions of the Group 1 family with water:

Metal	What you see
Lithium	A lot of fizz around the floating metal
Sodium	It shoots around on the surface of water
Potassium	It melts and the hydrogen bursts into flames
Rubidium	Sparks fly everywhere
Caesium	A violent explosion

reactivity increases

The increase in violence shows the metals are getting more **reactive**.
Reactivity increases as you go down Group 1.

Explaining the trend in reactivity

As you go down Group 1 the atoms get larger, because they add electron shells. They react in order to obtain a full outer shell. They do this by losing an electron. They become an ion with a charge of 1+.

outer electron lost during reaction — Li 2,1

this one is further from the nucleus so easier to lose — K 2,8,8,1

The larger the atom, the further the outer shell is from the positively charged nucleus. So the easier it is to lose an electron. So the more reactive the metal becomes.

Comparing Groups 1 and 2

The elements of Group 2 are quite similar to the alkali metals. For example magnesium and calcium react with water too – but much more slowly. Look at these trends:

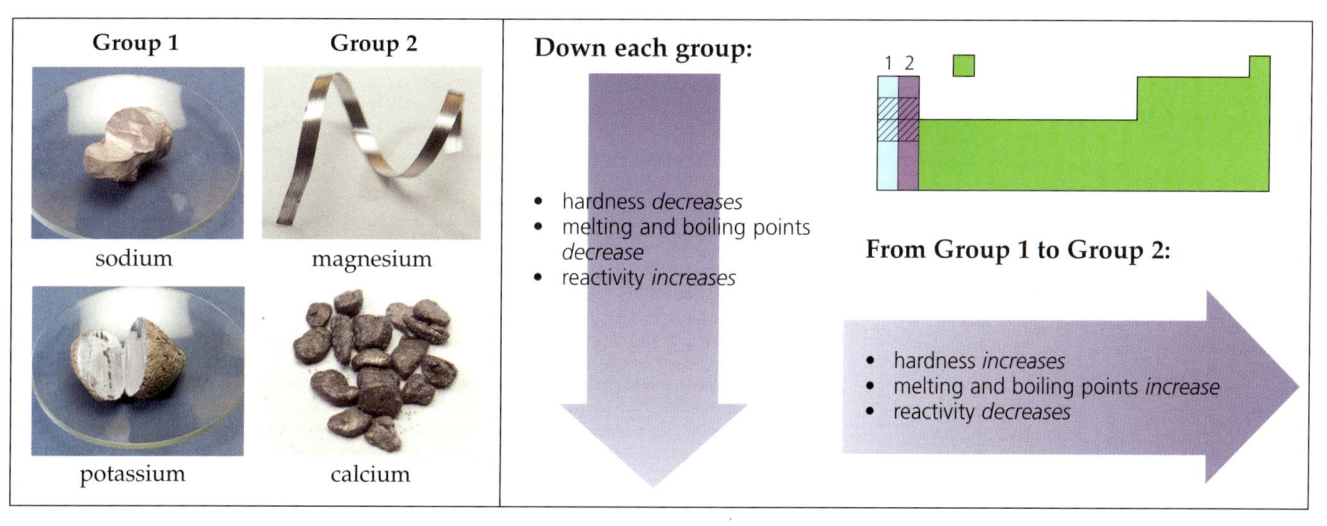

Group 1: sodium, potassium
Group 2: magnesium, calcium

Down each group:
- hardness *decreases*
- melting and boiling points *decrease*
- reactivity *increases*

From Group 1 to Group 2:
- hardness *increases*
- melting and boiling points *increase*
- reactivity *decreases*

When Group 2 metals react, they have to lose two electrons to obtain a full outer shell. This is more difficult than losing just one. And that is why they are less reactive than the Group 1 metals.

As you go down Group 2 the atoms add electron shells, as usual. So the outer shell gets further from the nucleus. That makes it easier to lose electrons from it. So reactivity increases down the group.

Questions

1. The Group 1 metals show a *trend* in melting points.
   a. What does that mean?
   b. Describe two other physical trends for the group.
2. Find one measurement that does *not* fit the trend exactly, in the table on the opposite page.
3. The Group 1 metals all have similar chemical properties. Why is this?
4. When a Group 1 metal reacts, what happens to the outer shell electron of its atoms?
5. a. Which is more strongly held, the outer electron in lithium or in sodium? Explain why you think so.
   b. Sodium is more reactive than lithium. Why?
6. Lithium is the first element in Group 1 and beryllium the first in Group 2. Which is more reactive? Why?

8.16 The non-metal groups in the periodic table

A reminder
Less than a quarter of the elements are non-metals. They lie on the right in the periodic table. Groups 7 and 0 are the two non-metal groups.

Group 0 – the noble gases
This group contains the elements helium, neon, argon, krypton, and xenon. These elements are all:

- non-metals.
- colourless gases (they occur naturally in air).
- **monatomic** – they exist as single atoms.
- unreactive. They do not normally react with anything. (That's why they are called **noble**.)

Why are they unreactive?
It's simple. Atoms react with each other to gain full outer shells of electrons. But the atoms of the noble gases have full shells already. **The noble gases are unreactive, and monatomic, because their atoms already have full outer electron shells.**

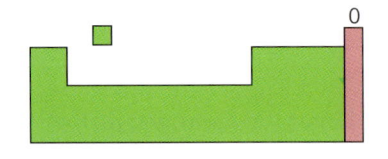

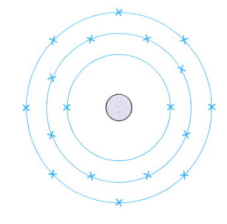

An argon atom: full shell. No need to gain, lose, or share electrons.

Trends within Group 0
The noble gases are not all identical, however. Look at this table.

Noble gas	Its atoms	A balloon full of this gas …	Boiling point /°C
Helium		rises quickly into the air	–269
Neon	the atoms increase in size and mass	rises slowly	–246
Argon		falls slowly	–186
Krypton		falls quickly	–152
Xenon		falls very quickly	–107

(the density of the gases increases ↓) (the boiling points increase ↓)

The gases get denser (or 'heavier') as you go down the group because their atoms are heavier. The increase in boiling points is a sign that the attraction between the atoms is increasing: it gets harder to separate them from each other to form a gas.

Uses of the noble gases
Because the noble gases are unreactive or **inert** they are safe to use.

- Helium is used in balloons and airships – it is much less dense than air.
- Argon is used in light bulbs, and as an inert gas around metals that are being welded. (Oxygen would react with the hot metals.)
- Neon is used in fluorescent lighting and advertising signs.
- Krypton is used in lasers which produce very intense beams of light.

Helium is lighter than air.

198

Group 7 – the halogens

This group of non-metals contains the elements fluorine, chlorine, bromine, and iodine. They have similar properties. For example they all:

- form coloured vapours. Fluorine is a pale yellow gas, chlorine a green gas, bromine forms a red vapour, and iodine a purple one.
- are poisonous.
- are brittle and crumbly in their solid form, and non-conductors of electricity (all typical properties of non-metals).
- form molecules containing a pair of atoms (**diatomic molecules**), for example Cl$_2$.
- have low boiling points – but boiling points change down the group as this table shows:

Halogen	Symbol	At room temperature (20 °C) it is …	Melts at … /°C	Boils at … /°C
Fluorine	F	a yellow gas	−220	−188
Chlorine	Cl	a green gas	−101	−35
Bromine	Br	a red liquid	−7	59
Iodine	I	a black solid	114	184

- are very reactive. For example they react vigorously with metals such as sodium to form compounds called **halides**.

Why do they have similar properties?

The halogens all have similar properties because their atoms all have the same number of electrons in their outer shells.

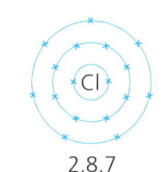

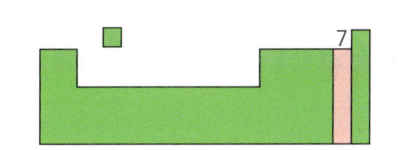

Because chlorine is poisonous, it was used as a weapon in World War I. This soldier is ready for a chlorine gas attack.

Why are they reactive?

They are reactive because their atoms are just one electron short of a full electron shell. They can gain this electron, and become stable, by bonding with atoms of other elements or with each other.

When they react with **metals,** an electron is transferred from a metal atom to a halogen atom. The metal atom becomes a positive ion, and the halogen atom a negative ion. For example when chlorine reacts with sodium, Na$^+$ and Cl$^-$ ions form. The ions arrange themselves to form an **ionic solid** with a giant structure.

a fluoride ion

With other **non-metals,** the halogen atoms *share* electrons to form **molecules** with a covalent bond. Pairs of halogen atoms also bond together covalently to give diatomic molecules.

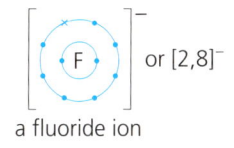

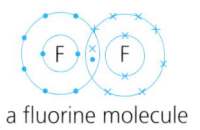

a fluorine molecule

Questions

1 Why do the members of Group 0 have similar properties?
2 Explain why the noble gases are unreactive.
3 What are the trends in density and boiling point for the noble gases?
4 Describe the appearance of the halogens.
5 Explain why the halogens are:
 a reactive b *not* monatomic
6 The fifth element in Group 7 is astatine. Would you expect it to be:
 a a gas, liquid, or solid? Why?
 b coloured or colourless? c harmful or harmless?

199

8.17 Reactions of the halogens

The trend in reactivity

Chlorine, bromine, and iodine are the three most common halogens. They react in a similar way – but there are differences. Look at their reactions with iron wool:

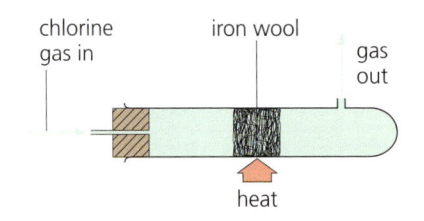

Hot iron wool glows brightly when chlorine passes over it. Brown smoke forms, and a brown solid is left behind. It is the salt iron(III) chloride.

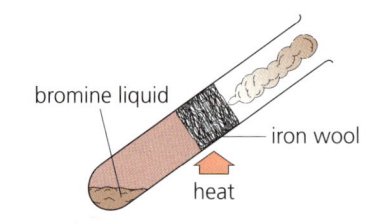

The iron glows less brightly when bromine is used. Brown smoke and a brown solid are formed. It is the salt iron(III) bromide.

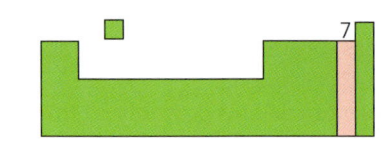

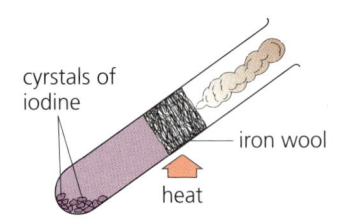

With iodine, the iron glows even less brightly. But once again, brown smoke and a brown solid are formed. It is the salt iron(III) iodide.

The trend is the same for all the halogen reactions:
Reactivity increases as you go up Group 7.

fluorine
chlorine
bromine
iodine

Their reactions with iron and other metals produce **ionic salts**. With chlorine the reaction is:

iron + chlorine ⟶ iron(III) chloride
$2Fe\ (s)\ +\ 3Cl_2\ (g)\ \longrightarrow\ 2FeCl_3\ (s)$

Explaining the trend

Halogen atoms need to *gain* an electron to obtain a full outer shell. An atom can attract an extra electron because of the positive charge on its nucleus. (Opposite charges attract.) But as atoms get larger their outer shells get further from the nucleus, so the force of attraction gets less. So the elements get less reactive.

Compare these chlorine and bromine atoms:

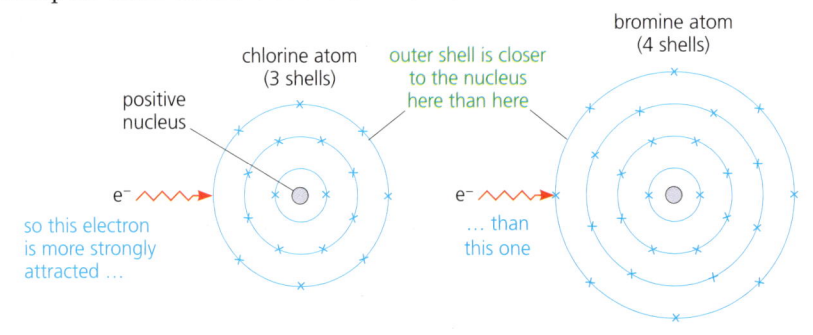

Chlorine reacts more violently with aluminium than with iron. Why?

200

Halogen displacement reactions

If you add this ...	to the colourless solution of this ...		
	potassium chloride (KCl)	potassium bromide (KBr)	potassium iodide (KI)
chlorine (Cl$_2$)		the solution goes a yellow-orange colour	the solution goes a red-brown colour
bromine water (Br$_2$)	no change observed		the solution goes a red-brown colour
iodine solution (I$_2$)	no change observed	no change observed	

Clearly chlorine has reacted with the potassium bromide solution:

potassium bromide + chlorine ⟶ potassium chloride + bromine
2KBr (aq) + Cl$_2$ (g) ⟶ 2KCl (aq) + Br$_2$ (aq)
colourless   yellow-orange

Chlorine has **displaced** the bromine from potassium bromide. It can do this because *it is more reactive than bromine*.
A more reactive halogen will always displace a less reactive halogen from solutions of its compounds.
Does this explain all the results in the table above? Check now.

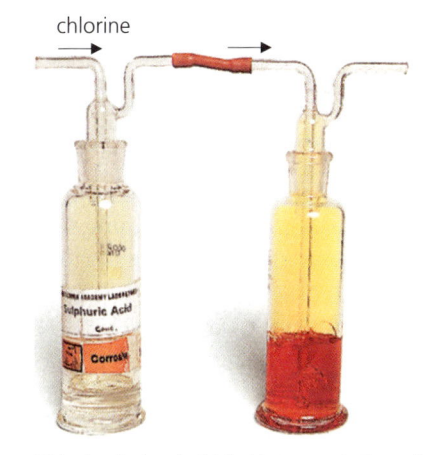

Displacement is a redox reaction

When chlorine displaces bromine in the reaction above, the chlorine atoms gain electrons to form chloride ions. So they are **reduced**. The bromide ions lose electrons to form atoms – they are **oxidized**. **So displacement is a redox reaction.**

2Br$^-$ + Cl$_2$ ⟶ Br$_2$ + 2Cl$^-$

Chlorine atoms are smaller than bromine atoms. So they attract electrons more strongly. This makes chlorine a stronger **oxidizing agent**.

Chlorine being bubbled into a solution of potassium bromide. The concentrated sulphuric acid is used to dry the chlorine.

Cleaning up with halogens

Chlorine is soluble in water. It reacts with it to form two acids:

Cl$_2$ (g) + H$_2$O (l) ⟶ HCl (aq) + HOCl (aq)
                            hydrochloric acid   hypochlorous acid

The solution is called **chlorine water**. It acts as a **bleach**. This is because the hypochlorous acid can lose its oxygen to other substances – it oxidizes them. Many coloured substances lose their colour when oxidized.

Like other bleaches, the solution also acts as a **sterilizing agent** – it kills bacteria. So it is used to sterilize our water supply at water treatment plants, and in swimming pools. Iodine also kills bacteria. **Tincture of iodine** is a solution of iodine in alcohol, and is used as an antiseptic.

Chlorine is used in swimming pools to kill bacteria.

Questions

1 Describe how chlorine reacts with iron wool.
2 What is the trend in the reactivity of the halogens?
3 Chlorine is more reactive than bromine. Why?
4 The fifth element in Group 7 is astatine. Would you expect it to be *more* or *less* reactive than iodine?
5 Bromine water is added to potassium iodide solution.
   a What would you expect to see?
   b Why does this reaction occur?
   c What is *oxidized* and what is *reduced* in the reaction?
6 Which halogen is used as: a an antiseptic? b a bleach?

201

8.18 Compounds of the halogens

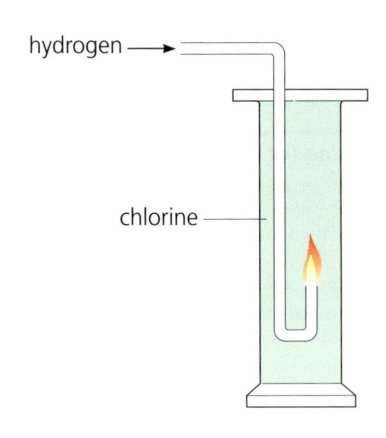

Hydrogen halides

Compounds of the halogens are called **halides**.
The halogens react with hydrogen to form gases – the hydrogen halides. For example chlorine reacts like this:

hydrogen + chlorine ⟶ hydrogen chloride
$H_2(g) + Cl_2(g) \longrightarrow 2HCl(g)$

The reaction must be carried out with care. A mixture of hydrogen and chlorine can explode in the presence of sunlight!

In all the hydrogen halides the atoms share electrons, forming **diatomic molecules**. Look at the diagram on the right.

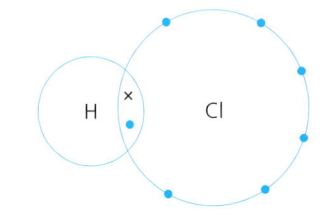

Two properties of hydrogen halides

Suppose you fill a test tube with hydrogen chloride gas, and close it with a bung. Then remove the bung in water containing universal indicator. This is what happens:

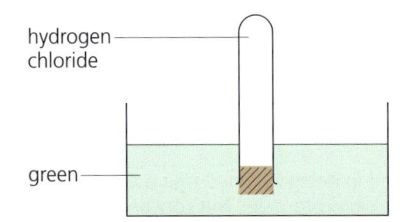

Before you remove the bung the water is green – so it is **neutral**.

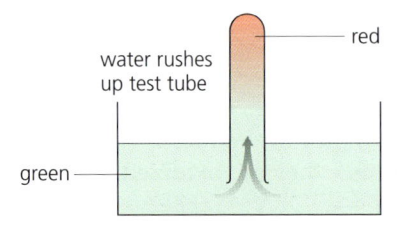

The water rushes up the test tube – and turns red as it goes.

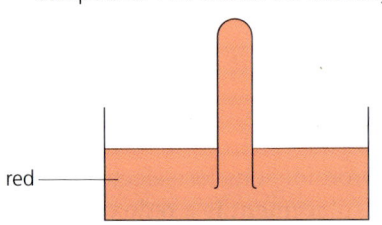

Soon all the water is red. It is no longer neutral. It is **acidic**.

Hydrogen bromide and hydrogen iodide give the same result.
So you can draw two conclusions:
1 Hydrogen halides are *very* soluble in water.
2 Their solutions are acidic. When dissolved in water –
   hydrogen chloride forms **hydrochloric acid** (HCl)
   hydrogen bromide forms **hydrobromic acid** (HBr)
   hydrogen iodide forms **hydriodic acid** (HI).

Sodium chloride

Sodium chloride is the most common halide. You eat it every day in food! It can be made by burning sodium in chlorine. But there is no need to make it in industry because vast amounts are found underground as **rock salt**, and in **sea water**.

Like all metal halides, sodium chloride is an ionic compound. It is made of sodium ions and chloride ions held in a giant lattice by strong ionic bonds. (That is why it is a solid at room temperature.)

The rock salt found in the UK is the starting point for a wide range of chemicals, including sodium hydroxide and chlorine. In winter it is sprinkled on roads to melt ice: it lowers the melting point of ice to below our usual winter temperatures.

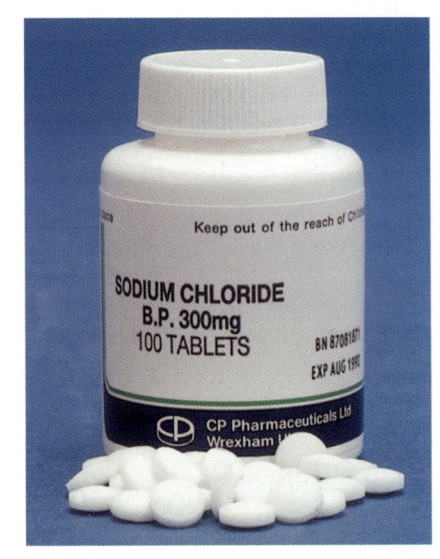

Hydrogen chloride is a covalent compound. This shows the bonding.

Your body loses sodium chloride in sweat. The result can be exhaustion. So people in hot climates often have to take salt tablets.

Silver halides

When aqueous solutions of sodium nitrate and sodium chloride are mixed, a white precipitate of silver chloride forms:

AgNO$_3$ (aq) + NaCl (aq) $\longrightarrow$ AgCl (s) + NaNO$_3$ (aq)

Sodium bromide and sodium iodide react in the same way, giving precipitates of silver bromide (cream) and silver iodide (yellow). After a while all three precipitates darken. Light causes them to break down, giving tiny crystals of silver. The reaction for silver chloride is:

silver chloride $\xrightarrow{\text{light}}$ silver + chlorine

2AgCl (s) $\xrightarrow{\text{light}}$ 2Ag (s) + Cl$_2$ (g)

The silver ions gain electrons to form atoms: they are **reduced**. As well as light, X-rays and radiation from radioactive substances also bring about the reaction. It is an example of a **photochemical** reaction. *Photo-* comes from the Greek word for light, and tells you that a reaction depends on light energy.

Silver bromide and silver iodide decompose in the same way. These reactions are important in photography. Photographic film and paper have a coating of silver chloride, bromide, or iodide in gelatine. Where light strikes it, the silver compound decomposes, giving a dark image. The rest is washed away during processing.

Fluorides

Fluoride compounds are usually toxic to humans. But it has been shown that very small quantities of fluoride in water can help prevent tooth decay. In some areas, depending on the rock, the water already has enough fluoride. In other areas it is added at water treatment plants in the form of compounds such as sodium fluoride. It is added to many brands of toothpaste too. (Check out the tube!)

Potassium iodide

Our bodies need tiny traces of iodine, to keep the thyroid gland working. Without it we would suffer from a disease called **goitre**. So to be on the safe side, tiny amounts of potassium iodide are often added to table salt.

CFCs

Compounds called **chlorofluorocarbons** were once widely made. They were cheap, stable, and very useful: for example as coolants in fridges, and propellants in aerosols. Then the problems emerged. Up in the atmosphere CFC molecules are broken down by UV light, releasing chlorine. This reacts with the ozone layer and destroys it. We need the ozone layer to protect us from the sun – so CFCs are now banned.

Fluoride added to the water supply helps ensure healthy teeth.

Questions

1 Describe how you could show that hydrogen halides are very soluble in water.
2 How does the pH change when hydrogen halides dissolve in water? Name the solutions that form.
3 Why is sodium chloride used on roads in winter?
4 What would you see when silver nitrate was added to a solution containing a chloride ion?
5 a What effect does light have on silver halides?
  b What is reduced in this reaction?
  c Explain how the reaction is used in photography.

8.19 The electrolysis of sodium chloride

A reminder

Sodium chloride is an ionic salt, Na$^+$Cl$^-$. So it can be decomposed by electrolysis. But first it must be:
- melted
- or dissolved in water

so that the ions are free to move.

The electrolysis of molten sodium chloride

If it is heated enough sodium chloride will melt, freeing the ions to move. (It melts at 801 °C.)

When the switch is closed, the positive sodium ions then move to the negative electrode. The negative chloride ions will move to the positive electrode.

At the negative electrode, the sodium ions gain electrons to form sodium atoms:

$$Na^+ + e^- \longrightarrow Na$$

At the positive electrode, chloride ions lose electrons to form chlorine atoms. Then two atoms join to form a chlorine molecule:

$$Cl^- - e^- \longrightarrow Cl$$
$$Cl + Cl \longrightarrow Cl_2$$

The overall equation for the decomposition is:

$$2NaCl\ (l) \longrightarrow 2Na\ (l) + Cl_2\ (g)$$

The electrolysis of molten sodium chloride gives sodium and chlorine.

The electrolysis of sodium chloride solution

The electrolysis of sodium chloride solution can be carried out in the lab using a **Hoffmann voltameter** as shown on the next page. This apparatus allows gases to be collected and tested.

During the experiment this is what you would observe:

At the positive electrode	At the negative electrode	In the solution
• Bubbles of gas form. • The gas is a pale green colour. • It turns pH paper white – it bleaches it.	• Bubbles of gas form. • The gas is colourless. • It burns with a pop when it is lit.	• pH paper is green at the start – the solution is neutral. • The pH paper turns blue – so the solution becomes alkaline.
The gas is **chlorine**.	The gas is **hydrogen**.	The solution is **sodium hydroxide**

The electrolysis of sodium chloride solution gives chlorine, hydrogen, and sodium hydroxide solution.
So don't do it if you want to get sodium!

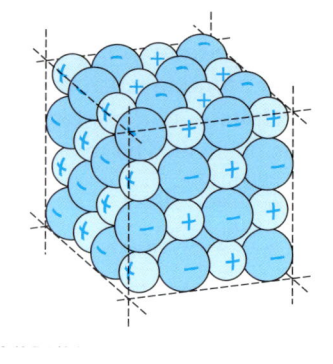

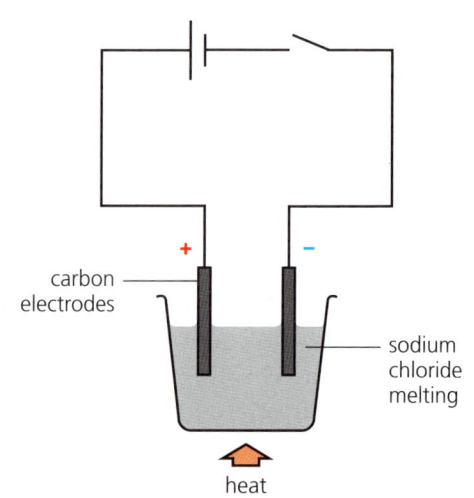

204

The Hoffmann voltameter

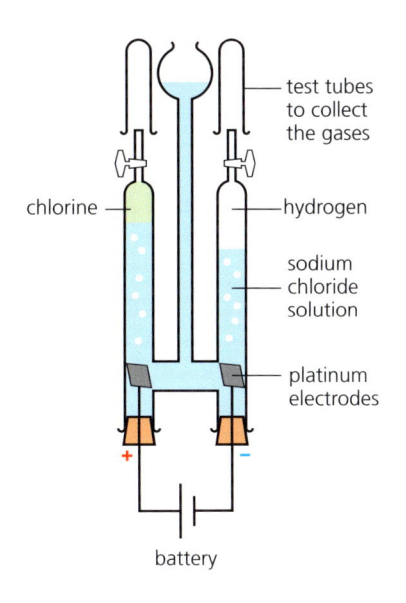

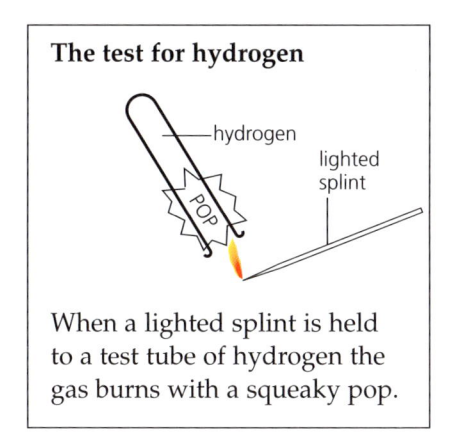

The test for chlorine

When damp pH paper is placed in chlorine the paper is bleached.

[Diagram: Hoffmann voltameter showing test tubes to collect the gases, chlorine and hydrogen gases collecting, sodium chloride solution, platinum electrodes, and battery]

The test for hydrogen

When a lighted splint is held to a test tube of hydrogen the gas burns with a squeaky pop.

Explaining the electrolysis of sodium chloride solution

In water, a tiny fraction of water molecules form ions, like this:

some water molecules → hydrogen ions + hydroxide ions
H_2O (l) → H^+ (aq) + OH^- (aq)

So let's look at what happens during the electrolysis.

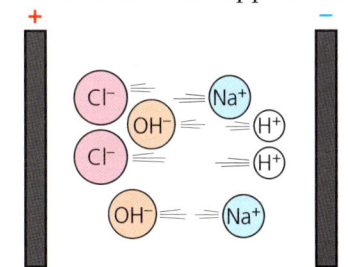

The solution contains Na^+ ions and Cl^- ions from the salt, and H^+ and OH^- ions from water. The positive ions go to the negative electrode and the negative ions to the positive electrode.

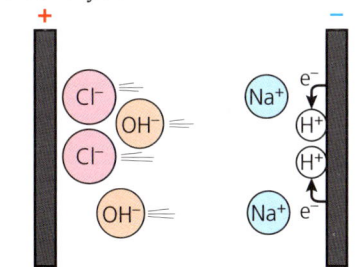

At the negative electrode it is the H^+ ions that accept electrons since sodium is more reactive than hydrogen:

$$2H^+ + 2e^- \rightarrow H_2$$

Hydrogen gas bubbles off.

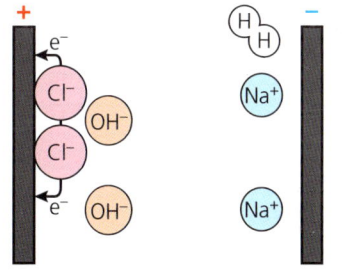

At the positive electrode the Cl^- ions give up electrons more readily than the OH^- ions do. Chlorine gas bubbles off:

$$2Cl^- - 2e^- \rightarrow Cl_2$$

The OH^- ions remain in solution.

When the hydrogen and chlorine bubble off, Na^+ and OH^- ions are left behind: a solution of sodium hydroxide is formed.

Questions

1 Which type of compound is sodium chloride?
2 What happens at:
 a the positive electrode
 b the negative electrode
 when molten sodium chloride is electrolysed?
3 How is the electrolysis different when the sodium chloride is dissolved in water?
4 When a solution of sodium chloride is electrolysed:
 a at which electrode is hydrogen released? Why?
 b what test shows the presence of this gas?
 c where has the hydrogen come from?
 d how would you show chlorine was produced?
 e what pH change occurs in the solution?
 f what causes this change in pH?

8.20 The chlor-alkali industry

An industry based on salt

A whole section of the chemical industry has developed around a single electrolysis reaction. It's called the **chlor-alkali** industry, and it is based on the electrolysis of a solution of sodium chloride, or **salt**.

Most of the salt in Britain comes from Cheshire, where it lies 200 metres below ground in seams 200 m thick. Some is mined as rock salt for gritting roads. But most is pumped out of the ground in a concentrated solution called **brine**, which is then electrolysed:

brine ⟶ sodium hydroxide + chlorine + hydrogen
$2NaCl\,(aq) + 2H_2O\,(l) \longrightarrow 2NaOH\,(aq) + Cl_2\,(g) + H_2\,(g)$

The products of the electrolysis are used in turn as raw materials for a whole range of other products.

Brine – the starting point for many new materials.

Getting the salt out from underground

First, three concentric pipes are buried 400 metres into the ground. Water is pumped down the second pipe to dissolve the salt. Air is pumped down the outer. The brine is forced up the inner pipe and into reservoirs, then piped to a purification plant when it is needed.

The brine is purified by adding sodium hydroxide to precipitate metal ions as hydroxides. (Most hydroxides are insoluble.) Sulphate ions are removed by adding barium ions to precipitate barium sulphate.

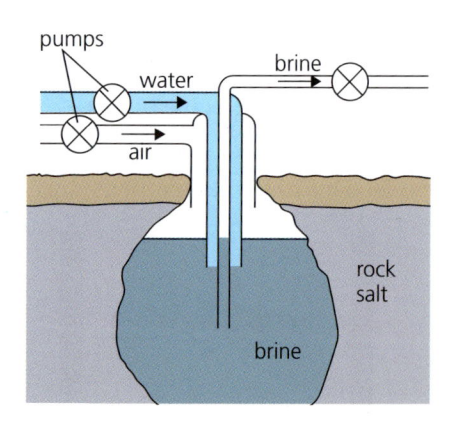

Getting salt out from underground.

Electrolysing the brine

The electrolysis is just like the one you can do in the lab, but on a much larger scale. The main thing is to make sure the hydrogen and chlorine are kept apart.

Different electrolysis cells can be used, but the **membrane cell** uses least electricity and is safest for the environment. The positive electrode is titanium and the negative is nickel. The cell has an **ion-exchange membrane** up the middle. This lets sodium ions through but keeps the gases apart. So the sodium ions can move freely to the negative electrode.

At the negative electrode   Hydrogen bubbles off:

$2H^+ + 2e^- \longrightarrow H_2$

At the positive electrode   Chlorine bubbles off:

$2Cl^- - 2e^- \longrightarrow Cl_2$

Na^+ and OH^- ions are left behind, which means a solution of sodium hydroxide forms. Some is evaporated to a more concentrated solution, and some evaporated completely to give solid sodium hydroxide.

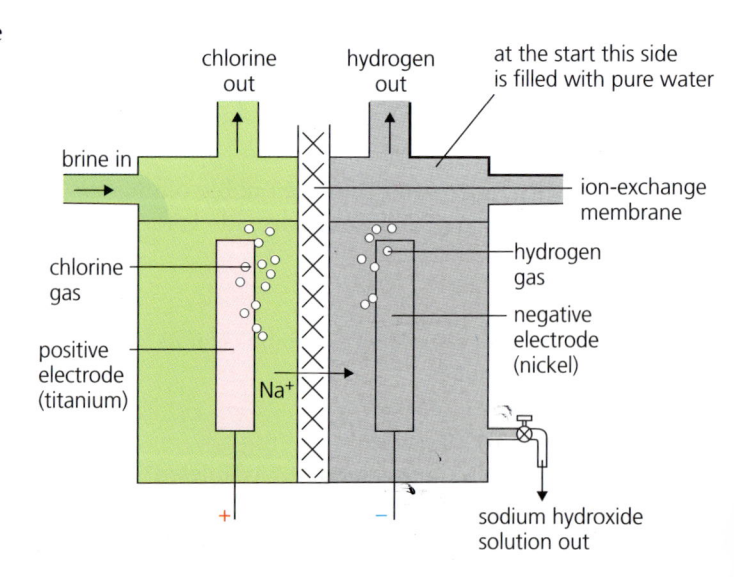

Chlorine is used as an anti-bacterial agent at water works.

Sodium hydroxide is extremely reactive and must be transported safely.

What the products are used for

A concentrated solution of sodium chloride is a very simple starting material. You could make it in your kitchen. But look what it leads to!

Electrolysis of brine

Chlorine, a poisonous yellow-green gas

Used for making . . .
- PVC (nearly $\frac{1}{3}$ of it used for this)
- solvents such as perchloroethylene for degreasing and dry-cleaning
- paints and dyestuffs
- bleaches, weedkillers, pesticides
- pharmaceuticals
- titanium dioxide, a white pigment used in paints, ceramics, cosmetics, and paper
- hydrogen chloride and hydrochloric acid

It is also used for . . .
- killing bacteria in tap water
- killing bacteria in swimming pools

Hydrogen, a colourless flammable gas

Used for . . .
- making nylon
- making hydrogen peroxide
- 'hardening' vegetable oils for margarines
- burning to make steam for other processes

Sodium hydroxide solution, alkaline and corrosive

Used for making . . .
- soaps
- detergents
- viscose (rayon) and other textiles
- paper (such as used in this book)

and in making many chemicals including dyestuffs and pharmaceuticals

Questions

1 Write a word equation for the electrolysis of brine.
2 Draw a diagram for the membrane cell used for it.
3 Why is the membrane needed?
4 List four uses for: **a** sodium hydroxide **b** brine
5 Where does the term 'chlor-alkali' come from?
6 Suppose your job is to keep a brine electrolysis plant running safely and smoothly. Try to think of four safety precautions you might need to take.

Questions for Module 8

1. This list shows a number of common chemical substances.
   - A iron
   - B carbon dioxide
   - C mercury
   - D hydrogen
   - E sodium chloride
   - F water

   a i Divide this list into two lists, one of elements and one of compounds.
   ii Using the idea of atoms, explain the difference between an element and a compound.
   b Divide the elements into metals and non-metals.
   c Give two further examples of:
      i metals
      ii non-metals
      iii compounds

2. In water (H_2O) the atoms are joined as shown in the diagram.

   Suggest how the atoms are joined in the following compounds:
   a sulphur dioxide SO_2
   b methane CH_4
   c methanol CH_4O
   d hydrogen peroxide H_2O_2
   e ammonia NH_3

3. For the following statements about the particles that make up the atom write:
   p if the answer is the **proton**
   e if the answer is the **electron**
   n if the answer is the **neutron**

   a the positively charged particle
   b the particle found with the proton in the nucleus
   c the particle that can be found in different numbers in two atoms of the same element
   d are held in shells around the nucleus
   e the negatively charged particle
   f the particle with negligible mass
   g the number of these particles is found by subtracting the atomic number from the mass number
   h the particles with no charge
   i the particles with the same mass as a neutron
   j the number of these particles is the atomic number

4. The atom for an element is shown here. $^{y}_{z}X$
   a What is represented by:
      i X? ii y? iii z?
   b How many neutrons are there in the following atoms:
      i $^{107}_{47}Ag$? ii $^{63}_{29}Cu$? iii $^{1}_{1}H$? iv $^{20}_{10}Ne$? v $^{238}_{92}U$?

5. Hydrogen, deuterium, and tritium are isotopes. Their structures are shown below.

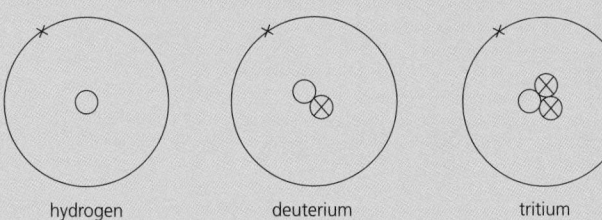

   hydrogen   deuterium   tritium

   a Copy and complete the following key:
      × represents _____
      ○ represents _____
      ⊗ represents _____
   b What are the mass numbers of hydrogen, deuterium, and tritium?
   c Copy and complete this statement:
      Isotopes of an element always contain the same number of _____ and _____ but different numbers of _____.
   d The average mass number of naturally-occurring hydrogen is 1.008. Which isotope is present in the highest proportion, in naturally-occurring hydrogen?

6. Copy and complete the following table for isotopes of some common elements.

Isotope	Name of element	Proton number	Mass number	Number of p	e	n
$^{16}_{8}O$	oxygen	8	16	8	8	8
$^{18}_{8}O$						
$^{12}_{6}C$						
$^{13}_{6}C$						
$^{24}_{12}Mg$						
$^{25}_{12}Mg$						
				17	17	18
				17	17	20

7 For each of the six elements aluminium (Al), boron (B), nitrogen (N), oxygen (O), phosphorus (P), sulphur (S), write down:
   a the period of the periodic table to which it belongs
   b its group number in the periodic table
   c its atomic number
   d the number of electrons in one atom
   e its electronic configuration
   f the number of outer-shell electrons in one atom
   Which of the above elements would you expect to have similar properties? Why?

8 Boron has two types of atom, shown below.

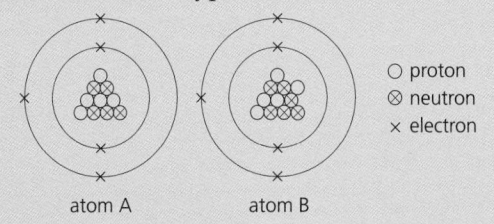

   atom A   atom B
   ○ proton
   ⊗ neutron
   × electron

   a What name is given to atoms like these shown?
   b Draw the shorthand form for each atom.
   c What is the mass number of atom A?
   d i What is the difference in mass number between atom B and atom A?
     ii Is atom B heavier or lighter than atom A?
   e i Give the electronic configuration of A and B.
     ii Comment on this.
   f 80% of the atoms in natural boron are atom B. Calculate the relative atomic mass of boron.

9 Gallium exists naturally as a mixture of two isotopes, gallium-69 and gallium-71. The atomic number of gallium is 31.
   a i How many neutrons are there in gallium-69?
     ii How many neutrons are there in gallium-71?
   The relative atomic mass for gallium is 69.8.
   b Comment on the proportion of each isotope in natural gallium.

10 The two metals sodium (atomic number 11) and magnesium (atomic number 12) are found next to each other in the periodic table.
   a Do the two metals have the same or different:
     i number of electron shells in their atoms?
     ii number of outer electrons?
   The relative atomic mass of sodium is 23.0.
   The relative atomic mass of magnesium is 24.3.
   b Which of the two elements may exist naturally as a single isotope? Explain.

11 Strontium, atomic number 38, is an element from the fifth period of the periodic table. It is a member of Group 2. Copy and complete the following: An atom of strontium therefore has:
   a .... electrons
   b .... shells of electrons
   c .... electrons in its outer shell

12 The following represents the electronic arrangement of an atom.

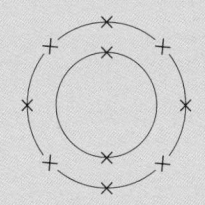

   a Write the electronic configuration for the atom.
   b i Which group of the periodic table does the atom belong to?
     ii What name is given to this group?
   c Name another element with a similar number of electrons in the outer shell.

13 Scientists have made many attempts to produce a model for the atom. Two such models were:
   Model A – All elements are composed of spherically shaped atoms which are indivisible and indestructible. All atoms of the same element are identical. Atoms of different elements have different masses.
   Model B – An atom is a solid sphere containing positive and negative charges. The negative charges are stuck on the positively charged material like currants on a bun.
   a Explain the meaning of the terms:
     i *indivisible* ii *indestructible*
   b What evidence is there that model B was proposed nearly 100 years after model A?
   c Taking carbon as an example sketch each element as it would look as described in the two models.
   d Using your knowledge of atomic structure, point out some of the inaccuracies in model A.
   e Which particle had not been discovered when model B was proposed?
   f Which model was proposed by the scientist Thomson and which by Dalton?
   g Explain in your own words what scientists mean by a model.

14 The table below describes the structure of several particles.

Particle	Electrons	Protons	Neutrons
A	12	12	12
B	12	12	14
C	10	12	12
D	10	8	8
E	9	9	10

  a Which three particles are neutral atoms?
  b Which particle is a negative ion? What is the charge on this ion?
  c Which particle is a positive ion? What is the charge on this ion?
  d Which two particles are isotopes?
  e Look through the chapter and identify the particles A to E.

15 This question is about the ionic bond formed between the metal lithium (atomic number 3) and the non-metal fluorine (atomic number 9).
  a How many electrons are there in a lithium atom? Draw a diagram to show its electron structure. (You can show the nucleus as a dark circle at the centre.)
  b How does a metal atom obtain a full outer shell of electrons?
  c Draw the structure of a lithium ion, and write the symbol for the ion.
  d How many electrons are there in a fluorine atom? Draw a diagram to show its electron structure.
  e How does a non-metal become a negative ion?
  f Draw the structure of a fluoride ion, and write a symbol for the ion.
  g Draw a diagram to show what happens when a lithium atom reacts with a fluorine atom.
  h Draw the arrangement of ions in the compound that forms when lithium and fluorine react together.
  i Write the name and formula for the compound in part **h**.
  j What type of structure will this compound have, giant or molecular?
  k Name two properties that this compound will have.

16 The following show the electronic arrangement for two elements, aluminium and nitrogen. These elements react to form an ionic compound, aluminium nitride.

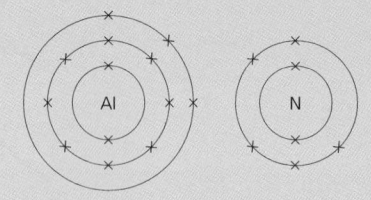

  a For each element answer these questions.
    i Does it gain or lose electrons to form an ion?
    ii How many electrons will be transferred?
    iii Is the ion formed positive or negative?
    iv What size charge will the ion have?
  b **i** Write the electronic configuration of each ion formed.
    ii What do you notice about these configurations? Explain your findings.

17 a The electronic configuration of a neon atom is (2,8). What is special about the outer shell of a neon atom?
  b The electronic configuration of a calcium atom is (2,8,8,2). What must happen to a calcium atom for it to achieve a noble gas structure?
  c Draw a diagram of an oxygen atom, showing its eight protons (p), eight neutrons (n), and eight electrons (e).
  d What happens to the outer-shell electrons of a calcium atom, when it reacts with an oxygen atom?
  e Name the compound that is formed when calcium and oxygen react together. What type of bonding does it contain?
  f Write the formula for the compound in **e**.
  g Would this compound have a high, or low, melting point? Explain your answer.

18 a Write down a formula for each of the following:
    i a nitrate ion
    ii a sulphate ion
    iii a carbonate ion
    iv a hydroxide ion
  b The metal strontium forms ions with the symbol Sr^{2+}. Write down the formula for each of the following:
    i strontium oxide
    ii strontium chloride
    iii strontium nitrate
    iv strontium sulphate

19 Nitrogen is in Group 5 of the periodic table, and its atomic number is 7. It exists as molecules, each containing two atoms.
   a Write the formula for nitrogen.
   b What type of bonding would you expect between the two atoms?
   c How many electrons does a nitrogen atom have in its outer shell?
   d How many electrons must each atom obtain a share of, in order to gain a full shell of 8 electrons?
   e Draw a diagram to show the bonding in a nitrogen molecule. You need show only the outer-shell electrons. (It may help you to look back at the bonding in an oxygen molecule.)
   f The bond in a nitrogen molecule is called a triple bond. Can you explain why?
   g Nitrogen (N_2), oxygen (O_2), chlorine (Cl_2) and hydrogen (H_2) all exist as diatomic molecules. What does *diatomic* mean?

20 This table gives information about properties of certain substances A to G.

Substance	Melting point /°C	Electrical conductivity solid	Electrical conductivity liquid	Solubility in water
A	−112	poor	poor	insoluble
B	680	poor	good	soluble
C	−70	poor	poor	insoluble
D	1495	good	good	insoluble
E	610	poor	good	soluble
F	1610	poor	poor	insoluble
G	660	good	good	insoluble

   a Which of the substances are metals? Give reasons for your choice.
   b Which of the substances are ionic compounds? Give reasons for your choice.
   c Two of the substances have very low melting points, compared with the rest. Explain why these could *not* be ionic compounds.
   d Two of the substances are molecular. Which ones are they?
   e Which substance is a giant covalent structure?

21 Silicon lies directly below carbon in Group 4 of the periodic table. The following table lists the melting and boiling points for silicon, carbon (diamond), and their oxides.

Substance	Symbols or formula	Melting point /°C	Boiling point /°C
Carbon	C	3730	4530
Silicon	Si	1410	2400
Carbon dioxide	CO_2	(sublimes at −78°C)	
Silicon dioxide	SiO_2	1610	2230

   a In which state are the two *elements* at room temperature (20°C)?
   b Is the structure of carbon (diamond) giant covalent or molecular?
   c What type of structure would you expect silicon to have? Give reasons.
   d In which state are the two oxides at room temperature?
   e What type of structure does carbon dioxide have?
   f Does silicon dioxide have the same structure as carbon dioxide? What is your evidence?

22 Hydrogen bromide is a compound of the two elements hydrogen and bromine. It has a melting point of −87°C and a boiling point of −67°C. Bromine is in Group 7 of the periodic table.
   a Is hydrogen bromide a solid, a liquid, or a gas at room temperature (20°C)?
   b Is it made of molecules, or does it have a giant structure? How can you tell?
   c What type of bond is formed between the hydrogen and bromine atoms in hydrogen bromide? Show this on a diagram.
   d Write a formula for hydrogen bromide.
   e Name two other compounds that would have bonding similar to hydrogen bromide.
   f Write formulae for these two compounds.

23 a Use the structures of diamond and graphite to explain why:
      i graphite is used for the 'lead' in pencils
      ii diamonds are used in cutting tools
      iii graphite is used in electrical circuits
   b Give reasons why:
      i copper is used in electrical wiring
      ii steel is used for domestic radiators
   c Ethanol is used as the solvent in perfume and aftershave, because it evaporates easily. What does that tell you about the bonding in it?

24 The elements of Group 0 are called the noble gases. They are all monatomic gases.
   a Name four of the noble gases.
   b i What is meant by *monatomic*?
     ii Explain why the noble gases, unlike all other gaseous elements, are monatomic.
   Reactive non-metals when they react are said to become *like noble gases*.
   c i Explain what is meant by this statement.
     ii What does this tell you about the reactivity of Group 7 ions?

25 Identify the following non-metal elements.
   a a colourless gas, used in balloons and airships
   b a green gas, which dissolves to form a bleaching solution
   c a colourless gas which produces a red light in advertising signs
   d a red liquid
   e a yellow gas that is so reactive that it is not used in school laboratories
   f a black solid, and needed by our thyroid gland

26 The elements of Group 7 have similar properties.
   a What name is given to this group of elements?
   Reactions with iron can be used to show the similar chemical properties.
   b i Describe how the reactions differ when iron reacts with chlorine, bromine, and iodine.
     ii Describe the trend in the reactivity of the three halogens.
   c i Suggest how fluorine would react with iron.
     ii Name the compound formed in this reaction.

27 Argon has atomic number 18, one more than chlorine. Explain why:
   a both chlorine and argon are gases at room temperature
   b chlorine is reactive and argon unreactive
   c chlorine forms diatomic molecules whereas argon is monatomic

28 The diagram shows some of the elements in Group 7 of the periodic table.

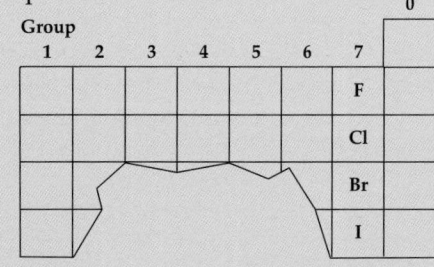

Group							0
1	2	3	4	5	6	7	
						F	
						Cl	
						Br	
						I	

   a What name is given to the elements in this group of the periodic table?
   Chlorine reacts explosively with hydrogen. The equation for the reaction is:

   hydrogen + chlorine ⟶ hydrogen chloride

   The reaction requires sunlight but does not require heat.
   b i Describe how fluorine might react with hydrogen.
     ii Write the word equation for the reaction.
   c i Describe how bromine might react with hydrogen.
     ii Write the word equation for the reaction.
   d Explain the trend in reactivity you have described above.

29 In the electrolysis of sodium chloride solution, hydrogen is formed at the negative electrode.
   a Name the gas formed at the positive electrode.
   b Give one important use of the gas.
   When this gas is passed through a solution of potassium bromide the solution changes colour.
   c Describe the colour change you would see.
   d Explain the cause of the colour change.

30 This question is about the electrolysis of molten lithium chloride. Lithium chloride is ionic, and contains lithium ions (Li$^+$) and chloride ions (Cl$^-$).
   a Copy the following diagram and use arrows to show which way:
     i the ions flow when the switch is closed
     ii the electrons flow in the wires

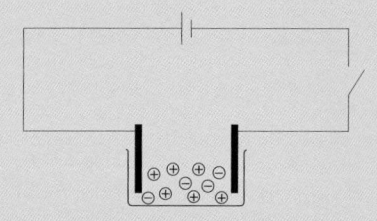

   b Write equations for the reaction at each electrode, and the overall reaction.
   c How would the products at the electrodes be different if the lithium chloride had been dissolved in water?

31 The table below shows four groups of 'substances'. This table was published by Antoine Lavoisier in 1789. Lavoisier thought these 'substances' were elements. The modern names of some of them are given in brackets.

Acid-making elements	Gas-like elements	Metallic elements	Earthy elements
sulphur	light	cobalt	lime
phosphorus	caloric	mercury	(calcium oxide)
charcoal	(heat)	copper	magnesia
(carbon)	oxygen	nickel	(magnesium oxide)
	azote	gold	barytes
	(nitrogen)	platina (platinum)	(barium sulphate)
	hydrogen	iron	argilla
		silver	(aluminium oxide)
		lead	sitex
		tin	(silicon dioxide)
		manganese	
		tungsten	
		zinc	

a Write a definition of:
  i an element   **ii** a compound
b Name a substance in the table that is a compound not an element.
c Why do you think Lavoisier thought this substance was an element?
d Name a 'substance' in the table that is neither an element nor a compound.
e Give two ways in which Mendeleev's periodic table published in 1869 is more useful than Lavoisier's four groups shown above.

32 Before the development of the periodic table, a scientist called Döbereiner discovered sets of three elements with similar properties, which he called **triads**. Some of these triads are shown as members of the same group, in the modern periodic table.
  a Copy and complete the following triads by inserting the missing element.
     chlorine (Cl) .... iodine (I)
     lithium (Li) .... potassium (K)
     calcium (Ca) .... barium (Ba)
  b What information about each of the elements was used by Döbereiner to formulate his *Law of Triads*?

33 Silver bromide is used in photography. In the presence of light it breaks down into silver.
  a Write an equation for the reaction.
  b Explain why the silver in the silver bromide has been reduced.
  c What colour is silver bromide?
  d What observation would show that the reaction has occurred?

34 When a piece of damp blue litmus paper is placed in a gas jar of chlorine, the litmus changes colour. First the blue colour turns red, then the red colour quickly disappears, leaving the paper white.
  a Explain why the paper turns red.
  b Explain why the red colour then disappears.

35 The British chlor-alkali industry is based in Cheshire. Write down three reasons why Runcorn is a good location for the electrolysis of brine.

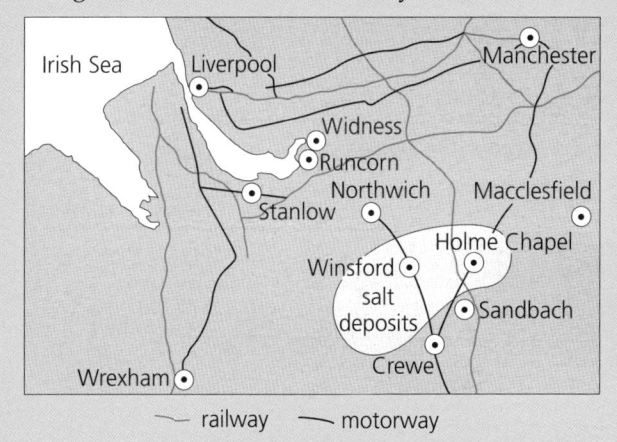

36 Hydrogen chloride is obtained when hydrogen reacts with chlorine. It is a molecular compound.
  a i What type of bond is formed between the hydrogen and chlorine atoms?
    ii Draw a diagram to show the bond.
  b Hydrogen chloride is very soluble in water.
    i Describe an experiment that shows this.
    ii Name the solution formed.
The solution exists as ions.

$$HCl\,(g) \longrightarrow H^+\,(aq) + Cl^-\,(aq)$$

When this solution is electrolysed:
  c i Name the gas formed at the positive electrode.
    ii Write the equation for the reaction.
  d i Name the gas formed at the negative electrode.
    ii Write the equation for the reaction.

Further topics

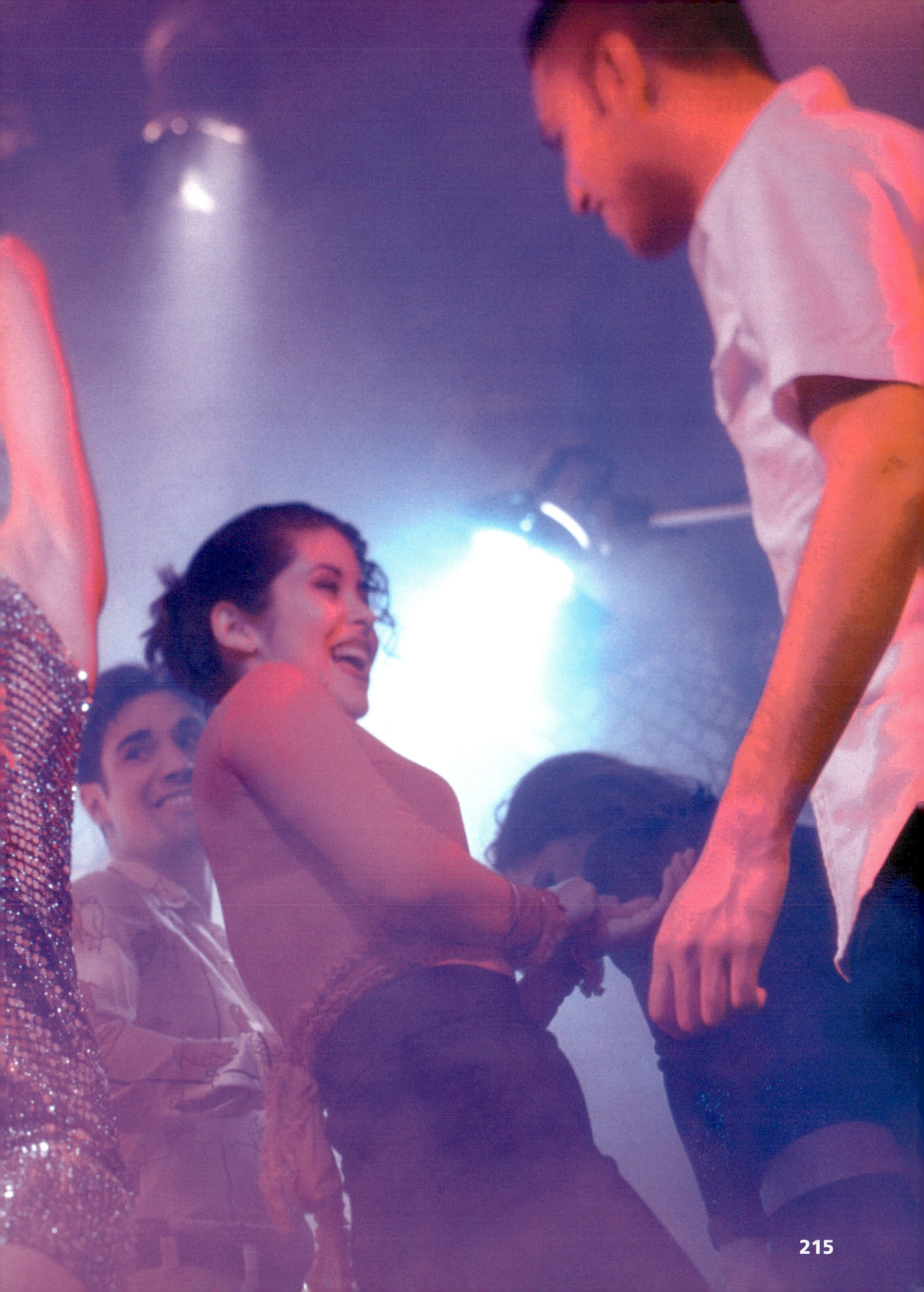

FT 1 Revision: solids, liquids, and gases

You know already that a substance can change from solid to liquid to gas. The individual *particles* of the substance are the same in each state. It is their *arrangement* that is different.

State	How the particles are arranged	Diagram of particles
Solid 	The particles in a solid are packed tightly in a fixed pattern. There are strong forces holding them together, so they cannot leave their positions. The only movements they make are tiny vibrations to and fro.	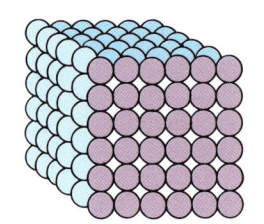
Liquid 	The particles in a liquid can move about and slide past each other. They are still close together but are not in a fixed pattern. The forces that hold them together are weaker than in a solid.	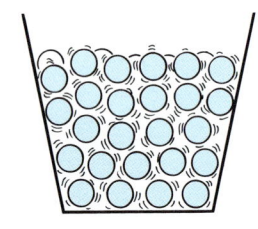
Gas 	The particles in a gas are far apart, and they move about very quickly. There are almost no forces holding them together. They collide with each other and bounce off in all directions.	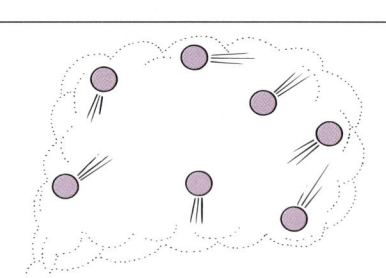

Changes of state

Melting When a solid is heated, its particles get more energy and vibrate more. This makes the solid **expand**. At the melting point, the particles vibrate so much that they break away from their positions. The solid becomes a liquid.

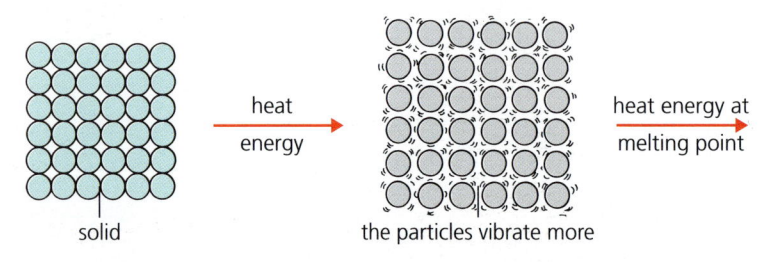

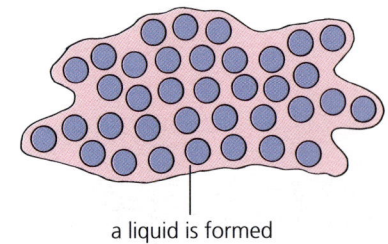

solid → heat energy → the particles vibrate more → heat energy at melting point → a liquid is formed

Boiling   When a liquid is heated, its particles get more energy and move faster. They bump into each other more often and bounce further apart. This makes the liquid expand. At the boiling point, the particles get enough energy to overcome the forces holding them together. They break away from the liquid and form a gas.

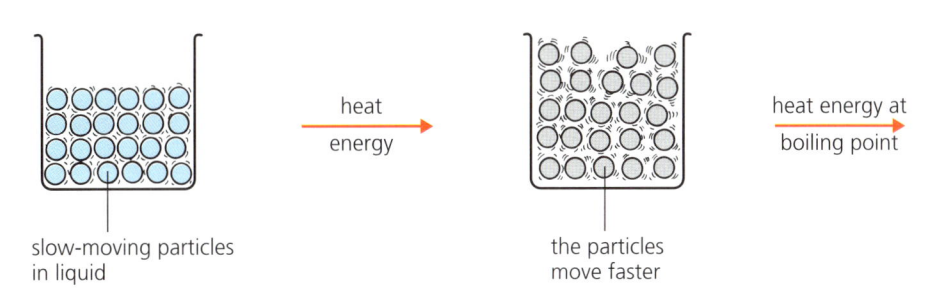

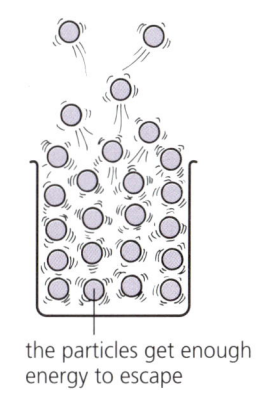

slow-moving particles in liquid → heat energy → the particles move faster → heat energy at boiling point → the particles get enough energy to escape

Evaporating   Some particles in a liquid have more energy than others. Even when a liquid is well below boiling point, *some* particles have enough energy to escape and form a gas. This is called **evaporation**. It is why puddles of rain dry up in the sunshine.

Condensing and solidifying   When a gas is cooled, the particles lose energy. They move more and more slowly. When they bump into each other, they do not have enough energy to bounce away again. They stay close together and a liquid forms. When the liquid is cooled, the particles slow down even more. Eventually they stop moving, except for tiny vibrations, and a solid forms.

Compressing a gas

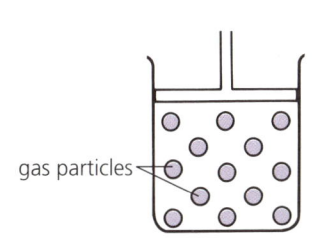

There is a lot of space between the particles in a gas. You can force the particles closer by pushing in the plunger.

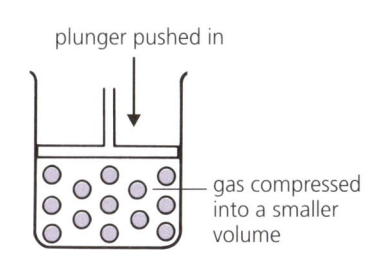

The gas gets squeezed or **compressed** into a smaller volume. We say gases are **compressible**.

If enough force is applied to the plunger, the particles get so close together that the gas turns into a liquid. But liquids and solids cannot be compressed because their particles are already close together.

These divers carry compressed air so that they can breathe under water.

Questions

1 Using the idea of particles, explain why:
   a it is easy to pour a liquid
   b a gas will completely fill any container
   c a solid expands when it is heated
2 Draw a diagram to show what happens to the particles in a liquid, when it boils.
3 Explain why a gas can be compressed into a smaller volume, but a solid can't.

217

FT 2 A closer look at gases

When you blow up a balloon, you fill it with air particles moving at speed. The particles knock against the sides of the balloon and exert **pressure** on it. The pressure is what keeps the balloon inflated. In the same way, *all* gases exert pressure. The pressure depends on the **temperature** of the gas and the **volume** it fills, as you will see below.

How gas pressure changes with temperature

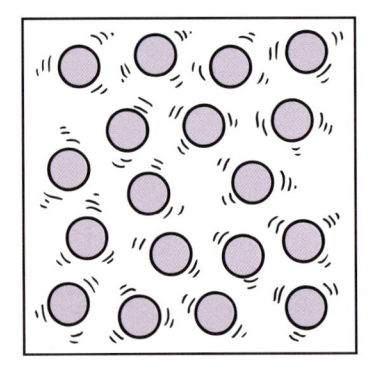

The particles in this gas are moving at speed. They hit the walls of the container and exert pressure on them. If the gas is heated . . .

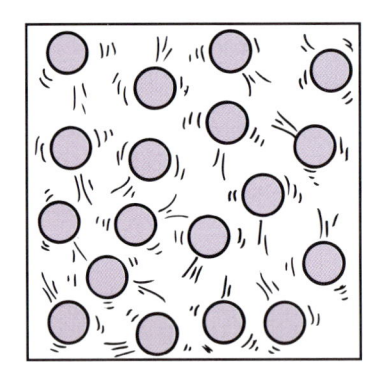

. . . the particles take in heat energy and move even faster. They hit the walls more often and with more force. So the gas pressure increases.

When you blow up a balloon the gas particles exert pressure on the balloon and make it inflate. The more you blow, the greater the pressure.

The same happens with all gases:

If the volume of a gas is kept constant, its pressure increases with temperature.

How gas pressure changes with volume

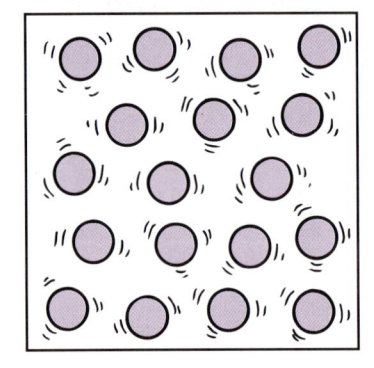

Here again is the gas from above. Its pressure is due to the particles colliding with the walls of the container.

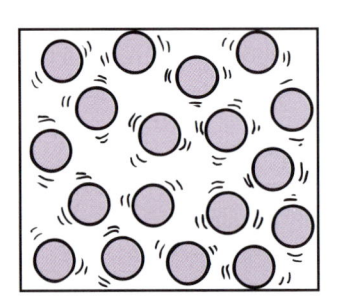

This time the gas is squeezed into a smaller volume. So the particles hit the walls more often. The gas pressure increases.

In a pressure cooker, water vapour is heated to well over 100 °C, so it reaches very high pressure. It is dangerous to open a pressure cooker without first letting it cool.

The same thing is true for all gases:

When a gas is squeezed into a smaller volume, its pressure increases.

218

How gas volume changes with temperature

Now let's see what happens if the gas pressure is kept constant, but the temperature changes:

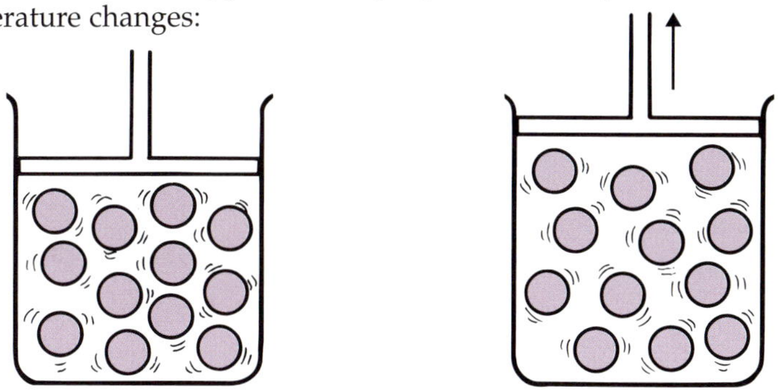

The plunger in this container can move in and out. When the gas is heated, the particles hit it more often . . .

. . . and with more energy, so it moves out. This means the pressure doesn't change. But now the gas fills a larger volume.

This shows that:

If the pressure of a gas is constant, its volume increases with temperature.

The diffusion of gases

The movement of different particles among each other, so that they become evenly mixed, is called **diffusion**. A particle of ammonia gas has about half the mass of a particle of hydrogen chloride gas. So will it diffuse faster? Let's see:

1. Cotton wool soaked in ammonia solution is put into one end of a long tube. It gives off ammonia gas.
2. *At the same time*, cotton wool soaked in hydrochloric acid is put into the other end of the tube. It gives off hydrogen chloride gas.
3. The gases diffuse along the tube. White smoke forms where they meet.

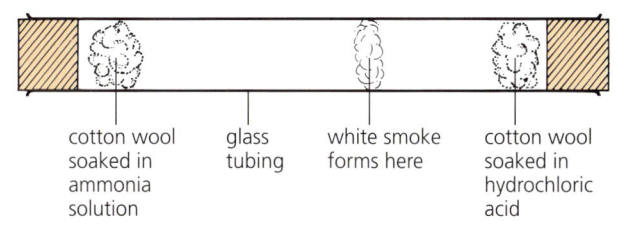

cotton wool soaked in ammonia solution | glass tubing | white smoke forms here | cotton wool soaked in hydrochloric acid

The sweet smell of freesias can fill a room. 'Smells' are caused by gas particles diffusing through the air. They dissolve in moisture in the lining of your nose. Then cells in the lining send a message to your brain.

The white smoke forms closer to the right-hand end of the tube. So the ammonia particles have travelled further than the hydrogen chloride particles, in the same length of time.

The lighter the particles of a gas, the faster the gas will diffuse.

Questions

1. What causes the pressure in a gas?
2. Why does a balloon burst if you keep on blowing?
3. A gas is in a sealed container. How do you think the pressure will change if the container is cooled? Explain your answer.
4. A gas flows from one container into a larger one. What happens to its pressure? Draw diagrams to explain.
5. Of all gases, hydrogen diffuses fastest. What can you tell from that?

FT 3 Composite materials

Often one material is combined with another to give a **composite** material with better properties. For example if you melt plastic film between sheets of glass, you get **safety glass**, which is shatterproof.

Wood, bone, ivory, and teeth are natural composite materials. Synthetic composite materials include concrete, reinforced concrete, glass-fibre-reinforced plastic or fibreglass, and plywood.

Bone

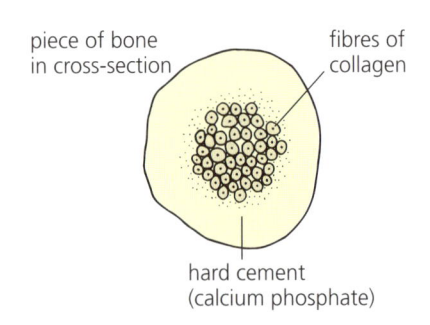

Bone is tough, flexible fibres of a protein called collagen, embedded in a hard cement of calcium phosphate. The cement contains the bone's living cells.

The result is hard, light, and slightly flexible, quite strong under tension and compression, supporting your flesh and whatever else you carry.

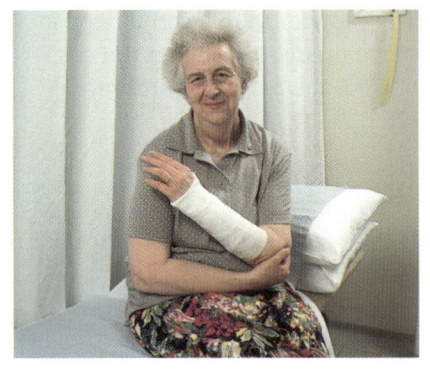

But calcium is continually lost from bones. If it's not absorbed again from food, the cement gets fragile and the bones break more easily.

Concrete

Concrete consists of **aggregate** (a mixture of stone chips and sand) bound together by **cement**. Cement is a mixture of calcium aluminate and calcium silicate, made by heating limestone with shale. For concrete, the cement is mixed with aggregate and water added.

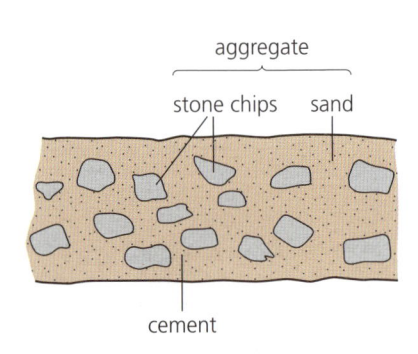

As the concrete 'sets', particles in the cement bond together forming a hard mass of crystals with water molecules trapped inside them. This is called **water of crystallization**.

The hard crystals and chips make the concrete strong under compression. But it is weak under tension – a concrete beam will crack if you place a heavy load on it.

To improve the tensile strength of concrete, steel rods or steel mesh is embedded in it. The result is called **reinforced concrete**. It is used for bridges and multi-storey buildings.

220

Reinforced plastic

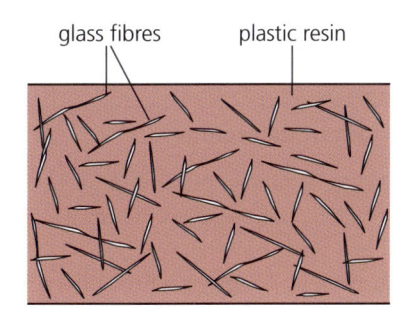

Glass-fibre-reinforced plastic is made by mixing glass fibres with a liquid plastic resin that sets hard. The result is . . .

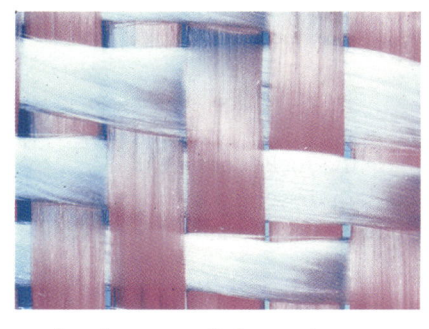

. . . hard, strong, light, resistant to corrosion and attack by chemicals, and does not shatter easily.

Glass-fibre-reinforced plastic is used for making boats, surfboards, and some car bodies.

Carbon fibres can be used instead, to give **carbon-fibre-reinforced plastic**. This is used for tennis rackets and golf clubs.

Composite fabrics

Most clothes these days are made of composite fabrics. These combine the hard-wearing and easy-care properties of synthetic fabrics with the texture and warmth of natural fabrics.

Cotton fabric is usually light and comfortable, but not very hard-wearing. It wrinkles when you wash it. Ironing cotton garments can be a big chore.

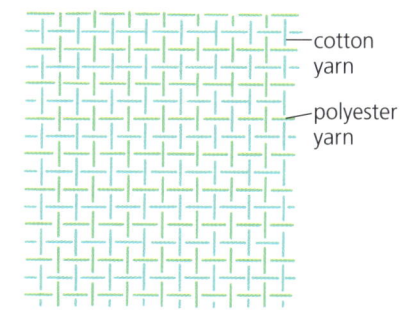

Cotton can be woven with polyester to give polyester cotton. This is easier to wash and dry, wears better, and needs less ironing. Good for shirts.

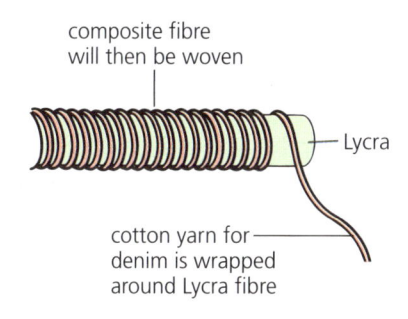

Denim is a tough cotton fabric. But combining the fibre with Lycra gives a tougher fabric with stretch. It keeps its shape well. Good for jeans.

Laminated materials

In some composite materials, the different materials are combined in sheets. These are called **laminates**. One example is safety glass. Another is **plywood**, where layers of different woods are glued together with the grain at right angles. In **plasterboard**, a layer of plaster is sandwiched between sheets of very strong paper. The cover of this book is a laminate of paper and a thin film of plastic.

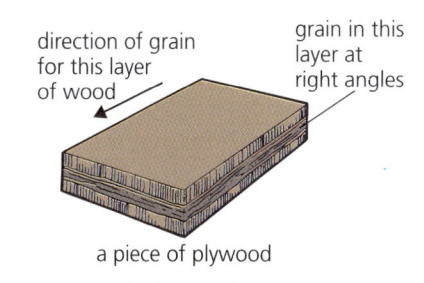

a piece of plywood

Questions

1 What is a composite material? Give two examples.
2 Explain why bone is strong under tension.
3 Would you expect plywood to bend easily? Why?
4 Name six composite materials used in your home.

FT 4 Soft and hard water

Scum alert!

If your tap water comes from an area where the rocks contain chalk, limestone, dolomite, or gypsum, it will contain dissolved calcium and magnesium compounds: sulphates and hydrogencarbonates. How can you tell? Soap will give you a clue.

Lather or scum?

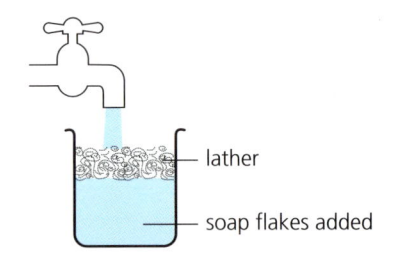

If soap lathers easily, it means the water contains very little calcium and magnesium compounds. It is **soft** water.

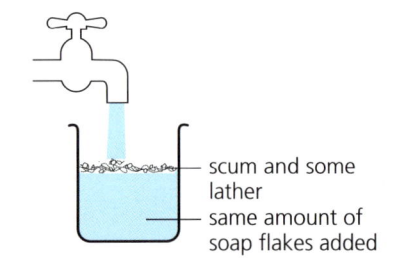

But a greyish scum and hardly any lather shows that larger amounts are present. The water is called **hard** water.

The scum forms because the calcium and magnesium compounds react with soap to give an insoluble product. For example:

calcium sulphate + sodium stearate ⟶ calcium stearate + sodium sulphate
                  (soap)   (scum)

Modern detergents

Modern detergents don't contain soap, so don't form a scum with hard water. But studies in America have shown that:

- hard water uses up more detergent than soft water
- hard water needs higher wash temperatures to get laundry clean
- fabrics tend to wear out up to 15% faster in hard water washes
- washing machines can wear out up to 30% faster in areas where water is very hard.

How water acquires hardness

Calcium hydrogencarbonate is the main cause of hard water. It forms when rain falls on rocks of **limestone** and **chalk**. These are made of calcium carbonate which is *not* soluble in water. But rainwater contains carbon dioxide dissolved from the air, which makes it acidic. So it reacts with the rocks to form calcium hydrogencarbonate which *is* soluble and ends up in our taps. The reaction is:

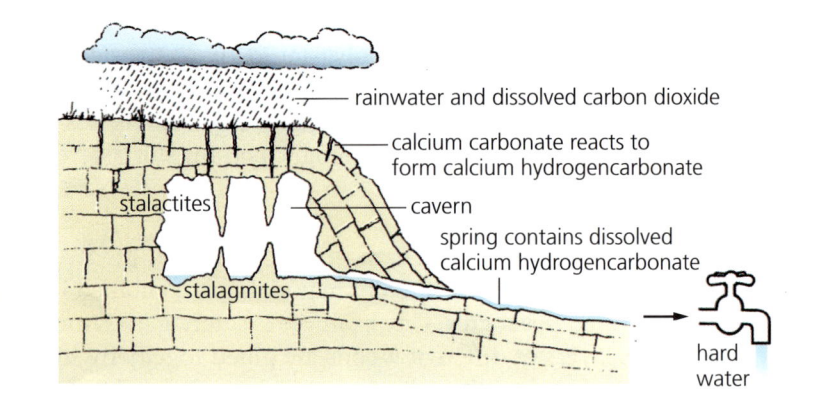

$$H_2O\ (l) + CO_2\ (g) + CaCO_3\ (s) \rightleftharpoons Ca(HCO_3)_2\ (aq)$$
  water   carbon dioxide   calcium carbonate   calcium hydrogencarbonate

In the same way, rainwater reacts with **dolomite** ($CaCO_3 \cdot MgCO_3$) and **gypsum** ($CaSO_4 \cdot 2H_2O$) to give the other compounds that make water hard. These reactions are part of the **weathering** process.

Temporary or permanent?

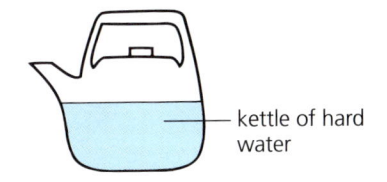

This water is hard: it contains calcium hydrogencarbonate. On heating, this compound breaks down to form calcium carbonate which is insoluble.

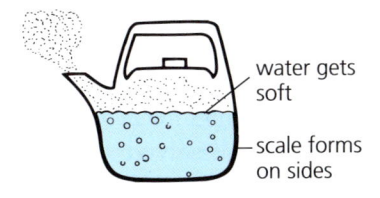

The new compound forms a hard **scale** on the kettle. But the water itself is now soft since the calcium hydrogencarbonate has been removed.

The same reaction occurs inside boilers and pipes in factories and central heating systems. Scale has blocked this pipe almost completely.

Since it can be removed just by boiling, the hardness caused by calcium hydrogencarbonate is called **temporary hardness**. Hardness caused by other compounds is called **permanent hardness** because boiling does not affect it. Even after boiling water with permanent hardness, you'll still get a scum with soap.

Hardness in water: good or bad?

Compare the advantages and disadvantages of hard and soft water. Which kind of water would *you* prefer?

Hard water	Soft water
Advantages <ul><li>The dissolved substances give it a pleasant taste.</li><li>The calcium compounds in it are good for bones and teeth.</li><li>There are fewer heart attacks in areas with hard water.</li><li>A coating of calcium carbonate inside pipes, boilers, and radiators helps to prevent corrosion.</li></ul> *Disadvantages* <ul><li>When the coating develops into a scale, it makes pipes, boilers, and radiators less efficient. It must be removed from time to time to prevent blockages.</li><li>Hard water uses up more soap than soft water does.</li><li>It leaves a messy scum, and is tougher on laundry.</li></ul>	*Advantages* <ul><li>It makes soap lather better and gets clothes cleaner.</li><li>Laundry uses less soap and can be done at lower temperatures.</li><li>It means less scum in your bath tub.</li><li>It leaves no scale in pipes or boilers.</li></ul> *Disadvantages* <ul><li>Soft water contains more *sodium* ions than hard water does. Sodium is linked to heart disease.</li><li>Soft water dissolves metals such as cadmium and lead from metal pipes. Lead is poisonous. Cadmium has been linked to hypertension. (But it also dissolves copper which your body needs.)</li></ul>

Questions

1 Hard water wastes soap. Explain why. Use an equation to help you.
2 Chalk is insoluble in water. Why does it dissolve in rainwater? Write an equation for the reaction.
3 What is the difference between *temporary* and *permanent* hardness?
4 Hardness in water costs money. Write down as many reasons as you can to explain why.

FT 5 Making hard water soft

Ways to soften hard water

Because of the problems it causes, hard water is often softened for use in factories, laundries, and homes. That means removing the dissolved calcium and magnesium ions. Below are ways to do this.

Boiling This removes temporary hardness, by causing calcium carbonate to precipitate. But it uses a lot of fuel which makes it expensive to do on a large scale – unless you need the hot water anyway. (You'll still end up with scale in the boiler!)

Distillation In distillation, the water is boiled and the steam collected, cooled, and condensed. Distilled water is pure water. *All* the dissolved substances have been left behind. Like boiling, it is an expensive option in terms of fuel. But it is essential for some purposes, for example for lab experiments and for making drugs.

Adding washing soda Washing soda is sodium carbonate, Na_2CO_3. It removes both temporary and permanent hardness by precipitating calcium carbonate. For example, it reacts with calcium sulphate like this:

$$Na_2CO_3\,(aq) + CaSO_4\,(aq) \longrightarrow CaCO_3\,(s) + Na_2SO_4\,(aq)$$
sodium carbonate + calcium sulphate → calcium carbonate + sodium sulphate

Bath salts contain sodium carbonate to soften the bath water. The sodium sulphate that forms does not affect soap. But the precipitate of calcium carbonate leaves a ring on your bath.

Ion exchange In ion exchange, unwanted ions are removed by replacing them with 'harmless' ions.

A typical ion exchanger is a container full of small beads. These are made of a special plastic called **ion exchange resin**, which has the 'harmless' ions weakly attached to it. They are usually sodium ions, as shown on the right. When hard water flows through, the calcium and magnesium ions in it switch places with the sodium ions and attach themselves to the resin. The sodium ions are carried away:

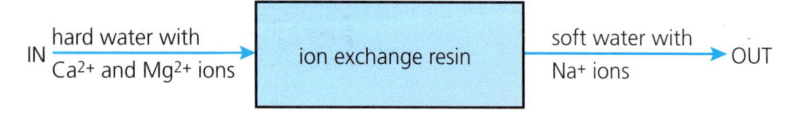

IN — hard water with Ca^{2+} and Mg^{2+} ions → [ion exchange resin] → soft water with Na^+ ions — OUT

After a time all the sodium ions have gone so no more hardness can be removed. The resin has to be **regenerated**. A concentrated solution of sodium chloride is poured into it. The sodium ions push the calcium and magnesium ions off the resin and the ion exchanger is ready for use again.

Other ions could be used instead of sodium for the resin. But sodium ions have one big advantage: salt is cheap!

It's advisable to use distilled water in steam irons. Why?

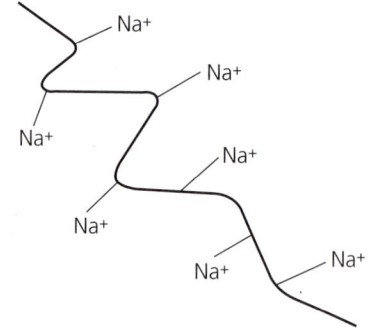

This is only part of a long molecule of resin. (Each bead contains many molecules.)

How well do these methods work?

A simple experiment will help you find out. Collect samples of water softened in different ways, and then compare them with distilled water using a lather test. Here's how to do it:
1 Measure a sample of water (say 20 cm^3) into a conical flask.
2 Add a little soap solution from a burette, and shake the flask carefully.
3 Add more soap solution, a little at a time, until a lather lasting at least half a minute is obtained.
4 Note the volume of soap solution used.

In the experiment in the photograph, water from the same tap was used for A, B, and C. For A it was left untreated. For B it was boiled. For C it was passed through an ion exchanger. D is distilled water. Here are the results:

Water sample	Volume of soap solution used (cm^3)
A (untreated tap water)	10
B (boiled)	7
C (ion exchanger)	0.5
D (distilled)	0.5

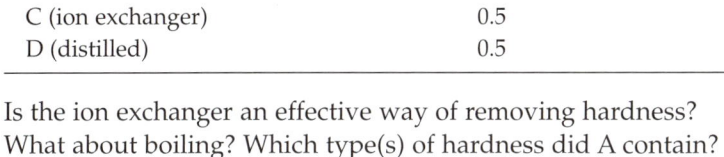

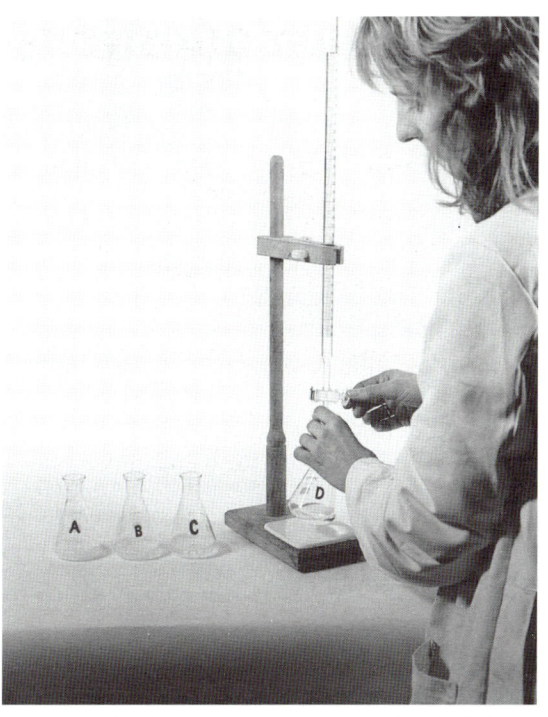

Is the ion exchanger an effective way of removing hardness? What about boiling? Which type(s) of hardness did A contain?

Comparing hardness in water samples, using the lather test.

A little more about ion exchangers

- Ion exchangers for water softening can be any size, from small ones designed for home use to large ones for factories.
- A typical ion exchanger for the home has two tanks. One contains the resin beads. The other contains a concentrated solution of sodium chloride (brine). A meter records how much water has gone through the resin. The brine is then automatically flushed through at an appropriate stage.
- Ion exchangers aren't used only for water softening. For example, they can be designed to remove nitrate ions, or to recover valuable metal ions from waste solution.
- They aren't always made of plastic either. Some are made of natural or synthetic compounds called **zeolites**. These are silicates in which some silicon atoms are replaced by aluminium atoms. They have an open porous structure.

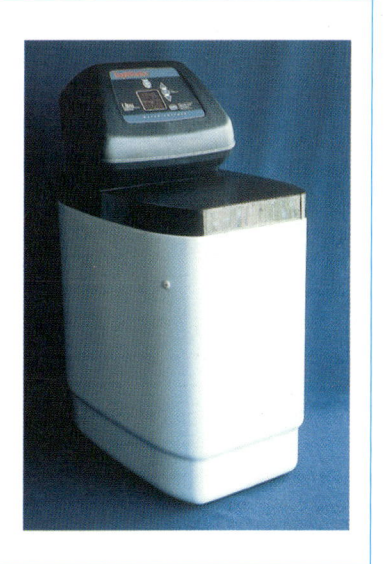

Questions

1 Which method of water softening would you expect to be: **a** cheapest? **b** most expensive?
2 Why is sodium carbonate used in bath salts?
3 **a** Explain how an ion exchanger works.
  b Suggest another ion to use instead of sodium.
4 Explain how the experiment above shows that:
  a the tap water contains both types of hardness
  b most of the hardness in it is permanent
  c ion exchange is just as effective as distillation in removing hardness

Exam-style questions

1. Aluminium is a widely used metal. It has some very useful properties.
   a. Why is aluminium the preferred metal for:
      i. overhead electricity cables?
      ii. window frames?
      (Give at least two reasons for each use.)
   Alloys of aluminium are often used rather than the pure metal.
   b. What is an alloy?
   c. Give two reasons for using alloys rather than pure metals.
   d. How is aluminium extracted from its ore?
   e. Why is aluminium quite an expensive metal?

2. Use ideas about *structure* and *bonding* to explain the following observations about some of the substances involved in the extraction of aluminium from aluminium oxide.
   a. Aluminium oxide has a *very high melting point*.
   b. Graphite electrodes *conduct electricity*.
   c. Cryolite *conducts electricity once it has become molten*.
   d. Oxygen is released as a *gas at room temperature*.
   e. Aluminium forms at the *negative electrode*.
   f. The aluminium obtained is both *malleable* and a *good conductor of electricity*.

3. Tin has been used as a metal for thousands of years. It is extracted in a similar way to iron, using carbon.
   a. Is the method of extraction consistent with tin being a reactive or an unreactive metal? Explain your answer.
   The main ore of tin is tin(IV) oxide, SnO_2.
   b. i. Write a word equation to show how tin is obtained from tin(IV) oxide.
      ii. What type of chemical reaction is this?
   Tin could also be obtained from an aqueous solution of one of its compounds, if a more reactive metal such as magnesium was added.
   c. i. Write a word equation to show how tin could be obtained from tin chloride.
      ii. What type of chemical reaction is this?
      iii. Why would this not be an economic way of extracting tin?
   d. Describe how you could purify tin using electrolysis.

4. Chromium is one of the transition metals.
   a. Where are the transition metals found in the periodic table?
   b. Describe the properties you would expect chromium to have. Include the following:
      i. electricity conductivity
      ii. melting point
      iii. hardness
      iv. appearance
      v. colour of compounds
   Iron is another transition metal used to make ships' hulls. Because iron rusts it is often protected by placing lumps of zinc in contact with the iron.
   c. i. What two things cause iron to rust?
      ii. What type of chemical reaction is rusting?
   d. Explain how zinc stops an iron hull rusting.
   e. What other methods are used to prevent iron from rusting?

5. Copper(II) sulphate can be made in the lab using the reaction between copper(II) oxide and sulphuric acid. An excess of copper oxide is used. The equation for the reaction is:

   $$CuO\,(s) + H_2SO_4\,(aq) \xrightarrow{heat} CuSO_4\,(aq) + H_2O\,(l)$$
   copper(II) oxide (black powder) + dilute sulphuric acid → copper(II) sulphate (blue solution) + water

   a. What would you observe that would indicate that the chemical reaction had taken place?
   b. i. Why is an excess of copper(II) oxide used?
      ii. How could you remove unreacted copper(II) oxide?
   c. Describe how you would obtain crystals of copper(II) sulphate from the blue solution.

6. Limestone is a rock which often contains fossils. Many of the fossils are the remains of marine animals.
   a. i. Explain how this type of limestone formed.
      ii. What type of rock is limestone?
      iii. What is the chemical name for the mineral in limestone?
   The same mineral is found in the *metamorphic* rock marble.
   b. i. How does limestone become marble?
      ii. Would you expect to find fossils in the marble? Explain your answer.

c Name one other rock which contains the same mineral as limestone and marble.

The mineral can be identified by dropping dilute acid on to a sample of the rock.

d i What observation tells you that the mineral is present?
  ii Write a word equation for the reaction.

When limestone is heated a decomposition reaction occurs.

e i What is the common name of the product obtained from the decomposition?
  ii Name one use for the new material.

7 Crude oil is separated into a range of products by fractional distillation. Six of the fractions are shown in the following table.

Fraction	Boiling range (°C)	Number of carbon atoms in each molecule
1	−40 to 40	1 to 4
2	40 to 100	4 to 8
3	100 to 160	6 to 10
4	160 to 250	10 to 16
5	250 to 300	16 to 20
6	300 to 350	20 to 25

a Which fraction:
  i contains gases at room temperature?
  ii contains molecules with 12 carbon atoms in them?
  iii could be used as petrol?
b Which of three 'middle' fractions, 3, 4, or 5, will be the most:
  i volatile?
  ii viscous?
  iii flammable?

Fraction 6 is fuel oil. It is cracked to make more useful products.

c i Why does fuel oil on its own have only limited use?
  ii Explain how cracking is carried out.
  iii In which two ways are the carbon chains different in the molecules before and after cracking?

8 Propane and propene are both hydrocarbons obtained from the fractional distillation and cracking of crude oil.
  a Draw the displayed formula of propane and propene.
  b Which (propane, propene, or both):
    i can be used to make polymers?
    ii contain only carbon and hydrogen atoms?
    iii have a chain length of 3?
    iv can only be obtained from cracking other chemicals?
    v burn well in oxygen?
    vi are gases at room temperature?
    vii will not react with bromine water?
    viii is a molecule containing 6 hydrogen atoms?
    ix can be used as a fuel?

9 The following equation represents the fermentation of glucose to form alcohol and carbon dioxide.

$$C_6H_{12}O_6 \;(aq) \longrightarrow 2C_2H_5OH \;(aq) + 2CO_2 \;(g)$$
glucose   alcohol   carbon dioxide

a What does (aq) stand for?
b What is added as a catalyst for this reaction?
c What does this catalyst contain that speeds up this reaction?
d Why must the temperature of the reaction be controlled carefully?
e Calculate the relative formula mass (M_r) for glucose.
  (A_r: C = 12, H = 1, O = 16.)
f What mass of alcohol could be obtained from 90 kg of glucose?

10 The reduction of silver compounds in the presence of light plays an important role in black and white photography. The 'black' areas are very small particles of silver.
   a i Write a word equation for the reduction of silver bromide (AgBr).
      ii Why is the reaction called a reduction?
      iii Why is this reaction called a photochemical reaction?

Controlling temperature is very important in the developing processes. Here are some typical results using fresh developer each time.

temperature of developer / °C	time needed to develop the film/s
40	155
38	201
36	248
34	300
32	345

   b Plot a graph of the results.
   c Identify a trend from these results.
   d Give a scientific reason that supports the trend.
   e Estimate the time needed to develop a film at 30 °C.
   f How could the reliability of the results be improved?
   g If the experiment is repeated without using fresh developer each time, the time needed at each temperature is longer than before. Give a scientific reason why this happens.
   h What would be the effect on the time of using a smaller particle size of the silver bromide? Explain your answer.

11 Copper(II) oxide reacts with hydrogen according to the equation:

$$CuO\,(s) + H_2\,(g) \xrightarrow{heat} Cu\,(s) + H_2O\,(g)$$

   a Explain why in this reaction copper(II) oxide is reduced.
   b Which substance is oxidized in this reaction?
   c Calculate the relative formula mass (M_r) for copper(II) oxide.
      (A_r: Cu = 64, H = 1, O = 16.)
   d What mass of copper would be obtained from 4 g of copper(II) oxide?
   e What volume of hydrogen would be required to bring about this reaction?

(The relative formula mass (M_r) in grams, of a gas, has a volume of 24 000 cm^3 at room temperature and pressure.)

   f If copper(I) oxide (Cu_2O) was used in place of copper(II) oxide, would more or less hydrogen be required to obtain the same mass of copper?

12 Ethene and bromine react in an addition reaction to form dibromoethane:

$$C_2H_4\,(g) + Br_2\,(l) \longrightarrow C_2H_4Br_2\,(l)$$
ethene   dibromoethane
colourless dark red   colourless

   a Draw a displayed formula for ethene. (It contains a carbon–carbon double bond.)
   b Draw a displayed formula of dibromoethane. (It contains only single bonds.)
   c Explain why it is called an addition reaction.
   d What colour change would tell you that a reaction had occurred?

An approximate value for the energy change in this reaction can be found using bond energies:

Bond	Bond energy / kJ
C–C	347
C=C	612
C–H	413
Br–Br	193
C–Br	290

   e Use the bond energies to calculate the energy:
      i   to break the bonds in ethene
      ii to break the bonds in bromine
      iii to make the bonds in dibromoethane
   f Calculate the energy difference in bond breaking and bond making.
   g Is this reaction exothermic or endothermic? Explain your answer.

13 Nitric acid is manufactured from ammonia. The first step involves forming nitrogen monoxide by passing ammonia and oxygen over a heated platinum gauze. The word equation is:

ammonia + oxygen ⟶ nitrogen monoxide + water

   a Which chemical is oxidized in this reaction?
   b What is the purpose of the platinum gauze? During the process the gauze glows red hot.
   c i Is the reaction exothermic or endothermic?
      ii Which has the stronger bonds, the reactants or the products? Explain your answer.

The nitrogen monoxide is reacted further with oxygen and water to form nitric acid.

   d i Write a word equation for this reaction.
      ii Which chemical is oxidized in this reaction?

Much of the nitric acid made is reacted with more ammonia to form a fertilizer. Copy and complete the equation using the chemical name for the fertilizer.

ammonia + nitric acid ⟶ _____ + water

14 Ammonia is produced by the Haber process.

$N_2 (g) + 3H_2 (g) \rightleftharpoons 2NH_3 (g)$

   a Name the two gases from which ammonia is made.
   b The reaction is reversible. What problem does this create when manufacturing ammonia?

The conditions for the manufacture are chosen to give a high percentage yield of ammonia. The yield is affected by temperature and pressure as the following table shows:

	100 atm pressure	200 atm pressure
400 °C	30% yield	40% yield
500 °C	15% yield	20% yield

   c What effect does increasing the pressure have on the yield of ammonia?
   d i What effect does increasing the temperature have on the yield of ammonia?
      ii Is the formation of ammonia exothermic or endothermic? Explain your answer.
   e Explain why the reaction is only economic if iron is also used in the process.

15 The formula for the compound iron(II) sulphate is $FeSO_4$.
   a Name the three elements in this compound.
   b What does the (II) refer to?
   c Calculate the relative formula mass (M_r) of this compound. (Show your working.)
      (A_r: Fe = 56, S = 32, O = 16.)

Iron(II) sulphate can be made from iron and dilute sulphuric acid in a conical flask. This is the equation for the reaction:

Fe (s) + H_2SO_4 (aq) ⟶ $FeSO_4$ (aq) + H_2 (g)

   d Draw a diagram of the apparatus used for this reaction.
   e What would you observe in the flask?
   f Calculate the mass of iron(II) sulphate that could be obtained from 5.6 g of iron. (Show your working.)
   g By how much would the flask and its contents lose mass during the reaction? Explain your answer.

16 When magnesium is added to hydrochloric acid, hydrogen is released according to this equation:

Mg (s) + 2HCl (aq) ⟶ $MgCl_2$ (aq) + H_2 (g)

   a What mass of hydrogen would be released if 24 g of magnesium were used?
      (A_r: Mg = 24, H = 1.)
   b What volume of gas is this?
      (The relative formula mass (M_r) in grams, of a gas, has a volume of 24 000 cm³ at room temperature and pressure.)
   c How much hydrogen would be collected if 0.024 g of magnesium were used?

17 *Weathering* of rock produces small bits of rock which may then be *eroded*.
   a Explain clearly what is meant by:
      i weathering
      ii erosion
   b Name the agents of erosion.
   c Explain how the shapes of the small bits of rock can be further changed during erosion.
   d The bits of rock are eventually *deposited*.
      i Explain what that means.
      ii Where would you expect to see only large pieces of rock? Explain your answer.

229

18 Mendeleev published his periodic table in 1869. The following shows how Groups 1 and 7 looked. These were the *outer* two groups in his table.

	Group 1		Group 7	
Period 1	H			
Period 2	Li		F	
Period 3	Na		Cl	
Period 4	K		Mn	
		Cu		Br
Period 5	Rb			
		Ag		I

Use information from this and the modern periodic table to answer these questions.

a Explain what is meant by Period 1, 2, etc.
b i How does Group 1 differ in the modern periodic table?
  ii Why is the modern table more appropriate for this group?
  iii What name is given to the Group 1 elements?
c i What name is given to the Group 7 elements?
  ii Why is Mn out of place in Group 7?
Below Mn is one of the several places where Mendeleev left a gap in his table.
d Suggest why Mendeleev left these gaps.
e Why is there no group to the right of Group 7 in Mendeleev's table?

19 The following table lists the atomic structures of two elements.

	Number of protons	Number of electrons	Number of neutrons
Element X	3	3	4
Element Y	9	9	10

a Draw diagrams to show the arrangement of the particles in an atom of each element.
b What is the mass number of each atom?
c i To which group of the periodic table does element X belong?
  ii To which group of the periodic table does element Y belong?
Element X reacts with Y to form a solid compound.
d Explain what happens to the outer electrons when the two elements react.

e What type of bond is formed between the two elements in the compound?
f Deduce the formula of the compound.
g What type of structure, giant or covalent, will the compound have?

20 Sodium chloride and hydrogen chloride have the formulae NaCl and HCl respectively. One of the chlorides is an ionic compound and the other is a covalent compound.
a i Which chloride contains an ionic bond?
  ii Draw a diagram to show the bonding.
b i Which chloride contains a covalent bond?
  ii Draw a diagram to show the bonding.
c i Which chloride has a molecular structure?
  ii Describe the type of structure for the other chloride.
d Is your answer to part c consistent with the fact that sodium chloride is a solid and hydrogen chloride is a gas at room temperature? Explain your answer.

21 Ammonium nitrate is a salt with the formula NH_4NO_3. It can be made in the laboratory. It is very soluble in water.
a i Name the acid and alkali used to make ammonium nitrate in the laboratory.
  ii What type of chemical reaction is this?
b The chemical reaction is *exothermic*.
  i Will the temperature of the liquid rise, or fall, during the reaction?
  ii Draw a simple energy diagram for this reaction.
c The ammonium nitrate solution contains ions.
  i What are *ions*?
  ii The formula of the nitrate ion is NO_3^-. What is the formula of the ammonium ion?
  iii Will the solution be a good, or bad, conductor of electricity?

22 The following table lists the melting points of the metals from Group 2 of the periodic table.

Element	Melting point/°C
beryllium	1278
magnesium	649
calcium	839
strontium	769
barium	725

a i What is the general trend in melting point as you go down Group 2?

ii One metal does not fit the general trend. Which one is it?
  iii Does the data support the idea that beryllium atoms are held together more strongly than barium atoms? Explain your answer.
 b The trend in chemical reactivity for the metals in Group 2 is the same as for the alkali metals.
   i Which of the metals listed above is the most reactive?
   ii Explain the trend in reactivity.
   iii Suggest a reason to explain why Group 2 metals are *less* reactive than Group 1 metals from the same period.
 c Barium reacts with water like this:
   Ba(s) + H$_2$O(l) $\rightarrow$ Ba(OH)$_2$(aq) + H$_2$(g)
   i Write a word equation for the reaction.
   ii What would you *see* when barium is added to water?
   iii What colour change would you see if universal indicator was added to the water before the barium was added? Explain your answer.

23 Chlorine is one of the products of the chor-alkali industry. Thousands of tonnes of it are made each year from sodium chloride.
 a What is the main source of the sodium chloride used in this industry, in the UK?
 The chlorine is obtained by electrolysis.
 b i Explain why the sodium chloride is dissolved in water before it is electrolysed.
   ii Suggest reasons why it is dissolved rather than melted.
   iii At which electrode is the chlorine obtained?
   iv Describe a chemical test for chlorine.
   v What effect does chlorine have on bacteria in drinking water?
 Chlorine is also used to make antiseptics. One of these has the formula C$_6$H$_3$OCl$_3$.
 c Calculate the percentage of chlorine in this compound, (A_r: C = 12, H = 1, O = 16, Cl = 35.5)

24 Ethanol is one product from the fermentation of sugar. The chemical reaction is shown by this equation:
 C$_6$H$_{12}$O$_6$(aq) $\rightarrow$ 2C$_2$H$_5$OH(l) + CO$_2$(g)
 a What does the *(aq)* represent in the equation?
 b Name the gas given off during fermentation.
 c What else must be present for reaction to occur?

 d Under what conditions does this reaction work best?
 e How is this reaction used by the baking industry?
 Starting at room temperature, the reaction was studied in the laboratory and three stages were observed. They were:
   Stage A: reaction started slowly
   Stage B: reaction rate increased rapidly
   Stage C: reaction slowed down
 f i Sketch a graph (rate of reaction against time) to show these three stages.
   ii Suggest a way of measuring the rate of this reaction.
 The reaction is known to be exothermic.
 g i Suggest a reason why the reaction rate increases as shown by stage B.
   ii Why does the reaction begin to slow down eventually?

25 a Copy and complete this table about the particles that make up the atom:

Particle	Mass	Charge
proton		
electron		
neutron		

 Most atoms exist in more than one form.
 b i What name is given to different atoms of the same element?
   ii How do the atoms differ?
 Some atoms are radioactive. Many of these emit alpha particles. An alpha particle is a helium nucleus, represented by: ^{4}He^{2+}
 Use the **periodic table** to answer these questions.
 c i How will the atomic number change when an atom emits an alpha particle?
   ii How will the mass number change when an atom emits an alpha particle?
   iii Describe the atom that is obtained when an atom of radon-222 emits an alpha particle. (Radon-222 has a mass number of 222.)
   iv Describe the atom that emits an alpha particle to produce an atom of thorium-232.

Multiple-choice questions

Module 5 Metals

1. Transition metals, like other metals, are good conductors of electricity. What further property makes them suitable for use as a structural material?
   A They corrode in air.
   B They form coloured compounds.
   C They are good conductors of heat.
   D They are hard, tough, and strong.

2. Two electrical circuits are shown below.

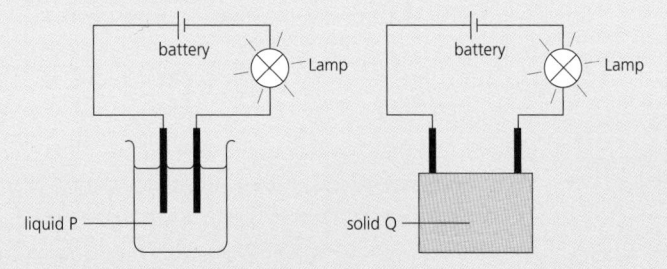

   The lamps are lit in both circuits. What could P and Q be?

	P	Q
A	bromine	silver
B	mercury	silver
C	naphtha	iron
D	molten sulphur	copper

3. Aluminium is used to make drinks cans. The following property makes it suitable for this:
   A it can be recycled
   B it is not corroded by acids
   C it has a low density
   D it is shiny and attractive

4. Carbon, aluminium oxide, and iron(III) oxide are mixed. The mixture is lit. What is reduced?
   A both aluminium oxide and iron(III) oxide
   B carbon dioxide from the air
   C iron(III) oxide only
   D aluminium oxide only

5. Nickel is between iron and copper in the reactivity series. Which of the following metals will react with nickel(II) oxide?
   A iron only
   B aluminium and iron
   C aluminium and copper
   D copper only

6. Part of the reactivity series is shown below.

   zinc
   tin
   lead
   silver

   In which of the following mixtures will there be *no* reaction?
   A lead with tin chloride solution
   B tin with lead nitrate solution
   C tin with silver nitrate solution
   D zinc with tin chloride solution

7. When molten lead bromide is electrolysed:
   A the number of lead ions in the beaker decreases
   B lead ions give up electrons to form lead
   C lead ions are oxidized to form lead
   D bromide ions are reduced to form bromine

8. The diagram shows a chemical cell. The voltage increases as the difference in reactivity of the two metals increases.

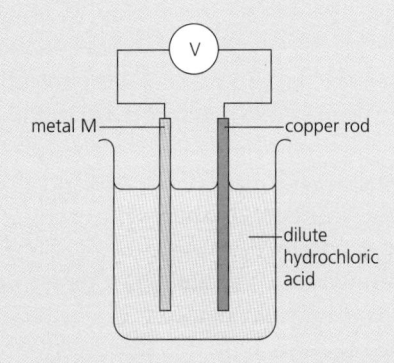

   The highest voltage would be observed if metal M was:
   A silver
   B magnesium
   C iron
   D zinc

9. Aqueous solutions of P and Q are mixed. The result is a solution with pH 7. Which could be the pH values of solutions P and Q?

	pH of solution P	pH of solution Q
A	3	5
B	3	12
C	7	11
D	8	13

10 Which **two** of the following reactants could be used to make ammonium chloride by neutralization?
   A sulphuric acid
   B ammonia solution
   C sodium chloride
   D hydrochloric acid
   E ammonium sulphate

11 Choose from the list the **two** substances that are needed to complete the word equation:
   ____ + copper oxide → copper sulphate + ____
   A hydrochloric acid
   B copper chloride
   C water
   D sulphuric acid
   E hydrogen

12 Which **two** of the following will produce hydrogen gas?
   A sodium + water
   B copper + hydrochloric acid
   C nickel oxide + sulphuric acid
   D sodium hydroxide + hydrochloric acid
   E zinc + sulphuric acid

13 Aluminium is used for overhead electricity cables for the national grid. What **two** properties of aluminium make it suitable for this purpose?
   A it has a low density
   B it is a good conductor of heat
   C it is shiny
   D it can be made into thin sheets
   E it is a good conductor of electricity

14 Lithium is one of the alkali metals. Which **two** of these properties does lithium have?
   A forms coloured compounds
   B has a low melting point
   C reacts with non-metals to form ionic compounds
   D has a high density
   E is chemically unreactive

15 Copper is purified using electricity. Which **two** of the following words describe this process?
   A neutralization
   B redox
   C displacement
   D electrolysis

Questions 16 to 20 refer to the extraction of iron in the **blast furnace**.

16 Which correctly describes the charge of raw materials used in the blast furnace?
   A iron ore, carbon monoxide
   B iron ore, coke, limestone
   C iron ore, coal, limestone
   D iron ore, carbon dioxide, air

17 Which substance is mainly responsible for reducing iron ore to iron?
   A oxygen
   B carbon
   C carbon dioxide
   D carbon monoxide

18 Iron can be extracted in the blast furnace because iron…
   A is below carbon in the reactivity series
   B has a low melting point
   C readily corrodes in the presence of air
   D is more reactive than aluminium

19 The waste gas from the blast furnace contains
   A unused air
   B carbon dioxide only
   C carbon monoxide
   D nitrogen and carbon dioxide

20 In the extraction of iron from iron ore in the blast furnace…
   A iron atoms are reduced to iron ions
   B iron ions lose electrons to form iron atoms
   C more oxygen is added to the iron ore
   D iron ions gain electrons to form iron atoms

21 In extracting aluminium from aluminium oxide, the aluminium oxide is dissolved in cryolite. This is because…
   A aluminium is too reactive on its own
   B it makes it easier to melt the aluminium oxide
   C no ions form unless cryolite is added
   D it prevents the electrodes being worn away

22 Electrolysis is used to extract several metals because…
   A many metals are more reactive than carbon
   B many metals are difficult to melt
   C many metals are less reactive than aluminium
   D it reduces the cost of extraction

233

Module 6 Earth materials

1. Crude oil is...
   A a renewable fuel
   B a fossil fuel
   C only found under the sea
   D always found with coal

2. What is the main substance that makes up natural gas?
   A kerosene
   B propane
   C methane
   D ethene

3. The diagram shows some apparatus set up to separate the fractions in petroleum (crude oil). There is a mistake in the set-up.

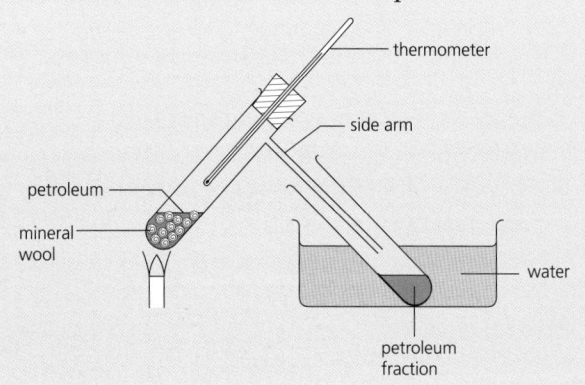

   What is the mistake?
   A the side arm is the wrong length
   B the fraction should be collected in a flask
   C there should be oil not water in the trough
   D the bulb of the thermometer is in the wrong position

4. The fractional distillation of crude oil gives the following products. Which one has the lowest boiling point?
   A kerosene
   B bitumen
   C naphtha
   D gasoline (petrol)

5. One of the following is *not* a fuel. Which one?
   A hydrogen
   B propane
   C oxygen
   D alcohol

6. Which substance produced when fossil fuels are burned does *not* cause damage to limestone buildings?
   A nitrogen oxides
   B carbon dioxide
   C carbon monoxide
   D sulphur dioxide

7. Which of the following results in more ultraviolet radiation reaching Earth's surface?
   A acid rain
   B destruction of the ozone layer
   C deforestation
   D excessive use of pesticides

8. Hexane is an alkane with the formula C_nH_{14}. What is the value of n?
   A 6
   B 7
   C 8
   D 10

9. Butane (C_4H_{10}) and butene (C_4H_8) can be distinguished using:
   A hydrogen gas
   B chlorine gas
   C bromine water
   D hydrogen bromide solution

10. Which row shows the correct polymer that results from polymerization of the given monomer?

234

11 Part of a polymer molecule is shown below.

$$-\underset{H}{\underset{|}{\overset{Cl}{\overset{|}{C}}}}-\underset{Cl}{\underset{|}{\overset{CH_3}{\overset{|}{C}}}}-\underset{H}{\underset{|}{\overset{Cl}{\overset{|}{C}}}}-\underset{Cl}{\underset{|}{\overset{CH_3}{\overset{|}{C}}}}-$$

Which of the following monomers can be used to make this polymer?

A $\quad\underset{CH_3}{\overset{Cl}{>}}C=C\underset{Cl}{\overset{H}{<}}$

B $\quad\underset{H}{\overset{Cl}{>}}C=C\underset{H}{\overset{CH_3}{<}}$

C $\quad\underset{Cl}{\overset{Cl}{>}}C=C\underset{H}{\overset{CH_3}{<}}$

D $\quad\underset{CH_3}{\overset{CH_3}{>}}C=C\underset{H}{\overset{Cl}{<}}$

12 The Earth's atmosphere changed when living things appeared. Which of the following helped to *reduce* the level of carbon dioxide in the atmosphere?
  A activity of volcanoes
  B the increase in green plants
  C the use of wood as a fuel
  D the development of the ozone layer

13 Today's atmosphere contains about 80% nitrogen. Which of the following caused the release of nitrogen into the atmosphere?
  A photosynthesis
  B the burning of fossil fuels
  C the activity of animals
  D the activity of denitrifying bacteria

14 The Earth's ozone layer…
  A developed from the oxygen in the atmosphere
  B is the cause of global warming
  C prevents photosynthesis
  D was caused by the use of CFCs

15 Limestone forms when quicklime when heated. The chemical name for quicklime is…
  A calcium carbonate
  B sodium carbonate
  C calcium oxide
  D calcium hydroxide

16 Slaked lime is made from quicklime by…
  A adding water
  B heating in a rotary kiln
  C thermal decomposition
  D mixing it with cement

17 Which of the following is **not** part of the layered structure of the Earth?
  A crust
  B mantle
  C magma
  D inner core

18 Which of the following is correct about the structure of the Earth?
  A the crust is less dense than the average density of the Earth
  B the inner core is liquid
  C the mantle is mainly iron and nickel
  D the mantle and core have the same density

19 The Earth's crust is broken into a number of large pieces called tectonic plates. Which of the following changes is associated with plates moving apart?
  A the formation of a subduction zone
  B the melting of rock to form magma
  C the formation of an oceanic ridge
  D the formation of mountain ranges

20 Magnetic reversal patterns…
  A show the strength of the magnetic field today
  B are found in intrusive igneous rock
  C show the direction of the Earth's magnetic field in the past
  D are found everywhere tectonic plates meet

21 Which is **not** true about magnetic reversal patterns?
  A found in sedimentary rock
  B occur parallel to oceanic ridges
  C occur in iron-rich minerals
  D the patterns are symmetrical about a central line

Revision and exam guidance

How to revise

There is no one method of revising which works for everyone. It is therefore important to discover the approach that suits you best. The following rules may serve as general guidelines.

Give yourself plenty of time

Leaving your revision until the last minute reduces your chances of success. There are very few people who can revise everything 'the night before' and still do well in an examination the next day. You need to plan your revision timetable for some weeks before the examinations start.

Plan your revision timetable

Plan your revision timetable well before the examinations start. Once you have done this, follow it – don't be sidetracked. Stick your timetable somewhere prominent where you will keep seeing it – or better still put several around your home!!!!

Relax

You will be working very hard revising. It is as important to give yourself some free time to relax as it is to work. So, build some leisure time into your revision timetable.

Ask others

Friends, relatives, teachers will be happy to help if you ask them. Go and talk to them if you are having any difficulties – don't just give up on something that is causing you a problem. And don't forget your parents too!

Find a quiet corner

Find the conditions in which you can revise most efficiently. Many people think they can revise in a noisy, busy atmosphere – most cannot! And don't try and revise in front of the television – it doesn't generally work. Revision in a distracting environment is very inefficient.

Use routemaps/checklists/pathways to help you

Use routemaps, checklists or other listing device to help you work your way logically through the material. When you have completed a topic, mark it off. You can also mark off topics you already feel confident about. That way you won't waste time revising unnecessarily.

Make short notes, use colours

As you read through your work or your textbooks make brief notes of the key ideas and facts as you go along. But be sure to concentrate on understanding the ideas rather than just memorizing the facts. Use colours and highlighters to help you.

Practise answering questions

As you finish revising each topic try answering some questions. At first you may need to refer to your notes or textbooks. As you gain confidence you will be able to attempt questions unaided, just as you will in the exam.

Give yourself a break

When you are working, work for perhaps an hour then reward yourself with a short break of 15 to 20 minutes while you have a coffee or cola; then go back for another period of revision.

Success in examinations

If the thought of an examination fills you with dread, then you are probably not fully prepared. Of course everybody will be a bit nervous about an important examination, but there is no reason why you should fall apart and fail to do yourself justice if you know what to expect before you go in.

If you have worked hard for two years and put in a lot of effort, there are some steps you can take to ensure that your examination results reflect this.

Be prepared

Make sure you have everything you need ready the night before, including pens, pencils, ruler, and calculator. Your teacher can tell you what mathematical skills you are expected to have, or they are listed in your syllabus. Make sure you are familiar with these well in advance.

Read carefully

Before you start, read the paper all the way through and make sure you know exactly what you have to do. Turn over all the pages. Then, when you are ready to start, read each question more than once to make sure it really does say what you think it says. Follow the instructions to the letter, as marks can only be awarded for precise answers to the question asked, not for any other information about the subject.

Plan your time

Work out how much time you should spend on each question, based on how many marks it has. Don't spend ages racking your brains to think of the answer to the last part of a question – go on and start the next question instead. You will probably get more marks for answering the beginning of all the questions than for finishing one or two questions completely, but not attempting later questions. If you have time left at the end, you can go back and think about the more difficult parts when you will be feeling more relaxed.

Present your work clearly

Remember that the examiner will have a large pile of papers to mark in a short space of time. Struggling to read your writing or to follow an illogical argument will not help, so write as clearly as you possibly can in the time available and think through what you are going to write before you start. The examiner is trying to award you marks – make it easy to find them.

If you are drawing diagrams, draw them clearly and simply, using single lines where appropriate. Label the diagrams and make sure the label lines point exactly to the relevant structures.

Stay calm

If you find a question you have no idea about, do not panic! Breathe slowly and deeply and have another look. It's likely that if you stay calm and think clearly, it will start to make more sense, or at least you may be able to answer part of it. If not, then don't agonise about it but concentrate first on the questions you *can* answer.

Module summaries/checklists

Photocopy the lists of topics below and put a cross against those that are *not* included in your examination specification. (Your teacher should be able to tell you which they are.) Tick off the other items on the list as you revise. The spread numbers in brackets tell you where to find more information.

Module 5 Metals

- ❏ The position of the metals in the periodic table (5.01)
- ❏ The properties of Group 1 (5.02)
- ❏ The general properties of the transition elements (5.03)
- ❏ Comparing how different metals react (5.04)
- ❏ The reactivity series, based on their reactions (5.05)
- ❏ What happens when a metal is:
  heated with the oxide of a less reactive metal (5.05)
  placed in a solution of a compound of a less reactive metal (5.05)
- ❏ The six most common elements in Earth's crust (5.06)
- ❏ The definition of a metal ore (5.06)
- ❏ How the method of extraction depends on the metal's reactivity (5.07)
- ❏ Why extraction means the ore is reduced (5.07)
- ❏ Reduction is loss of oxygen (5.07)
- ❏ Reduction is also gain of electrons (5.07)
- ❏ Oxidation is gain of oxygen (5.07)
- ❏ Oxidation is also loss of electrons (5.07)
- ❏ How oxidation and reduction always occur together (5.07)
- ❏ What a redox reaction is (5.07)
- ❏ Using OIL RIG to identify what's oxidized and what's reduced (5.07)
- ❏ How the properties of metals dictate their uses (5.08)
- ❏ How a metal is made more useful by turning it into an alloy (5.08)
- ❏ Some common alloys and their uses (5.08)
- ❏ How iron is extracted in the blast furnace (5.09)
- ❏ What cast iron is, and what it is used for (5.09)
- ❏ How iron is turned into steels (5.09)
- ❏ What an electrolyte is (5.10)
- ❏ What happens to the ions during electrolysis (5.10)
- ❏ How to write ionic equations to show what happens at electrodes (5.10)
- ❏ How metal and non-metal ions behave during electrolysis (5.10)
- ❏ Electrolysis as a redox reaction (5.10)
- ❏ The general rules for the electrolysis of aqueous solutions (5.10)
- ❏ How aluminium is mined and extracted (5.11)
- ❏ Details of the electrolysis of aluminium (5.11)
- ❏ Properties and uses of aluminium (5.11)
- ❏ How electrolysis is used to purify copper (5.12)
- ❏ How to electroplate a metal object (5.12)
- ❏ What corrosion is (5.13)
- ❏ How both air and water are needed for rusting (5.13)
- ❏ Ways to prevent rusting (5.13)
- ❏ What sacrificial protection is (5.13)
- ❏ Why corrosion is limited, for aluminium (5.13)
- ❏ How metals changed the world (5.14)
- ❏ Common acids and their formulae (5.15)
- ❏ Strong and weak acids, with examples (5.15)
- ❏ Common alkalis and their formulae (5.15)
- ❏ Strong and weak alkalis, with examples (5.15)
- ❏ How acids and alkalis affect litmus paper (5.15)
- ❏ Some examples of neutral substances (5.15)
- ❏ The pH scale and what the numbers tell you (5.15)
- ❏ What universal indicator paper is (5.15)
- ❏ The reactions of acids (5.16)
- ❏ Using transition metal oxides and hydroxides (bases) to make salts (5.16)
- ❏ What makes a solution acidic (5.17)
- ❏ What makes a solution alkaline (5.17)
- ❏ What goes on during a neutralization reaction (5.17)
- ❏ Using indicator to tell when neutralization is complete (5.17)
- ❏ How to carry out a titration (5.17)
- ❏ Examples of neutralization outside the lab (5.18)
- ❏ Causes and effects of acid rain (5.18)

Module 6 Earth materials

- [ ] The uses of limestone, an important raw material (6.01)
- [ ] The problems associated with quarrying (6.02)
- [ ] Why oil is called a fossil fuel (6.03)
- [ ] Where oil is found and how it is extracted (6.03)
- [ ] What organic compounds are (6.03)
- [ ] What hydrocarbons are (6.03)
- [ ] Oil as a mixture of hydrocarbons (6.03)
- [ ] Why oil is a non-renewable resource (6.03)
- [ ] What refining means, and how it is carried out (6.04)
- [ ] Examples of the fractions produced in refining (6.04)
- [ ] What catalytic cracking is, and why it is useful (6.05)
- [ ] Ethene as a starting point for different plastics (6.05)
- [ ] The general formula of the alkanes (6.06)
- [ ] The names and structures of the first four alkanes (6.06)
- [ ] The properties of the alkanes (6.06)
- [ ] The general formula of the alkenes (6.06)
- [ ] The names and formulae of the three simplest alkenes (6.06)
- [ ] The bonding in ethene (6.06)
- [ ] The properties of the alkenes 6.06)
- [ ] What unsaturated compounds are (6.06)
- [ ] Testing for unsaturation using bromine water (6.06)
- [ ] What the terms monomer, polymer, and polymerization mean (6.07)
- [ ] How to write a chemical equation for an addition polymerization (6.07)
- [ ] Examples of plastics obtained from oil (6.07)
- [ ] The different structures and properties of thermosoftening plastics and thermosetting plastics (6.07)
- [ ] The problems of disposing of plastics (6.08)
- [ ] The impact of oil on the environment (6.09)
- [ ] The main pollutants produced by burning fossil fuels (6.09)
- [ ] Why carbon monoxide is poisonous (6.09)
- [ ] How carbon dioxide contributes to global warming (6.10)
- [ ] What a greenhouse gas is (6.10)
- [ ] The effects of global warming (6.10)
- [ ] Why countries are reluctant to fight against global warming (6.10)
- [ ] What the atmosphere is, and the layers that form it (6.11)
- [ ] What the composition of air is (6.11)
- [ ] How our atmosphere is different from Earth's earliest atmosphere, and why (6.11)
- [ ] What nitrifying, denitrifying, and anaerobic bacteria are (6.11)
- [ ] How the atmosphere remains in balance (6.12)
- [ ] The discovery of oxygen (6.12)
- [ ] How to measure the percentage of oxygen in air (6.12)
- [ ] What the ozone layer is, and how it protects us (6.13)
- [ ] How ozone is being depleted, and the cause (6.13)
- [ ] What CFCs are (6.13)
- [ ] The oceans as a solution (6.14)
- [ ] Why the oceans remain in balance (6.14)
- [ ] The role of the oceans in removing carbon dioxide (6.14)
- [ ] The oceans and global warming (6.14)
- [ ] The structure of planet Earth (6.15)
- [ ] How temperature, pressure, and density increase with distance below Earth's surface (6.15)
- [ ] The continental and oceanic crusts (6.15)
- [ ] The three types of rock in Earth's crust (6.15)
- [ ] What weathering is (6.16)
- [ ] Physical and chemical weathering, with examples (6.16)
- [ ] What erosion and transport are (6.16)
- [ ] How and why sediment is deposited (6.16)
- [ ] What you can learn from examining sediment (6.17)
- [ ] How sedimentary rock is formed (6.17)
- [ ] What fossils are, and how they provide useful information (6.18)
- [ ] How igneous rock is formed (6.19)
- [ ] How to tell a rock is igneous (6.19)
- [ ] Igneous rock from volcanic ash (6.19)
- [ ] The difference between magma and lava (6.19)
- [ ] The effect of cooling on crystal size, for igneous rock (6.19)
- [ ] How metamorphic rock is formed (6.20)
- [ ] Where metamorphic rock is usually found (6.20)
- [ ] The difference between contact and regional metamorphism (6.20)
- [ ] How rock gets weathered and eroded and eventually turned into new rock, in the rock cycle (6.21)
- [ ] How to sketch the rock cycle (6.21)
- [ ] The link between rock type and depth of burial (6.21)
- [ ] Examples of clues gained from studying rocks (6.21)
- [ ] What tectonic plates are, and how fast they move (6.22)
- [ ] What causes plates to move (6.23)
- [ ] What sea floor spreading is (6.23)
- [ ] The three ways plates can move (6.23)
- [ ] The meaning of constructive boundary, destructive boundary, conservative boundary and subduction (6.23)
- [ ] Why plate movements give rise to earthquakes and volcanic eruptions (6.24)

- ❏ How faults and folds result from plate movements (6.24)
- ❏ How fold mountains are formed by plate movements (6.24)
- ❏ Wegener's theory of continental drift, and evidence he used to support it (6.25)
- ❏ Ways of predicting earthquakes (6.26)
- ❏ Ways of predicting volcanic eruptions (6.26)

Module 7 Patterns of chemical change

- ❏ A_r and M_r (7.17)
- ❏ What the Law of Constant Composition says (7.18)
- ❏ How to calculate the percentage composition of a compound (7.18)
- ❏ What the formula of a compound tells you (7.19)
- ❏ What the empirical formula of a compound tells you (7.19)
- ❏ How to find the empirical formula by experiment (7.19)
- ❏ How to write a balanced chemical equation (7.20)
- ❏ How to calculate the masses of reactants and products, from a chemical equation (7.21)
- ❏ What the Law of Conservation of Mass states (7.21)
- ❏ How to calculate gas volumes from chemical equations (7.22)
- ❏ How to calculate amounts in electrolysis reactions (7.23)
- ❏ What rate of reaction means (7.01)
- ❏ Measuring the rate of a reaction that produces a gas (7.02)
- ❏ Using a graph to show how the rate of a reaction varies (7.02)
- ❏ How rate changes with concentration (7.03)
- ❏ How rate changes with temperature (7.03)
- ❏ How rate changes with surface area (7.04)
- ❏ What a catalyst is (7.04)
- ❏ Explaining rates using collision theory (7.05)
- ❏ How a catalyst affects activation energy (7.06)
- ❏ How different reactions need different catalysts (7.06)
- ❏ Examples of catalysts used in industry (7.06)
- ❏ What enzymes are (7.07)
- ❏ How enzymes work – the mechanism (7.07)
- ❏ The optimum conditions for enzymes (7.07)
- ❏ Traditional use of enzymes: baking, brewing and other examples (7.08)
- ❏ Modern uses of enzymes (7.09)
- ❏ The difference between batch and continuous processes (7.09)
- ❏ What exothermic and endothermic reactions are (7.10)
- ❏ How to draw simple energy diagrams for reactions (7.10)
- ❏ How bonds must be broken and new bonds formed, in a reaction (7.11)
- ❏ Why some reactions are exothermic and others endothermic (7.11)
- ❏ What bond energy is (7.11)
- ❏ What the activation energy for a reaction is (7.11)
- ❏ What a reversible reaction is, with examples (7.12)
- ❏ What dynamic equilibrium means (7.12)

- ❏ What Le Chatelier's principle states (7.13)
- ❏ Applying Le Chatelier's principle to improve the yield of ammonia (7.13)
- ❏ How ammonia is made in industry by the Haber process (7.14)
- ❏ What the chosen conditions for making ammonia are, and why (7.14)
- ❏ The elements plants need (7.15)
- ❏ What fertilizers are, with examples (7.15)
- ❏ How nitric acid is made in industry (7.15)
- ❏ Why fertilizers are important (7.16)
- ❏ Problems caused by fertilizers (7.16)

Module 8 Structures and bonding

- ❏ The definition of an element (8.01)
- ❏ What the periodic table is, and how it is set out and labelled (8.01)
- ❏ What a compound is (8.01)
- ❏ The structure of an atom (8.01, 8.06)
- ❏ Atomic number and mass number, and how they are linked (8.02)
- ❏ How to draw diagrams to represent the atoms of the first twenty elements (8.02)
- ❏ What isotopes are, and examples (8.03)
- ❏ What A_r is (8.03)
- ❏ What a radioisotope is (8.03)
- ❏ How electrons are arranged in shells (8.04)
- ❏ Past ideas about atoms and atomic structure (8.05, 8.06)
- ❏ Why elements form compounds (8.07)
- ❏ What an ion is, with examples (8.07)
- ❏ What a compound ion is, with examples (8.08)
- ❏ How to show an ionic bond on a diagram (8.10)
- ❏ The structure of solid sodium chloride (8.10)
- ❏ How to write the formula of an ionic compound (8.09)
- ❏ The general properties of ionic solids (8.10)
- ❏ What a covalent bond is, with examples (8.11)
- ❏ How to show a covalent bond on a diagram (8.11)
- ❏ Properties of molecular substances (8.12)
- ❏ Two giant covalent structures: diamond and graphite (8.12)
- ❏ How atoms are held together in a metal (8.12)
- ❏ The general properties of metals (8.12)
- ❏ How the periodic table developed through the work of many chemists (8.14)
- ❏ Mendeleev and the periodic table (8.14)
- ❏ The trends within Group 1, and why they occur (8.15)
- ❏ Why Group 0 elements are unreactive and monatomic (8.16)
- ❏ Properties and trends for Group 7 (8.16)
- ❏ Why Group 7 elements are reactive and diatomic (8.16)
- ❏ The trend in reactivity within Group 7 (8.17)
- ❏ The displacement reactions of the halogens (8.17)
- ❏ How hydrogen chloride is made (8.18)
- ❏ The properties of hydrogen chloride (8.18)
- ❏ Compounds of the halogens (8.18)
- ❏ Where sodium chloride is found, and how it is extracted (8.18, 8.20)
- ❏ The electrolysis of sodium chloride (8.13)
- ❏ What the products of the electrolysis are used for (8.13)

Glossary

acid rain rain in which acidic gases such as nitrogen dioxide and sulphur dioxide are dissolved; the gases are produced when fossil fuels are burned

acidic solution a solution that has a pH lower than 7; an acidic solution contains H^+ ions

activation energy the energy required for a reaction to start

addition reaction reaction in which a molecule adds across the double bond of an alkene

alkali metals the Group 1 elements of the periodic table

alkaline earth metals the Group 2 elements of the periodic table

alkaline solution a solution that has a pH higher than 7; alkaline solutions contain OH^- ions

alkanes a family of saturated hydrocarbons with the general formula C_nH_{2n+2}

alkenes a family of unsaturated hydrocarbons with the general formula C_nH_{2n}; alkene molecules contain a carbon–carbon double bond

allotropes different forms of the same element, for example diamond and graphite are allotropes of carbon

alloy mixture containing at least one metal; alloys of metals are made to improve the properties of the individual metals

anaerobic bacteria bacteria that respire without oxygen

atmosphere the mixture of gases surrounding the Earth

atomic number the number of protons in the atoms of an element

atoms the particles that make up elements; atoms contain protons, neutrons, and electrons

bacteria microorganisms, some of which can cause disease; others are responsible for the breakdown of dead plant and animal material

balanced equation a chemical equation in which the numbers of each type of atom are the same on each side of the equation

basaltic crust the crust forming the floor of an ocean basin

base a chemical that neutralizes an acid to form a salt and water

batch process a process in which the reactants are mixed to react, then the process is stopped to collect the products

battery a series of electrical cells which convert chemical energy into electrical energy

biodegradable a biodegradable substance will decay naturally in the soil

blast furnace the chemical engineering plant that produces iron in industry

boiling the physical change from a liquid to a gas, that happens at the boiling point; bubbles form throughout the body of the liquid

boiling point the temperature at which a substance boils

bond energy the energy required to break a bond, or the energy released when the bond is formed

bonding the way that atoms are joined in a molecule/compound; the three main types of bonding are ionic, covalent, and metallic

brewing the fermentation process by which alcoholic drinks are made

brine the industrial name for a solution of sodium chloride in water

burette a piece of laboratory equipment for delivering and measuring volumes of liquid

burial (of rocks) layers of sedimentary rock are forced into the Earth's crust by other layers on top

burning an exothermic chemical reaction in which the reactant combines with oxygen to form an oxide

cast iron the product from the blast furnace; it contains a high percentage of carbon which makes the iron brittle

catalyst a substance that speeds up chemical reactions without being used up in the process

cell (biological) the building blocks for animals and plants

cell (electrical) a device that converts chemical energy to electrical energy

cement a chemical agent that binds other materials together; building cement is made from limestone

centrifuging separating solids and liquids by rapid spinning; the more dense material falls to the bottom

ceramic a hard, chemically resistant material made by heating clay in a kiln

chalk calcium carbonate

charge a charged particle has the ability to attract or repel another charged particle; like charges repel, opposite charges attract

chemical change a reaction, in which a new chemical substance is made

chlor-alkali industry industry based on the electrolysis of salt

chlorofluorocarbons (CFCs) organic compounds containing halogens, which used to be used as coolants and refrigerants; they destroy the ozone in the atmosphere

chromatography separating a mixture using a solvent and a support such as paper; the paper absorbs each part of the mixture differently

coke a source of carbon for the blast furnace, made by heating coal

combustion another name for burning

compound fertilizer a fertilizer that provides nitrogen, potassium, and phosphorus

compound ion an ion containing more than one element

compound a substance in which two or more elements are chemically combined

concentration describes the quantitiy of one substance dissolved in another

condensation the physical change in which a cooled gas turns into a liquid

condenser a piece of laboratory equipment used to cool a gas rapidly and turn it into a liquid

conductor a substance that allows heat or electricity to pas through it easily

conglomerate a rock made up of small stones cemented together

constructive boundary zone on the Earth's crust where the plates are moving apart and new crust is being formed

continental crust the parts of the Earth's crust that form the continents

continental drift (crustal movement) the slow movement of the continents over millions of years

continuous flow process a process in which the reactants flow into the system continuously, and the products flow out

convection current transfer of heat energy in gases and liquids in which hot areas rise through the cooler areas because they are less dense

core (Earth) the centre of the Earth, consisting mainly of the metals iron and nickel; it is made of two parts, the inner core which is solid and the outer core which is liquid

corrosion the wearing away of a metal from the surface inwards; the most common form of corrosion is rusting

covalent bond a chemical bond formed when two atoms share electrons in joining together

covalent compound a compound made up of molecules

cracking the chemical process in which long-chain hydrocarbon molecules are broken down to shorter, more useful molecules

cross-linking the chemical bonds that hold polymer chains together in a three-dimensional lattice; plastics with cross-linking are thermosetting plastics

crude oil the fossil fuel formed from the remains of tiny plants and animals that lived in the sea

crust (Earth) the solid outer part of the Earth, to a depth of about 65 km

crystallization crystals forming from a saturated solution as it cools

decomposition reaction a chemical reaction involving the breaking down of a reactant into smaller substances

denature a permanent change in the properties of a protein that takes place when the protein is heated, or under other extreme conditions

denitrifying bacteria bacteria that convert nitrates in the soil back to nitrogen in the air

density a measure of how 'heavy' something is – its mass per unit volume; the density of water is 1 g/cm^3

deposition the dropping of earth material being carried by water to form sediments

destructive boundary the parts of the Earth's crust where the plates are converging and one plate is dragged down or subducted

detergent a chemical that cleans

diatomic describes a molecule consisting of two atoms joined together by a covalent bond

displacement reaction a chemical reaction in which a more reactive element takes the place of a less reactive element in a compound

dissolving the process in which a soluble substance forms a solution

distillation separating a liquid from a solution or mixture by the process of evaporating and condensing

double bond a covalent bond in which two atoms share two pairs of electrons

ductility the ability of a metal to be drawn into a wire

dynamic equilibrium describes an equilibrium in which the situation is not static; the forward and backward reactions take place at the same rate

earthquake vibration in the Earth's crust caused by rock giving way under tension or pressure

economic process a process that is profitable to carry out

electrode the electrical conductors used to transfer electric current into and out of an electrolyte

electrolysis the process of splitting up a compound by passing an electric current through it

electrolyte a liquid through which the current is passed in an electrolysis reactions

electron shells the different energy levels where electrons are found

electronic configuration how the electrons in an atom are arranged

electrons the negatively charged particles in an atom

electroplating the coating of one metal with another using electricity

electrostatic charge a stationary electric charge, for example on an ion

element a substance that cannot be split into anything simpler by a chemical reaction

empirical formula a formula that shows the simplest ratio in which atoms combine in a compound

endothermic describes a chemical reaction that takes in energy from the surroundings

environment the surroundings

enzymes biological catalysts that speed up reactions in cells

equation a representation of a chemical reaction (reactants to products) using words or symbols

equilibrium when a reversible reaction is in balance; the forward and back reactions proceed at the same rate

erosion the picking up of material from the land by water, waves, wind or glaciers

eutrophication the result of excessive amounts of nutrients in lakes and rivers, causing oxygen deficiency and damage to plant and animal life

evaporation the physical change in which a liquid turns to a gas at a temperature below the boiling point

exothermic describes a chemical reaction that gives energy out to the surroundings

extract to remove a metal from the ore in which it occurs

extrusive rock igneous rock formed as lava from a volcano cooled quickly in the open

fault (in rock) a crack in rock layers caused by earth movements

fermentation the process by which enzymes in yeast break down sugars to form alcohol (ethanol) and carbon dioxide

fertilizers chemicals that provide the nutrients for plants to grow well

filtering separating solids from liquids by pouring through filter paper or another medium

filtrate the liquid obtained from filtration (from which the solid has been removed)

flammable describes a substance that burns easily

fold (in rock) layers of rock turned over on themselves as a result of earth movements

formula the symbol for a compound, made up of the symbols of its elements in the ratio in which they combine

fossil fuels fuels obtained from the remains of living things that died millions of years ago

fossil the ancient remains of a living thing that have left an impression in a rock

fractional distillation the process of separating two or more liquids which have different boiling points

fractions different parts of a mixture with different boiling point ranges, obtained from fractional distillation

freezing the physical change from a liquid to a solid, that happens at the melting point

fuel a chemical that is burned to release energy

galvanized iron iron coated with zinc, to prevent it rusting

giant structure a continuous two- or three-dimensional network of strong bonds

global warming an increase in the average temperature of the Earth's atmosphere

grain (in metal) a crystal of metal

greenhouse gas a gas such as methane or carbon dioxide that contributes to global warming

group a vertical column of the periodic table

243

Haber process the industrial process for making ammonia from nitrogen and hydrogen

halogens the Group 7 elements of the periodic table

hard water water that contains calcium and magnesium ions

hydrocarbon a compound containing the elements carbon and hydrogen only

hydrogenation adding hydrogen to the double bonds of an unsaturated compound to make it saturated

hydrolysis the breaking down of a compound by reaction with water

igneous rock rock formed from cooling of magma

immiscible liquids liquids that do not mix together

immobilized trapped, unable to move

incomplete combustion burning of fuels in a limited supply of oxygen, producing carbon (soot) and carbon monoxide

indicator a chemical that shows by its colour whether a substance is acidic or alkaline

inert describes a substance that does not react except under extreme conditions

insoluble describes a chemical that does not dissolve in a solvent

insulator a poor conductor of heat or electricity

intermolecular forces forces between molecules

intrusive rock igneous rock formed as magma inside the Earth cooled slowly

ion a charged atom or group of atoms formed by gaining or losing electrons

ionic bond a chemical bond formed when one atom transfers electrons to another atom

ionic compound a compound made up of ions, joined by ionic bonds

ionosphere an ionized region of the atmosphere, above the mesosphere

isotopes different forms of the same element; isotopes have the same number of protons and electrons but different numbers of neutrons

lattice a regular arrangement of particles

lava molten rock that flows from a volcano

lithosphere the crust and the outer mantle of the Earth; it is broken inot plates

magma molten rock

malleability the ability of a metal to be hammered into shape

mantle the part of the Earth between the crust and the core

mass number the total number of protons and neutrons in an atom

mass spectrometer an instrument used to determine accurately the masses of atoms and molecules

melting point the temperature at which a solid substance melts

melting the physical change from a solid to a liquid

mesosphere the region of the atmosphere between ths stratosphere and ionosphere

metal an element to the left of the zig-zag line of the periodic table, that has characteristic metallic properties

metallic bond the bond that holds metal atoms together

metamorphic rock rock changed by pressure and/or heat

minerals inorganic compounds that occur naturally in the Earth; rocks are made up of different minerals

molecular substance a substance made up of molecules

molecule a unit of two or more atoms held together by covalent bonds

monatomic describes elements existing as single atoms, e.g. the noble gases

monomers small molecules that join together to form polymers

native describes a metal that is found in the earth as the element

neutral (electrical) describes particles that have no charge

neutral (solutions/liquids) describes a solution or liquid that is neither acidic nor alkaline, with a pH of 7

neutralization the chemical reaction between an acid and a base to form a salt and water

neutron the neutral particle that occurs in the nuclei of atoms some hydrogens have nuclei

nitrifying bacteria bacteria that convert nitrogen from the atmosphere to nitrates in the soil

nitrogenous fertilizer fertilizer that provides nitrogen for plants, in the form of nitrate or ammonium salts

noble gases the Group 0 elements of the periodic table

non-metal an element to the right of the zig-zag line in the periodic table, that does not have characteristic metallic properties

non-renewable resources resources that cannot be replaced, such as fossil fuels

nucleus the centre part of the atom

oceanic crust the crust forming the floor of an ocean basin

oceanic ridge a ridge formed as plates move apart and new rock erupts from underwater volcanoes

ore rock from which it is economic to extract a metal

organic compound a compound containing carbon, which originated in living things

oxidation a chemical reaction in which a substance gains oxygen, loses hydrogen, or loses electrons

oxide a compound formed between an element and oxygen

ozone layer the layer in the atmosphere which protects us from UV radiation

period a horizontal row of the periodic table

periodic table a table that places the elements in order of atomic number, and has similar elements together in columns called groups

pH scale a scale measuring the strength of acids and alkalis

photochemical reaction reaction that depends on light energy to start it off

photosynthesis the process in which green plants convert carbon dioxide and water to glucose

physical change a change in which no new chemical substance is made

physical properties properties such as density and melting point that relate to the substance (and not its reactions)

pipette a piece of laboratory equipment for accurately delivering a known volume of liquid

plastic a synthetic polymer

plate tectonics the theory that the Earth's lithosphere is divided into large rigid slabs called plates, which move around

pollutant substance that causes pollution

pollution harmful or poisonous substances released into the environment

polymer a compound made up of large molecules, formed by polymerization

polymerization the process in which many monomer units are joined together to form a polymer

precipitate an insoluble chemical produced during a chemical reaction

product a chemical made in a chemical reaction

protein long chain molecule made up of amino acids.

proton number the number of protons in the atoms of an element, also known as the atomic number

proton the positive particle that occurs in the nuclei of all atoms

quicklime calcium oxide

radioactive isotope (radioisotope) atoms of an element that emit alpha, beta, or gamma radiation

rate of reaction the speed of a reaction

reactant a starting chemical for a chemical reaction

reactive describes a substance that tends to react easily

reactivity describes how readily a substance reacts

reactivity series the metals listed in order of reactivity

recycling to reuse resources such as glass and plastics

redox reaction a chemical reaction in which one substance is oxidized and another is reduced

reducing agent a substance used to carry out reduction

reduction a chemical reaction in which a substance loses oxygen, gains hydrogen, or gains electrons

refining separating crude oil into fractions containing molecules of similar size

relative atomic mass (A_r) the average mass of the atoms of an element, measured relative to the mass of carbon-12

relative formula mass (M_r) the mass of a molecule or formula unit of a substance, found by adding together the relative atomic masses of the atoms in the formula

renewable resource a resource that will not run out, such as wind energy

respiration the reaction that takes place in our body cells to give us energy

reversible reaction a reaction that can go both ways, i.e. reactants ⇌ products

rock cycle a sequence of events that shows how rocks are formed and changed

rusting the corrosion reaction between iron, air, and water

sacrificial protection allowing one metal to corrode in order to protect another metal

salt a chemical formed in a neutralization reaction

saturated compound an organic compound in which the carbon atoms form only single covalent bonds

sea-floor spreading the formation of new ocean floor caused by plates moving apart

sedimentary rock rock formed from deposition, burial, and cementation of sediments

separating funnel a piece of laboratory equipment for separating two immiscible liquids

slaked lime calcium hydroxide

solubility how much of a solute will dissolve in 100 grams of a solvent

soluble describes a chemical that dissolves in a solvent

solute the chemical substance that dissolves in the solvent when making a solution

solution a mixture, the result of a solute dissolving in a solvent

solvent the liquid in which the solute is dissolved to make a solution

sonorous makes a ringing noise when hit

stable describes a substance that does not tend to react easily

state symbols symbols used to show the physical states of reactants and products in chemical equations (g = gas, l = liquid, s = solid, aq = aqueous)

stratosphere a layer of atmosphere above the troposphere extending to about 50 km above the Earth's surface

subduction how an oceanic plate is forced under a continental plate, when the two plates converge

sublimation the physical change in which a solid changes directly to a gas on heating

tectonic plate the Earth's lithosphere is cracked into large sections called tectonic plates

tensile strength a measurement of a material's resistance to breaking when stretched (under tension)

thermal decomposition the breaking down of a compound by heating it

Thermit process the chemical reaction between iron oxide and aluminium, which is highly exothermic

thermosetting describes a plastic that sets permanently when made

thermosoftening describes a plastic that repeatedly softens on heating and hardens on cooling

titration a chemical analysis technique used to find the concentration of a solution

toxic poisonous

transition metals the elements that occupy the middle block of the periodic table

trend a gradual change in properties within a group of the periodic table

universal indicator an indicator that shows colour changes across the full range of pH

unsaturated compound an organic compound in which some of the carbon atoms form double covalent bonds

viscosity a measure of how runny a liquid is; the lower the viscosity, the more runny and easy to pour the liquid is

volatile a measure of how easily a liquid turns into a vapour; a volatile liquid forms a vapour at low temperature

volcano an area of the Earth's crust where magma gets through to the surface

water of crystallization the water that is locked into a hydrated compound, for example hydrated copper sulphate $CuSO_4.5H_2O$

weathering the breaking down of rock, for example by the action of air, water, heat, cold, and plant roots

X-rays electromagnetic rays with a shorter wavelength than light

yield (in chemical reactions) the amount of a product actually obtained in a laboratory or industrial chemical reaction

Appendix 1   Separating solids from mixtures

Suppose you need only one substance from a mixture. How do you separate it from the mixture? Here are some methods.

How to separate a solid from a liquid

By filtering   Chalk can be separated from water by filtering the suspension through filter paper. The chalk gets trapped in the filter paper while the water passes through. The chalk is called the **residue**. The water is called the **filtrate**.

Other suspensions can be separated in the same way.

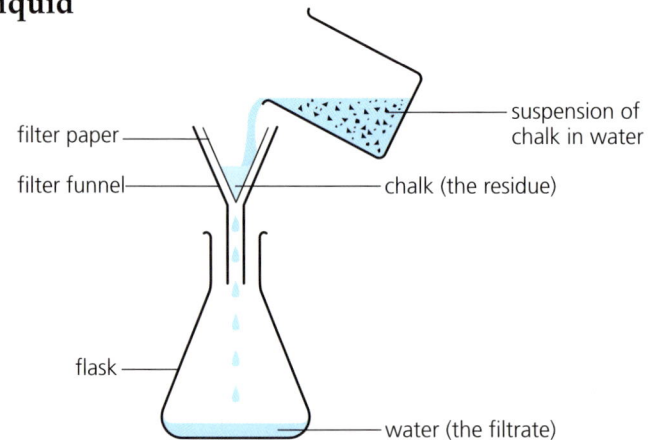

By centrifuging   A centrifuge is used to separate *small* amounts of suspension. In a centrifuge, test tubes of suspension are spun round very fast, so that the solid gets flung to the bottom:

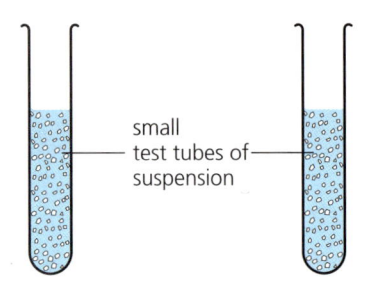

Before centrifuging, the solid is mixed all through the liquid.

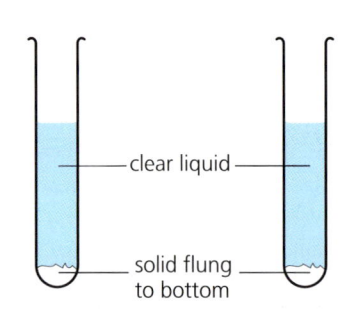

After centrifuging, all the solid has collected at the bottom.

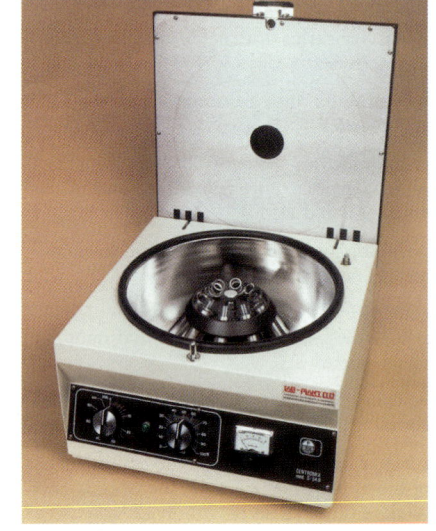

A centrifuge.

The liquid can be **decanted** (poured out) from the test tubes, or removed with a small pipette. The solid is left behind.

By evaporating the solvent   If the mixture is a *solution*, the solid cannot be separated by filtering or centrifuging. This is because it is spread all through the solvent in tiny particles. Instead, the solution is heated so that the solvent evaporates, leaving the solid behind. Salt is obtained from its solution by this method:

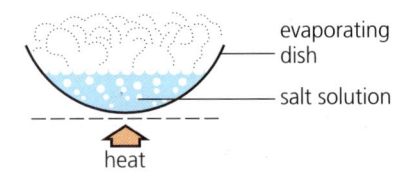

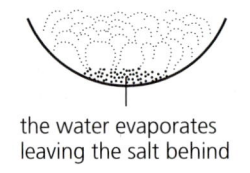

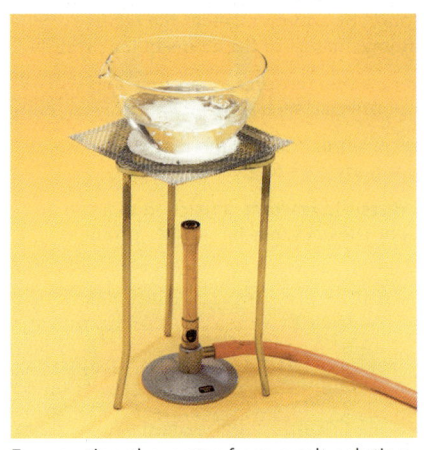

Evaporating the water from a salt solution.

246

By crystallizing You can separate many solids from solution by letting them form crystals. Copper(II) sulphate is an example:

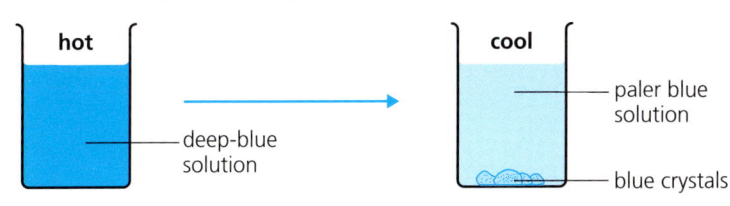

This is a saturated solution of copper(II) sulphate in water at 70 °C. If it is cooled to 20 °C ...

... crystals begin to appear, because the compound is less *soluble* at 20 °C than at 70 °C.

The process is called **crystallization**. It is carried out like this:

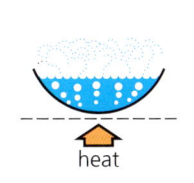

1 A solution of copper(II) sulphate is heated, to get rid of some water. As the water evaporates, the solution becomes more concentrated.

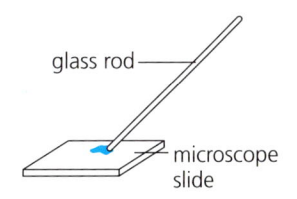

2 The solution can be checked to see if it is ready by placing one drop on a microscope slide. Crystals should form quickly on the cool glass.

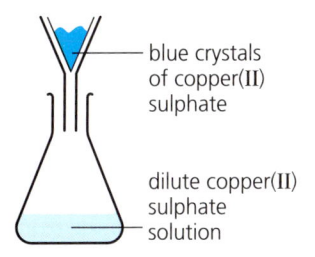

3 Then the solution is left to cool and crystallize. The crystals are removed by filtering, rinsed with water, and dried with filter paper.

How to separate a mixture of two solids

By dissolving one of them A mixture of salt and sand can be separated like this:

1 Water is added to the mixture, and it is stirred. The salt dissolves.
2 The mixture is then filtered. The sand is trapped in the filter paper, but the salt solution passes through.
3 The sand is rinsed with water and dried in an oven.
4 The salt solution is evaporated to dryness.

This method works because salt is soluble in water but sand is not. Water could *not* be used to separate salt and sugar, because it dissolves both. You could use ethanol, because it dissolves sugar but not salt. Ethanol is inflammable, so it should be evaporated from the sugar solution over a water bath, as shown on the right.

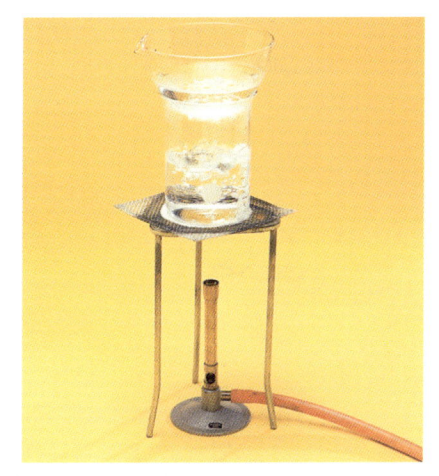

Evaporating ethanol from a sugar solution, over a water bath.

Questions

1 What is: a filtrate? a residue?
2 Describe two ways of separating the solid from the liquid in a suspension.
3 Sugar *cannot* be separated from sugar solution by filtering. Explain why.
4 What happens when a solution is evaporated?
5 Describe how you would crystallize potassium nitrate from its aqueous solution.
6 How would you separate salt and sugar? Mention any special safety precaution you would take.

247

Appendix 2 Distillation and chromatography

How to separate the solvent from a solution

By simple distillation This is a way of getting pure solvent out of a solution. The apparatus is shown on the right. It could be used to obtain pure water from salt water, for example. This is what happens:

1 The solution is heated in the flask. It boils, and steam rises into the condenser. The salt is left behind.
2 The condenser is cold, so the steam condenses to water in it.
3 The water drips into the beaker. It is completely pure. It is called **distilled water.**

This method could also be used to obtain pure water from sea water, or from ink.

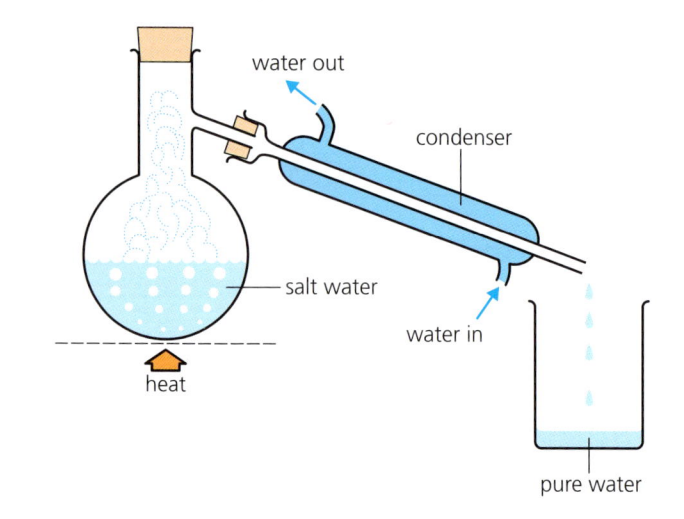

How to separate two liquids

Using a separating funnel If two liquids are **immiscible**, they can be separated with a separating funnel. For example when a mixture of oil and water is poured into the funnel, the oil floats to the top, as shown on the right. When the tap is opened, the water runs out. The tap is closed again when all the water has gone.

By fractional distillation If two liquids are **miscible**, they must be separated by fractional distillation. The apparatus is shown below. It could be used to separate a mixture of ethanol and water, for example. These are the steps:

1 The mixture is heated. At about 78 °C, the ethanol begins to boil. Some water evaporates too, so a mixture of ethanol vapour and water vapour rises up the column.
2 The vapours condense on the glass beads on the column, making them hot.
3 When the beads reach about 78 °C, ethanol vapour no longer condenses on them. Only the water vapour does. The water drips back into the flask, while the ethanol vapour is forced into the condenser.
4 There it condenses. Liquid ethanol drips into the beaker.
5 Eventually, the thermometer reading rises above 78 °C. This is a sign that all the ethanol has been separated, so heating can be stopped.

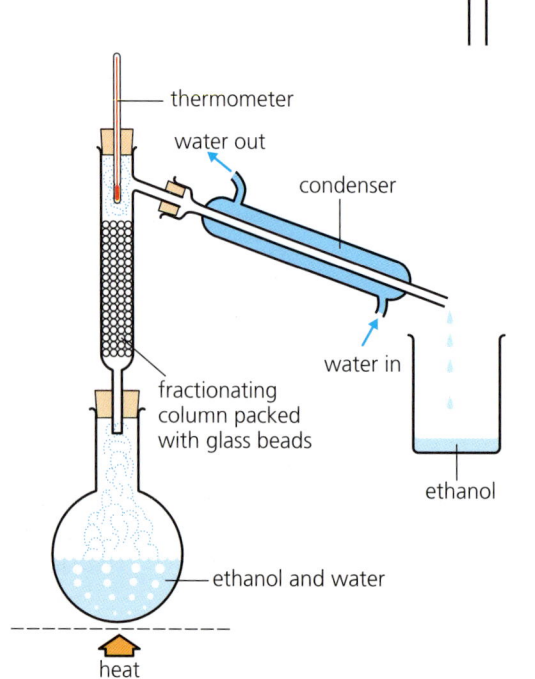

How to separate a mixture of coloured substances

Paper chromatography This method can be used to separate a mixture of coloured substances. For example it will separate the coloured substances in black ink:

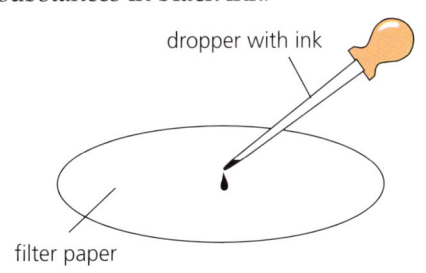

1 A small drop of black ink is placed at the centre of a piece of filter paper, and allowed to dry. Three or four more drops are added on the same spot.

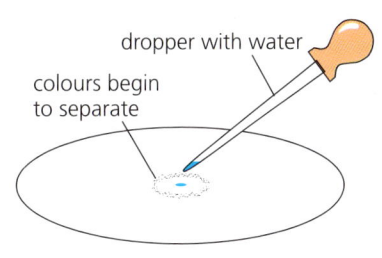

2 Water is then dripped on to the ink spot, one drop at a time. The ink slowly spreads out and separates into rings of different colours.

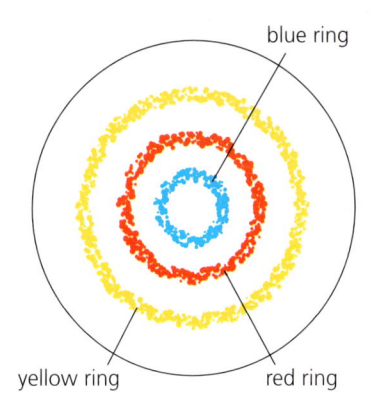

3 Suppose there are three rings – yellow, red, and blue. This shows that the ink contains three substances, coloured yellow, red, and blue.

The filter paper with its coloured rings is called a **chromatogram**. In the example above, the outside ring is yellow. This shows that the yellow substance is the most soluble in water. When the water drips on to the ink spot, it dissolves the yellow substance the most easily, and carries it furthest away. Can you tell which substance is the least soluble?

Paper chromatography is often used to **identify** the substances in a mixture. For example mixture X is thought to contain the substances A, B, C, and D, which are soluble in propanone. The mixture could be checked like this:

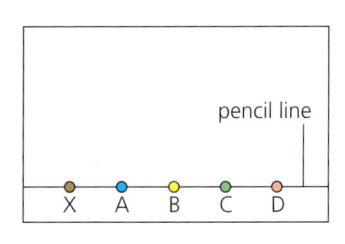

1 Concentrated solution of X, A, B, C, and D are made up in propanone. A spot of each is placed on a line, on a sheet of filter paper, and labelled.

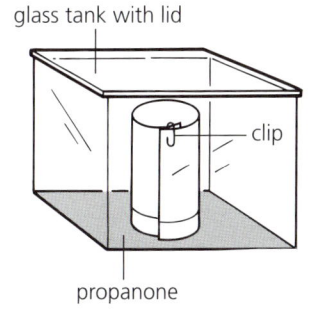

2 The paper is stood in a little propanone, in a covered glass tank. The solvent rises up the paper; when it's near the top, the paper is taken out again.

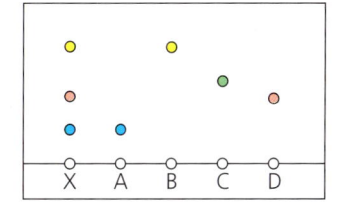

3 X has separated into three spots. Two are at the same height as A and B, so X must contain substances A and B. Does it also contain C and D?

Questions

1 How would you obtain pure water from ink? Draw the apparatus you would use, and explain how the method works.
2 Why are condensers so called? What is the reason for running cold water through them?
3 Water and turpentine are immiscible. How would you separate a mixture of the two?
4 Explain how fractional distillation works.
5 In the chromatogram above, how can you tell that X does not contain substance C?

249

Appendix 3 Working with gases in the lab

Preparing gases in the lab

The usual method is to displace the gas from a solid or solution, using apparatus like that shown here. The table gives some examples.

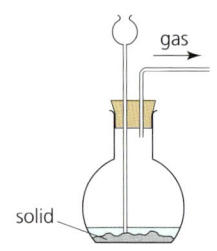

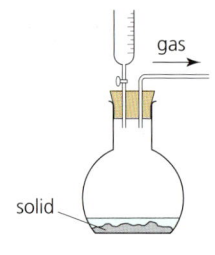

Apparatus for preparing gases. The dropping funnel in the right-hand apparatus is used for adding concentrated acids.

To make...	Place in flask...	Add...	Reaction
carbon dioxide	calcium carbonate (marble chips)	dilute hydrochloric acid	$CaCO_3 (s) + 2HCl (aq) \longrightarrow CaCl_2 (aq) + H_2O (l) + CO_2 (g)$
chlorine	manganese(IV) oxide (as an oxidizing agent)	conc hydrochloric acid	$2HCl (aq) + [O] \longrightarrow H_2O (l) + Cl_2 (g)$
hydrogen	pieces of zinc	dilute hydrochloric acid	$Zn (s) + 2HCl (aq) \longrightarrow ZnCl_2 (aq) + H_2 (g)$
hydrogen chloride	sodium chloride	conc sulphuric acid	$NaCl (s) + H_2SO_4 (aq) \longrightarrow NaHSO_4 (aq) + HCl (g)$
oxygen	manganese(IV) oxide (as a catalyst)	hydrogen peroxide	$2H_2O_2 (aq) \longrightarrow 2H_2O (l) + O_2 (g)$

But to make ammonia, you just heat any ammonium compound with any strong alkali.

Collecting gases in the lab

Here are four ways to collect the gas you've prepared:

Method	upward displacement of air	downward displacement of air	over water	gas syringe
Use when...	gas is heavier than air	gas is lighter than air	gas is sparingly soluble in water	you want to measure the volume
Apparatus	gas jar	gas jar	gas jar / water	gas syringe
Examples	carbon dioxide, CO_2 chlorine, Cl_2 sulphur dioxide, SO_2 hydrogen chloride, HCl	ammonia, NH_3 hydrogen, H_2	carbon dioxide, CO_2 hydrogen, H_2 oxygen, O_2	any gas

Tests for gases

You have a sample of gas. You suspect it is carbon dioxide, but you're not sure. So you need to do a test. Below are some tests for common gases. Each is based on particular properties of the gas, including its appearance and sometimes its smell.

Ammonia, NH_3
Properties	Ammonia is a colourless alkaline gas with a pungent smell.
Test	Carefully smell the gas, and hold damp indicator paper in it.
Result	Recognizable odour, and indicator paper turns blue.

Carbon dioxide, CO_2
Properties	Carbon dioxide is a colourless, weakly acidic gas. It reacts with **limewater** (a solution of calcium hydroxide in water) to give a white precipitate of calcium carbonate: $$CO_2\,(g) + Ca(OH)_2\,(aq) \longrightarrow CaCO_3\,(s) + H_2O\,(l)$$
Test	Bubble the gas through limewater.
Result	Limewater turns cloudy or milky.

Chlorine, Cl_2
Properties	Chlorine is a green poisonous gas which bleaches dyes.
Test	Hold damp indicator paper in the gas, *in a fume cupboard*.
Result	Indicator paper turns white.

Hydrogen, H_2
Properties	Hydrogen is a colourless gas which combines violently with oxygen when ignited.
Test	Collect the gas in a tube and hold a lighted splint to it.
Result	The gas burns with a squeaky pop.

Hydrogen chloride, HCl
Properties	Hydrogen chloride is a colourless, acidic gas with a choking smell. It neutralizes ammonia to form the salt ammonium chloride: $$HCl\,(g) + NH_3\,(g) \longrightarrow NH_4Cl\,(s)$$
Test	Hold a rod dipped in concentrated ammonia solution in the gas, *in a fume cupboard*.
Result	A white 'smoke' of ammonium chloride forms.

Oxygen, O_2
Properties	Oxygen is a colourless gas. Fuels burn much more readily in it than in air.
Test	Collect the gas in a test tube and hold a glowing splint to it.
Result	The splint immediately bursts into flame.

Sulphur dioxide, SO_2
Properties	Sulphur dioxide is a colourless acidic gas with a choking smell. It reacts with acidified potassium dichromate solution, reducing the orange dichromate ion, $Cr_2O_7^{2-}$, to the green chromium ion Cr^{3+}: $$3SO_2\,(g) + 2H^+\,(aq) + Cr_2O_7^{2-}\,(aq) \longrightarrow 3SO_4^{2-}\,(aq) + H_2O\,(l) + 2Cr^{3+}\,(aq)$$ from the acid / orange solution / green solution
Test	Hold a piece of filter paper soaked in acidified potassium dichromate solution in the gas, in a fume cupboard.
Result	The paper turns from orange to green.

Appendix 4 Testing for ions in the lab

You have a solid. You think it's sodium iodide, but you're not sure. Time for some detective work! Below are three methods of testing. Which would you use to test for sodium and iodide ions?

Method 1: Form a precipitate

If the salt is soluble in water, make a solution. Then try to precipitate an insoluble compound from it. You can use this method to test for any ion that forms an insoluble compound.

Testing for halide ions (Cl⁻, Br⁻, I⁻)
- Take a small amount of solution.
- Add an equal volume of dilute nitric acid.
- Then add silver nitrate solution.
- Silver halides are insoluble, so if halide ions are present a precipitate will form.

Precipitate	Indicates presence of ...	Reaction taking place
white	chloride ions, Cl⁻	$Ag^+ (aq) + Cl^- (aq) \longrightarrow AgCl (s)$
cream	bromide ions, Br⁻	$Ag^+ (aq) + Br^- (aq) \longrightarrow AgBr (s)$
yellow	iodide ions, I⁻	$Ag^+ (aq) + I^- (aq) \longrightarrow AgI (s)$

Note that iodide ions can also be identified using acidified lead(II) nitrate solution. A deep yellow precipitate of lead(II) iodide forms.

Testing for sulphate ions, SO_4^{2-}
- Take a small amount of solution.
- Add an equal volume of dilute hydrochloric acid.
- Then add barium chloride solution.
- Barium sulphate is insoluble, so if sulphate ions are present a precipitate will form.

Precipitate	Indicates presence of ...	Reaction taking place
white	sulphate ions, SO_4^{2-}	$Ba^{2+}(aq) + SO_4^{2-} (aq) \longrightarrow BaSO_4 (s)$

Testing for metal ions (Cu^{2+}, Fe^{2+}, Fe^{3+}, Al^{3+}, Zn^{2+}, Ca^{2+})
- Take a small volume of solution.
- Add a few drops of dilute sodium hydroxide solution.
- If a *coloured* precipitate forms, it indicates the presence of Cu^{2+}, Fe^{2+} or Fe^{3+} ions. These form insoluble coloured hydroxides:

Precipitate	Indicates presence of ...	Reaction taking place
pale blue*	copper ions, Cu^{2+}	$Cu^{2+} (aq) + 2OH^- (aq) \longrightarrow Cu(OH)_2 (s)$
green	iron(II) ions, Fe^{2+}	$Fe^{2+} (aq) + 2OH^- (aq) \longrightarrow Fe(OH)_2 (s)$
red-brown	iron(III) ions, Fe^{3+}	$Fe^{3+} (aq) + 3OH^- (aq) \longrightarrow Fe(OH)_3 (s)$

* This will redissolve in excess ammonia solution to form a deep blue solution.

- But a *white* precipitate indicates the presence of Al^{3+}, Zn^{2+}, or Ca^{2+} ions. They form insoluble white hydroxides.
- To find out which of these ions it is, divide the solution containing the precipitate into two equal volumes.

- To one, add double the volume of sodium hydroxide solution. To the other, add double the volume of ammonium hydroxide solution. Check the result against this table:

If the precipitate ...	It indicates the presence of ...
dissolves again in sodium hydroxide solution	Al^{3+} ions
dissolves in both sodium and ammonium hydroxide solutions	Zn^{2+} ions
dissolves in neither	Ca^{2+} ions

Method 2: Carry out a reaction to produce a gas you can test

You can identify carbonate, ammonium, or nitrate ions in an unknown substance by carrying out a reaction that produces carbon dioxide or ammonia.

Carbonate ions (CO_3^{2-})
- Take a small amount of solid or solution.
- Add a little dilute hydrochloric acid.
- If the mixture bubbles and gives off a gas that turns limewater milky, the unknown substance contained carbonate ions.

The reaction that takes place is:
$$2H^+ (aq) + CO_3^{2-} (aq) \longrightarrow CO_2 (g) + H_2O (l)$$
from the acid carbon dioxide
(turns limewater milky)

Ammonium ions (NH_4^+)
- Take a small amount of solid or solution.
- Add a little dilute sodium hydroxide solution and heat gently.
- If ammonia gas is given off, the unknown substance contained ammonium ions. (Ammonia gas has a pungent smell and turns red litmus blue.)

The reaction that takes place is:
$$NH_4^+ (aq) + OH^- (aq) \longrightarrow NH_3 (g) + H_2O (l)$$
from sodium ammonia
hydroxide

Nitrate ions (NO_3^-)
- Take a small amount of solid or solution.
- Add a little sodium hydroxide solution.
- Add some aluminium powder or Devarda's alloy. Heat gently.
- If ammonia gas is given off, the unknown substance contained nitrate ions.

The reaction that takes place is:
$$8Al (s) + 3NO_3^- (aq) + 5OH^- (aq) + 2H_2O (l) \longrightarrow 3NH_3 (g) + 8AlO_2^- (aq)$$
from sodium
hydroxide

Method 3: Do a flame test

You can use a flame test to test for some metal ions.
- Make a paste of the solid in concentrated hydrochloric acid.
- Dip the end of a flame test wire (nichrome wire) in the paste, and then hold it in a Bunsen flame.

Colour of flame ...	Indicates presence of ...
yellow	sodium ions, Na^+
lilac	potassium ions, K^+
brick-red	calcium ions, Ca^{2+}
green-blue	copper ions, Cu^{2+}

Appendix 5 Maths toolkit

Units

For GCSE chemistry, you need to be familiar with these units:

Measurement	Unit	Symbol
Mass	gram	g
Volume	cubic centimetre	cm^3
	litre	l
Time	second	s
Energy	joule	J
Rate of reaction	volume per second or	cm^3/s
	mass loss per second	g/s

1 litre = 1000 cm^3
1 litre = 1000 ml (millilitres)
so 1 cm^3 is the same as 1 ml

You should give units for all measurements you write down.

Ratio

A ratio is the **simplest relative value** of two numbers. For example:

1st number	2nd number	Ratio
6	2	3 : 1
8	2	4 : 1
10	6	5 : 3

Percentages

The whole of something is 100%. In a mixture of two substances A and B, if 20% of the mixture is A, then 80% will be B. Out of every 100 parts (for example out of 100 grams or 100 kilograms) 20 will be A and 80 will be B.

Percentages can be converted to ratios and fractions. Look at these examples for mixtures of two substances A and B.

Substance A	Substance B	Ratio A : B	Fraction of A in total
20%	80%	1 : 4	1/5
25%	75%	1 : 3	1/4
50%	50%	1 : 1	1/2
60%	40%	3 : 2	3/5

To write a percentage as a fraction:

$10\% = \frac{10}{100} = \frac{1}{10}$ $20\% = \frac{20}{100} = \frac{2}{10}$ or $\frac{1}{5}$ and so on

Graphs for rates of reaction

When drawing graphs for changes in mass or volume in rate experiments you should make sure that:

Axes
- mass or volume is on the vertical axis
- time is on the horizontal axis
- the axes are labelled and the correct units used
- the scales you choose suit the range of values you want to plot. The maximum value should lie near the end of the axis.

Points
- experimental results are accurately placed using × or •.

Lines
- you draw **best fit lines**. The line does not have to go through all the points. Look at the examples below.

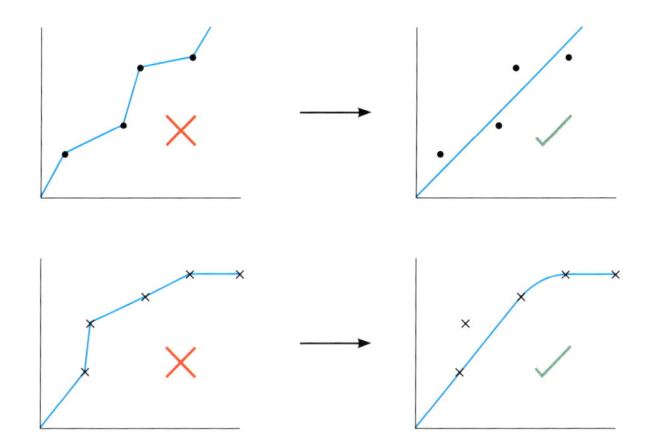

Information from the graph
The *slope* or *gradient* of the graph is a measure of the rate of the reaction. The table of units on the opposite page shows units for rate.

To find the slope:
- choose two widely spaced points on the straight line portion of the line.
- draw a right-angled triangle using these points as corners.
- find the increase along the vertical line (A) and the horizontal line (B) using the scales on the axes.

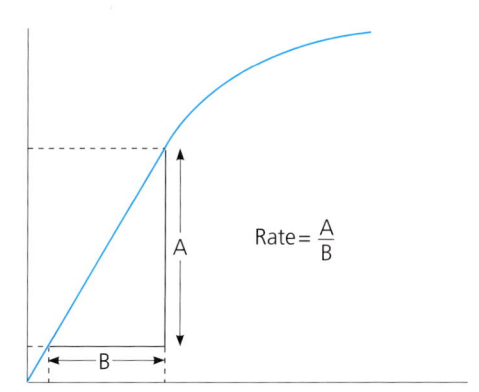

$$\text{Rate} = \frac{A}{B}$$

- then divide A by B to find the slope.
- The rate is the slope with the correct units added.

255

Appendix 6   The periodic table and atomic masses

Group	1	2											
1						$^{1}_{1}$H Hydrogen							
2	$^{7}_{3}$Li lithium	$^{9}_{4}$Be beryllium											
3	$^{23}_{11}$Na sodium	$^{24}_{12}$Mg magnesium											
4	$^{39}_{19}$K potassium	$^{40}_{20}$Ca calcium	$^{45}_{21}$Sc scandium	$^{48}_{22}$Ti titanium	$^{51}_{23}$V vanadium	$^{52}_{24}$Cr chromium	$^{55}_{25}$Mn manganese	$^{56}_{26}$Fe iron	$^{59}_{27}$Co cobalt	$^{59}_{28}$Ni nickel	$^{64}_{29}$Cu copper	$^{65}_{30}$Zn zinc
5	$^{85}_{37}$Rb rubidium	$^{88}_{38}$Sr strontium	$^{89}_{39}$Y yttrium	$^{91}_{40}$Zr zirconium	$^{93}_{41}$Nb niobium	$^{96}_{42}$Mo molybdenum	$^{99}_{43}$Tc technetium	$^{101}_{44}$Ru ruthenium	$^{103}_{45}$Rh rhodium	$^{106}_{46}$Pd palladium	$^{108}_{47}$Ag silver	$^{112}_{48}$Cd cadmium
6	$^{133}_{55}$Cs caesium	$^{137}_{56}$Ba barium	$^{139}_{57}$La lanthanum	$^{178.5}_{72}$Hf hafnium	$^{181}_{73}$Ta tantalum	$^{184}_{74}$W tungsten	$^{186}_{75}$Re rhenium	$^{190}_{76}$Os osmium	$^{192}_{77}$Ir iridium	$^{195}_{78}$Pt platinum	$^{197}_{79}$Au gold	$^{201}_{80}$Hg mercury
7	$^{223}_{87}$Fr francium	$^{226}_{88}$Ra radium	$^{227}_{89}$Ac actinium									

$^{140}_{58}$Ce cerium	$^{141}_{59}$PR praseodymium	$^{144}_{60}$Nd neodymium	$^{147}_{61}$Pm promethium	$^{150}_{62}$Sm samarium	$^{152}_{63}$Eu europium	$^{157}_{64}$Gd gadolinium	$^{159}_{65}$Tb terbium
$^{232}_{90}$TH thorium	$^{231}_{91}$Pa protactinium	$^{238}_{92}$U uranium	$^{237}_{93}$Np neptunium	$^{244}_{94}$Pu plutonium	$^{243}_{95}$Am americium	$^{247}_{96}$Cm curium	$^{247}_{97}$Bk berkelium

Relative atomic masses based on internationally agreed figures

Element	Symbol	Atomic number	Relative atomic mass
actinium	Ac	89	
aluminium	Al	13	26.9815
americium	Am	95	
antimony	Sb	51	121.75
argon	Ar	18	39.948
arsenic	As	33	74.9216
astatine	At	85	
barium	Ba	56	137.34
berkelium	Bk	97	
beryllium	Be	4	9.0122
bismuth	Bi	83	208.980
boron	B	5	10.811
bromine	Br	35	79.909
cadmium	Cd	48	112.40
caesium	Cs	55	132.905
calcium	Ca	20	40.08
californium	Cf	98	
carbon	C	6	12.01115
cerium	Ce	58	140.12
chlorine	Cl	17	35.453
chromium	Cr	24	51.996
cobalt	Co	27	58.9332
copper	Cu	29	63.54
curium	Cm	96	
dysprosium	Dy	66	162.50
einsteinium	Es	99	

Element	Symbol	Atomic number	Relative atomic mass
erbium	Er	68	167.26
europium	Eu	63	151.96
fermium	Fm	100	
fluorine	F	9	18.9984
francium	Fr	87	
gadolinium	Gd	64	157.25
gallium	Ga	31	69.72
germanium	Ge	32	72.59
gold	Au	79	196.967
hafnium	Hf	72	178.49
helium	He	2	4.0026
holmium	Ho	67	164.930
hydrogen	H	1	1.00797
indium	In	49	114.82
iodine	I	53	126.9044
iridium	Ir	77	192.2
iron	Fe	26	55.847
krypton	Kr	36	83.80
lanthanum	La	57	138.91
lawrencium	Lw	103	
lead	Pb	82	207.19
lithium	Li	3	6.939
lutetium	Lu	71	174.97
magnesium	Mg	12	24.312
manganese	Mn	25	54.9380
mendelevium	Md	101	

Group

3	4	5	6	7	0
					$^{4}_{2}$He helium
$^{11}_{5}$B boron	$^{12}_{6}$C carbon	$^{14}_{7}$N nitrogen	$^{16}_{8}$O oxygen	$^{19}_{9}$F fluorine	$^{20}_{10}$Ne neon
$^{27}_{13}$Al aluminium	$^{28}_{14}$Si silicon	$^{31}_{15}$P phosphorus	$^{32}_{16}$S sulphur	$^{35.5}_{17}$Cl chlorine	$^{40}_{18}$Ar argon
$^{70}_{31}$Ga gallium	$^{73}_{32}$Ge germanium	$^{75}_{33}$As arsenic	$^{79}_{34}$Se selenium	$^{80}_{35}$Br bromine	$^{84}_{36}$Kr krypton
$^{115}_{49}$In indium	$^{119}_{50}$Sn tin	$^{122}_{51}$Sb antimony	$^{128}_{52}$Te tellurium	$^{127}_{53}$I iodine	$^{131}_{54}$Xe xenon
$^{204}_{81}$Tl thallium	$^{207}_{82}$Pb lead	$^{209}_{83}$Bi bismuth	$^{210}_{84}$Po polonium	$^{210}_{85}$At astatine	$^{222}_{86}$Rn radon

| $^{162}_{66}$Dy dysprosium | $^{165}_{67}$Ho holmium | $^{167}_{68}$Er erbium | $^{169}_{69}$Tm thulium | $^{173}_{70}$Yb ytterbium | $^{175}_{71}$Lu lutetium |
| $^{251}_{98}$Cf californium | $^{252}_{99}$Es einsteinium | $^{257}_{100}$Fm fermium | $^{258}_{101}$Md mendelevium | $^{259}_{102}$No nobelium | $^{260}_{103}$Lw lawrencium |

Approximate atomic masses for calculations

Element	Symbol	Atomic mass for calculations
aluminium	Al	27
bromine	Br	80
calcium	Ca	40
carbon	C	12
chlorine	Cl	35.5
copper	Cu	64
helium	He	4
hydrogen	H	1
iodine	I	127
iron	Fe	56
lead	Pb	207
lithium	Li	7
magnesium	Mg	24
manganese	Mn	55
neon	Ne	20
nitrogen	N	14
oxygen	O	16
phosphorus	P	31
potassium	K	39
silicon	Si	28
silver	Ag	108
sodium	Na	23
sulphur	S	32
zinc	Zn	65

Element	Symbol	Atomic number	Relative atomic mass
mercury	Hg	80	200.59
molybdenum	Mo	42	95.94
neodymium	Nd	60	144.24
neon	Ne	10	20.179
neptunium	Np	93	
nickel	Ni	28	58.71
niobium	Nb	41	92.906
nitrogen	N	7	14.0067
nobelium	No	102	
osmium	Os	76	190.2
oxygen	O	8	15.9994
palladium	Pd	46	106.4
phosphorus	P	15	30.9738
platinum	Pt	78	195.09
plutonium	Pu	94	
polonium	Po	84	
potassium	K	19	39.102
praseodymium	Pr	59	140.907
promethium	Pm	61	
protactinium	Pa	91	
radium	Ra	88	
radon	Rn	86	
rhenium	Re	75	186.2
rhodium	Rh	45	102.905
rubidium	Rb	37	85.47
ruthenium	Ru	44	101.07

Element	Symbol	Atomic number	Relative atomic mass
samarium	Sm	62	150.35
scandium	Sc	21	44.956
selenium	Se	34	78.96
silicon	Si	14	28.086
silver	Ag	47	107.868
sodium	Na	11	22.9898
strontium	Sr	38	87.62
sulphur	S	16	32.064
tantalum	Ta	73	180.948
technetium	Tc	43	
tellurium	Te	52	127.60
terbium	Tb	65	158.924
thallium	Tl	81	204.37
thorium	Th	90	232.038
thulium	Tm	69	168.934
tin	Sn	50	118.69
titanium	Ti	22	47.90
tungsten	W	74	183.85
uranium	U	92	238.03
vanadium	V	23	50.942
xenon	Xe	54	131.30
ytterbium	Yb	70	173.04
yttrium	Y	39	88.905
zinc	Zn	30	65.37
zirconium	Zr	40	91.22

Appendix 7   Safety first!

The chemistry lab can be a dangerous place. But these rules will help you avoid accidents. How many of them do you follow already?

- Follow instructions precisely.
- Always wear your safety glasses when working in the lab. If chemicals get in your eyes they could blind you.
- Tie hair back.
- Don't fool around with Bunsen burners.
- Stand up when working with liquids.
- Don't splash chemicals around.
- Mop up spills carefully with a cloth for that purpose.
- If you have to smell a substance, do so carefully. Do NOT smell poisonous gases such as chlorine.
- Do NOT taste anything in the lab.
- Handle chemicals with the respect they deserve: especially acids, alkalis, and poisonous and flammable substances.
- Use a fume cupboard for experiments involving poisonous or highly reactive substances.
- Handle glassware with care.
- Find out NOW what to do in case of fire, or chemicals in your eye, or other kinds of accident. Talk the safety drill over with your teacher.
- Know where the fire extinguisher and fire blankets are, and how to use them.
- Know the symbols for chemical hazards. They're given below. Watch out for them on bottles and other chemical containers.

Toxic

These substances can kill. They may act when you swallow them, or breathe them in, or absorb them through your skin.
Examples: chlorine gas
mercury

Harmful

These are similar to toxic substances but less dangerous.
Examples: copper(II) sulphate
lead oxide

Highly flammable

These substances catch fire easily. They pose a fire risk.
Examples: hydrogen
ethanol

Oxidizing

These provide oxygen which allows other substance to burn more fiercely.
Examples: hydrogen peroxide
potassium manganate(VII)

Corrosive

These attack and destroy living tissue, including eyes and skin.
Examples: conc sulphuric acid
bromine

Irritant

These are not corrosive but can cause reddening or blistering of skin.
Examples: calcium chloride
zinc sulphate

Numerical answers

Module 5 Metals
17 c i 990 kg   ii 0.25%

Module 7 Patterns of chemical change

2 c i 14 cm³ per minute   ii 9 cm³ per minute   iii 8 cm³ per minute
  e 40 cm³   f 5 minutes   g 8 cm³ per minute
11 a drop of 4 °C for ammonium nitrate, rise of 20 °C for calcium chloride
   d i drop of 8 °C for ammonium nitrate, rise of 40 °C for calcium chloride
      ii drop of 2 °C for ammonium nitrate, rise of 10 °C for calcium chloride
      iii drop of 4 °C for ammonium nitrate, rise of 20 °C for calcium chloride
12 c 55.6 kJ
13 c i 2148 kJ   ii 2354 kJ   iii 206 kJ
14 b 4   c 416 kJ   e 30 kJ
15 c i 2220 kJ   ii 2801 kJ   d 581 kJ
20 a 92   b 685   c 227   d 278   e 60
22 a there is 80% of copper in each sample
23 a 64 g   b MnO_2
24 a 2.4 g   b 1.6 g   c MgO
25 a Zn_3P_2 b 27%
28 20.1 g of mercury, 1.6 g of oxygen
29 a 32 g   b sulphur   c 11 g of iron(II) sulphide, 6 g of sulphur   d 17.5 g
30 b i 160   ii 112 g   iii 224 g   iv 224 tonnes
31 b 74   c 24 000 cm³   d 1 g
32 a 10 g   b 2000 cm³
33 a 1   b 25 cm³   c 75 cm³   d 50 cm³
34 a 19   b 456 dm³
35 a 84   c i 24 000 cm³   ii 2400 cm³

Module 8 Structures and bonding

4 b i 60   ii 34   iii 0   iv 146
8 c 10   f 10.8
9 a i 38   ii 40

Exam-style questions

 9 e 180   f 46 kg
10 e 390 to 400 s
11 c 80   d 3.2 g   e 1200 cm³
12 e i 2264 kJ   ii 193 kJ   iii 2579 kJ f 122 kJ
15 c 152   f 15.2 g   g 0.2 g
16 a 2 g   b 24 000 cm³   c 24 cm³
23 53.9%

Multiple-choice questions

Module 5 Metals

1 D	2 B	3 B	4 C	5 B	6 A	7 A
8 B	9 B	10 B, D	11 C, D	12 A, E	13 A, E	14 B, C
15 B, D	16 B	17 D	18 A	19 D	20 D	21 B
22 A

Module 5 Metals

1 B	2 C	3 D	4 D	5 C	6 C	7 B
8 A	9 C	10 D	11 A	12 B	13 D	14 A
15 C	16 A	17 C	18 A	19 C	20 C	21 A

Index

acid rain 45, 70
acids 38–44
   reactions 40–1, 42–3
activation energy 124, 135
addition polymerization 66
addition reaction 65
aggregate 55, 57, 200
air 74
alchemists 176
algae 145
alkali metals 11, 12–13, 196–7
   compounds 13
alkaline earth metals 11, 197
alkalis 38–9, 44
   reaction with acids 40, 42–3
alkanes 64
alkenes 63, 65
allotropes, of carbon 191
alloys 15, 24, 25, 31
alpha particles 178, 179
Alps 99
alumina 30, 31
aluminium
   and acid rain 45
   anodized 35
   and corrosion 35
   in Earth's crust 20
   extraction 22, 30–1
   properties 31
   uses 24, 25
aluminium oxide 30, 35
   electrolysis 22, 23
   ore 21
amino acids 75
ammonia
   in atmosphere 75
   bonding 189
   and cleaners 44
   decomposition 134
   diffusion 219
   dynamic equilibrium 137
   and fertilizers 140, 143
   manufacture 138–9, 140–1
   and nitric acid manufacture 143
   solution 39, 43
ammonium
   chloride 133, 136
   ion 185
   ions, in soil 142, 143
   nitrate 143
   sulphate 143
amylase 126, 130
anaerobic bacteria 75
Andes 99, 101
anhydrous copper(II) sulphate 136
anodized aluminium 35
anti-charm 179
anti-strange 179
anti-up 179
A_r 146, 173
argon 196

Aristotle 176
atmosphere 74–9
atomic mass units 170
atomic number 170, 171
   and periodic table 195
atoms 168, 170–1, 176–9
Avogadro 156

bacteria 75, 130
Bakelite 67
baking soda 44
balanced equation 152
barium hydroxide 133
barium sulphate 57
barytes, quarrying 57
basalt 83, 90, 94, 95, 103
batch process 131
bauxite 21, 30
Becker 179
Becquerel 178
bee sting 44
beer 129
benzene 70
biodegradable 69
biological detergents 130
biological weathering 84
bitumen 61
black smokers 130
blast furnace 26
bleach 118, 201
blue baby syndrome 145
Bohr, Niels 174, 279
boiling 217
boiling point 186
bond energy 134, 135
bone 220
Bothe 179
Boyle, Robert 177
brass 25
bread 128
breccia 95
brine 206
bromine 199, 200–1
   isotopes 173
   water, test for unsaturation 65
bronze 25
Bronze Age 36
Brown, Robert 177
butane 62, 64

cadmium 14
caesium 12–13, 196
calamine lotion 44
calcium
   and bone 220
   in Earth's crust 20
   properties 197
   reaction with water 16, 17
calcium carbonate 41, 44, 45, 54–5, 80, 92, 120–1
   see also chalk, limestone, marble

calcium fluoride 57
calcium hydrogencarbonate 84, 222–3
calcium hydroxide 39, 44, 45, 54
calcium oxide 44, 54, 55, 132
calcium silicate 27
Calor gas 64
camping gas 64
cancer, skin 78, 79
carbohydrase 130
carbon
   allotropes 191
   cycle 76
   dating 172
   isotopes 172
   in reactivity series 18
   as standard atom 146
carbon dioxide
   in atmosphere 74, 75, 76
   and fermentation 128
   and global warming 70, 72–3
   and oceans 81
   in rainwater 45, 84
carbon monoxide 27, 70–1, 74, 75
carbonate ion 185
carbonates, reaction with acids 41
carbon–carbon double bond 63, 65, 66
carbon-fibre-reinforced plastic 221
carbonic acid 45, 84
Carlisle, William 37
cast iron 27
catalase 126
catalysts 124–5, 126
   in ammonia manufacture 139
   and reaction rate 121, 123
   transition metals as 15
catalytic cracking 63
cathode rays 178
cellulase 130
cement 45, 54, 55, 220
cementation 87, 95
Chadwick, James 179
chain reaction 78
chalk 44
   and hard water 222
charm 179
cheese 129
chemical equation 152
chemical weathering 84
chemists 177
chlor-alkali industry 206–7
chloride ion, electron shells 181, 182
chlorides, of metals 17
chlorine 199, 200–1
   bonding 188
   isotopes 146, 173
   and ozone layer 78
   production and uses 206–7
   reactions 13, 134, 180, 200–1
   test 205
   water 201
chlorofluorocarbons (CFCs) 78–9, 203

chromium 14
    plating 33, 35
chymosin 129
citric acid 38, 133
coal 58, 71
collisions 122, 123, 124
combustion 132, 135
common salt 182
    see also sodium chloride
composite fabrics 221
composite materials 220–1
compound 10, 169, 180
compound ions 185
compression 217
concentration, and reaction rate 118, 122
concrete 45, 54, 55, 114, 220
condensing 217
conglomerate 87
conservative boundary 99
constructive boundary 98
contact metamorphism 93
continental crust 83
continental drift 102
continental plate 99
continents 102
continuous flow process 131
converging plates 98, 99
coolants 78
copper 14–15, 20, 36
    displacement reaction 19
    extraction 22
    in granite 91
    ions 185
    reaction with oxygen 16
    purification 32–3
    structure 192
    uses 24
copper(II) bromide, electrolysis 29
copper(II) oxide, displacement reaction 18
copper(II) sulphate 19, 136
core, Earth 82
corrosion of metals 34
corrosive 38
covalent bond 188
covalent compounds 189, 192
covalent substances 190–1
cracking 62–3, 65, 140
cross-bedding 86
crude oil 58–63, 70–1
    refining 60–1, 63
    uses 59, 61
crust, Earth 20, 82, 83
cryolite 31
crystals 186, 190
    in rock 91, 95
cupronickel 25
curds 129
Curie, Marie 178

Dalton, John 148, 177
Davy, Humphry 37
decay, radioactive 172
Democritus 176
denatured 127
denitrifying bacteria 75
deposition 85

destructive boundary 99
detergents 222
deuterium 172
diamond 168, 191
diesel oil 61
diffusion, gases 219
displacement reaction 19, 201
distilled water 224
diverging plates 98, 101
Döbereiner, Johann 194
dolomite 222
double bond 63, 65, 188
drinking water 145
ductile 10, 193
dynamic equilibrium 137

Earth
    atmosphere 74–5
    crust 20
    structure 82–3
earthquakes 96, 97, 99, 100, 104
eka-aluminium 195
eka-boron 195
eka-silicon 195
electrode 28
electrolysis 22, 28–9, 133
    of aluminium oxide 30–1
    calculations 158
    of copper(II) bromide 29
    of copper(II) sulphate 32–3
    and the discovery of metals 37
    of lead bromide 28
    of potassium iodide 29
    of sodium chloride 29, 204–5, 206–7
electrolyte 28
electron shells 174, 179
electronic configuration 175, 197
electrons 168, 170, 178
electroplating 33
elementary particles 179
elements 10, 168, 177
    in Earth's crust 20
    metals found as 22
elixir of life 176
empirical formula 150
endothermic reactions 133, 134, 136
energy changes in reactions 135
energy level
    diagram 132, 133
    of electron shells 174, 175
environment, and crude oil 70–1
enzymes 121, 126–31
    action 127
    as catalysts 126–7
    and food 127, 128–9
    immobilized 131
    new uses 130–1
equation 152
equilibrium 137, 138–9
erosion 84, 94
ethane 62, 63, 64
ethanoic acid 38, 133
ethanol 128, 129
ethene 63, 65, 66
eutrophication 145
evaporation 186, 217

exhaust, car 124
exothermic reactions 18, 132, 134, 136
expansion 216
extraction
    of aluminium 30–1
    of metals 22–3
extremophiles 130
extrusive igneous rock 90, 91, 94

fault 100
fermentation 128–9
fertilizers 140, 142–5
field-ion microscope 177
flammable 60, 61
fluorides 203
fluorine 199, 200
fluorspar, quarrying 57
fold mountains 101
folds 100
foliation 92
foreshock 104
formula 150, 169
fossil fuels 58
    and global warming 72–3
fossils 87, 88–9, 102
fractional distillation 60–1
fractions 60, 62
francium 12–13
freeze–thaw weathering 84
fuel oil 61
fuels, from crude oil 59, 61, 70

galena 37
gallium 36
galvanized iron 35
gas 58, 61, 64, 216, 218–19
gas syringe 116
gasoline 61
genetically modified bacteria 130
giant covalent structure 190–1
giant ionic structure 182, 183
giant structure 186
glass 55
glass-fibre-reinforced plastic 221
global positioning system (GPS) 104, 105
global warming 70, 72–3, 76, 81
glucose 76, 135
gneiss 93
goitre 203
gold 16, 20, 22, 34, 36–7
    and alchemists 176
    in granite 91
    occurrence 21
    properties 10, 14
graded bedding 86
granite 83, 89, 90, 94
graphite 191
gritstone 89
groundwater 145
group, in periodic table 11, 168
Group 1 11, 12–13, 196
Group 2 11, 197
Group 7 199, 200–1
Gulf Stream 72
gypsum 55, 222

Haber, Fritz 140
Haber process 15, 140, 144
haematite 26
haemoglobin 71
half equation 158
half-life 172
halogens 199, 200–1
    compounds 202–3
hard water 222–5
Harman, Joe 78
helium 196
high density polythene 68
high-grade metamorphic rock 93
Himalayas 99, 101
Hofmann voltameter 204–5
hydrated crystals 55
hydriodic acid 202
hydrobromic acid 202
hydrocarbons 59, 60, 62–3, 64–5
hydrochloric acid 17, 38, 202
hydrogen
    and ammonia manufacture 140
    bonding 188
    and energy 135
    ions 42
    isotopes 172
    production and uses 206–7
    reaction with chlorine 134
    in reactivity series 18
    test 205
hydrogen chloride 202, 219
hydrogen halides 202
hydrogen peroxide 121, 126
hydrogenation of vegetable oils 15
hydrogencarbonate ion 185
hydrothermal solution 91
hydroxide ion 42, 185
hydroxides
    of alkali metals 12
    of metals 17
    reaction with acids 41

igneous rock 83, 90–1
immobilized enzyme 131
incomplete combustion 71
indigestion 44
Industrial Revolution 36, 72
inert 196
infrared cameras 105
intermolecular forces 190
intrusive igneous rock 90, 94
invertase 130
iodine 190, 199, 200–1
ion 181, 184–5
ion exchange, and hard water 224, 225
ion-exchange membrane 206
ionic bond 182
ionic compounds 13, 29, 182, 184–7, 192
iron 36, 37
    as catalyst 121, 125, 139, 140, 141
    competition reactions 18–19
    and corrosion 34
    in Earth's core 82
    in Earth's crust 20
    extraction 22, 26–7
    ions 185
    ore 22, 26

    properties and uses 10, 14–15, 24
    reactions 16, 17, 133, 155
    in rock 103
iron(III) oxide, reduction 27
Iron Age 36
island arcs 101
isomerase 130
isotopes 146, 172–3

joule 132

kerosene 61
kilojoule 132
krypton 196

lactase 131
lactic acid 129
Lactobacillus acidophilus 129
lactose 129
laminated materials 221
lattice 190
    ionic 186
lava 90, 95
Lavoisier, Antoine 77
Law of Conservation of Mass 154
Law of Constant Composition 148
Law of Octaves 194
Law of Triads 194
Le Chatelier's principle 138
lead 20, 22, 36, 37
    in granite 91
    and pollution 70
    reaction with hydrochloric acid 17
    uses 24
lead bromide, electrolysis 28
lead(II) chloride 137
life, beginning of 75
lightning 142
limestone 45, 54–5
    in blast furnace 27
    chemical weathering 84
    formation in oceans 80
    and hard water 222
    in Peak District 89
    quarrying 56–7
    uses 54
limewater 128
lipase 126, 130
liquid 216
lithium 12–13, 196
lithosphere 82, 90
litmus 38
liver 126
low density polythene 68
low-grade metamorphic rock 93
lubricating oil 61

magma 90, 94
magnesium
    burning 153
    in Earth's crust 20
    properties 197
    reaction with acid 17, 40, 116–17, 118, 122–3
    reaction with oxygen 16
    reaction with water 16, 17
magnesium chloride, electron shells 183

magnesium oxide 187
    electron shells 183
magnetism, in ocean floor 103
malleable 10, 193
manganese, on ocean floor 81
manganese steel 25
manganese(IV) oxide, as catalyst 121, 126
mantle 82
marble 92
marble chips 120
mass number 170
mass spectrometer 146
melamine 67
melting 216
melting point 186
membrane cell 206
Mendeleev, Dmitri 195
mercury 14, 20, 168, 176
metal oxides 22, 41
metallic bond 192
metals 8–37
    history 36–7
    and periodic table 10–11, 168
    positive ions 184
    properties 10, 24, 193
    reaction with acids 40
    reaction with halogens 199
    in reactivity series 18
    structure 192
metamorphic rock 83, 92–3
metamorphism 92
methane 62, 63, 72
    in atmosphere 75
    bonding 189
    and energy 135
methanoic acid 38
Midgely, Thomas 78
mild steel 24, 27
mining 21
model 179
molecular substances 188, 190–1, 192
molecule 188
monatomic 198
Montreal Protocol 78
Mount Pinatubo 105
mountain ranges 96, 99
M_r 147
Müller, Erwin 177

naphtha 61, 63
National Parks Authority 57
native 20, 22
natural gas (North Sea Gas) 64, 71
negative ion 181, 184
neon 196
neutral 39
neutralization 40–1, 42–3, 132
neutrons 179
Newlands, John 194
Nicholson, Anthony 37
nickel 14, 15, 25, 33, 82
nitrate
    ion 185
    in soil 76, 142, 143
    in water 145
nitric acid 38
    manufacture 125, 143

nitrifying bacteria 75, 76
nitrogen
    and ammonia manufacture 140
    in atmosphere 74, 75, 76
    cycle 75
    fertilizer 142, 143
nitrogen-fixing bacteria 142
nitrogen oxides *see* oxides of nitrogen
nitrogenous fertilizers 142
noble gases 180, 198
non-metals
    negative ions 184
    and periodic table 168
    properties 198
    reaction with halogens 199
non-renewable resource 22
nucleus 168, 170, 179

ocean plate 98, 99
ocean ridge 96
ocean trench 96, 99, 101
oceanic crust 83
oceans 80–1
oil *see* crude oil
oil trap 58
optimum temperature, enzymes 127
optimum yield 141
ores 21, 36, 37
organic compounds 59
organic farming 145
overturn 100
oxides of metals 16, 17
oxides of nitrogen, and acid rain 45, 70–1
oxygen
    in atmosphere 74, 75, 76
    bonding 188
    discovery 77
    in Earth's crust 20
    reaction with alkali metals 13
ozone 75, 78

paints 34
Pangaea 102, 103
paraffin waxes 61
particles 216–17
Peak District 56–7, 89
pentane 71
percentage composition 148–9
period, in periodic table 11, 168
periodic table 10–11, 168, 169
    development 194–5
    and electronic configuration 175, 197
    and properties of elements 196–9, 200
permanent hardness 223
pH scale 39, 44
pH, and enzymes 127, 130, 131
phenolphthalein 43
phosphorus, fertilizer 142
photochemical reaction 203
photography 203
photosynthesis 75, 76, 133
physical weathering 84
phytoplankton 80, 81
plankton 78, 80
plasterboard 221
plastics 66–9

plate tectonics *see* tectonic plates
platinum 14, 20, 22
platinum, as catalyst 124
platinum–rhodium gauze catalyst 125
plywood 221
pollution 70–1, 145
poly(chloroethene) (polyvinyl chloride, PVC) 66, 67
poly(phenylethene) (poly(styrene)) 66, 67
poly(propene) 66, 67
polymerization 66–7, 133
polymers 65, 66–9
polythene 66, 67, 68–9, 133
positive ion 181, 184
potassium 34
    in Earth's crust 20
    fertilizer 142
    properties 12–13, 196
    reaction with water 17
potassium hydroxide solution 39
potassium iodide 29, 203
pressure
    and equilibrium 139, 141
    and gases 218–19
Priestley, Joseph 77
product 152
propane 62, 64
propene 63, 65
protease 126, 130
proteins 75, 76, 126
proton number 170
protons 170, 179
Proust, Joseph Louis 177

quarrying, limestone 56–7
quicklime 44, 54, 132

radiation 172
radioactive decay, in Earth 82
radioactive isotopes (radioisotopes) 172
radioactivity 12, 178
radon 104
rate of reaction 114–23
reactant 152
reactivity
    of metals 16–17
    series 18
recycling
    metals 22
    plastics 69
redox reaction 23, 28, 201
reducing agent 18
reduction, of metal oxides 22, 27
refining crude oil 60–1, 63
regional metamorphism 93
reinforced concrete 220
reinforced plastic 221
relative atomic mass 146, 147, 173
relative formula mass 147
rennet 129
reservoir rock 58
respiration 132
reversible reactions 136–41
rhodium 124
ripening of fruit 127
rock

    cycle 94–5
    formation 83, 87, 90, 92–3
    salt 21, 202
    types 83
root nodules 142
rubidium 12–13, 196
rusting 34, 114
Rutherford, Ernest 179, 195

sacrificial protection 35
safety glass 220
salt 206
    and alchemists 176
    preparation 43
    see also sodium chloride
salts 40
    in oceans 80
San Andreas fault 99, 104
sandstone 87, 94
saturated hydrocarbons 64
scale 44, 223
scanning electron microscope 177
scarce metals 20
Scheele, Karl 77
schist 93
sea floor spreading 98
sediment 83, 86
sedimentary rock 83, 86–7, 88–9, 94
seismometer 104, 105
selenium 36
shale 57, 87, 89
sherbet 133
silica 20, 90
silicon, in Earth's crust 20
siltstone 87
silver 20, 22, 36
    in granite 91
    ions 185
silver chloride, precipitation 114, 132
silver halides 203
silver nitrate, displacement reaction 19
skin cancer 78, 79
slag 27
slaked lime 44, 45, 54, 55
slate 92, 93, 94
smokeless fuel 157
sodium 168
    atom 170
    in Earth's crust 20
    ion, electron shells 181, 182
    properties 10, 12–13, 14, 196
    reaction with chlorine 180
    reaction with oxygen 16
    reaction with water 16, 17
    uses 25
sodium carbonate 133, 224
sodium chloride 13, 202
    electrolysis 29, 204–5, 206–7
    ionic bonding 180–2
    in oceans 80
    ore 21
    production and uses 206–7
    properties 187
    structure 186
sodium hydrogencarbonate 44, 133
sodium hydroxide 39, 41, 44

sodium oxide 13
sodium thiosulphate 119
soft water 222–5
solder 25
solid 216
stable 180
stainless steel 24, 25
starch 76
starter culture 129
state symbols 152
steel, and corrosion 15, 27, 34
sterilizing agent 201
Stone Age 36
stonewashed denim 130, 131
strainmeter 104
strange 179
stratosphere 74
strong acids and alkalis 38, 39
subduction 99, 101
sulphate ion 185
sulphur 133, 155, 176
sulphur dioxide
   and acid rain 45, 72, 70–1
   pollution 74
   and sulphuric acid manufacture 125
   and volcanoes 105
sulphur trioxide 125
sulphuric acid 38, 125
Sumner, James 128
surface area, and reaction rate 120–1, 123
surface catalyst 124
sustainable development 57
symbols, for elements 168

tannic acid 38
tectonic plates 95, 96–103
temperature
   and enzymes 127, 130
   and equilibrium 138, 141
   and gases 218–19
   and reaction rate 119, 123
temporary hardness 223
thermal decomposition 62, 136
Thermit process 18
thermosetting plastics 67
thermosoftening plastics 67
Thomson, J. J. 178
tiltmeter 104, 105
tin 20, 36, 37
   in granite 91
   plating 33, 35
   uses 25
tincture of iodine 201
titanium 25, 36
transform motion 99
transition metals
   as catalysts 121
   ions 185
   and periodic table 11
   properties 14–15
   salts 15
transport, rock 84, 85, 94
tritium 172
troposphere 74
tuff 91

ultraviolet radiation 78
universal indicator 39
unreactive 180
unsaturated hydrocarbons 65, 66
up 179
uranium 82, 178
urease 128

vanadium pentoxide 125
vinegar 38
viscous 60, 61
volatile 60
volcanic ash 91
volcanoes 94, 95, 96, 97, 99, 101, 105
volume, of gases 156, 218–19

washing soda 224
water
   bonding 189
   of crystallization 136, 220
   cycle 76
   in electrolysis 29
   soft and hard 222–3
   vapour, in atmosphere 74, 75, 76, 80
weak acids and alkalis 38, 39
weathering 84, 94, 222
Wegener, Alfred 102–3
whey 129, 131
wine 129
world population 144
wort 129

xenon 196

yeast 128–9
yoghurt 129

zeolites 225
zinc 14, 22
   in granite 91
   ions 185
   reaction with acid 17, 115
   uses 25
   reaction with water 17
zinc carbonate 44
ZMS-5 125